for your very
fine contribution
to the Symposium.
David J. Findley
October 9, 1978

APPLIED TIME SERIES ANALYSIS

ACADEMIC PRESS RAPID MANUSCRIPT REPRODUCTION

Proceedings of the First Applied Time Series
Symposium held in Tulsa, Oklahoma,
May 14–15, 1976

APPLIED TIME SERIES ANALYSIS

EDITED BY

DAVID F. FINDLEY

Division of Mathematical Sciences
The University of Tulsa
Tulsa, Oklahoma

ACADEMIC PRESS **New York San Francisco London** **1978**
A Subsidiary of Harcourt Brace Jovanovich, Publishers

ACADEMIC PRESS, INC.
111 Fifth Avenue, New York, New York 10003

United Kingdom Edition published by
ACADEMIC PRESS, INC. (LONDON) LTD.
24/28 Oval Road, London NW1 7DX

Applied Time Series Symposium, 1st Tulsa, Okla., 1976.
Applied time series analysis.

Sponsored by the University of Tulsa.
1. Time-series analysis–Congresses. I. Findley, David F. II. Tulsa, Okla. University. III. Title.
QA280.A66 1976 519.2'32 78-9007
ISBN 0-12-257250-5

PRINTED IN THE UNITED STATES OF AMERICA

CONTENTS

List of Contributors *vii*
Preface *ix*

Time Series Analysis and Control through Parametric Models 1
Hirotugu Akaike

Non-Linear Time Series Modeling 25
C. W. J. Granger and A. Andersen

On G-Spectral Estimation 39
H. L. Gray, A. G. Houston, and F. W. Morgan

Multivariate Autoregression Estimation Using Residuals 139
Richard H. Jones

Two-Dimensional Recursive Filtering in Theory and Practice 163
James H. Justice

Adaptive Processing of Seismic Data 225
Stanley J. Laster

A Comparative Case Study of Several Spectral Estimators 245
Don McIntire

Application of Homomorphic Filtering to Seismic Data Processing 261
A. Oppenheim and J. Tribolet

Waves Propagating in Random Media as Statistical Time Series 287
Enders A. Robinson

Estimating the Intensity of a Poisson Process 325
G. S. Watson

LIST OF CONTRIBUTORS

Numbers in parentheses indicate the pages on which the authors' contributions begin.

Hirotugu Akaike (1), Institute of Statistical Mathematics, 4-6-7 Minami-Azabu, Minato-Ku, Tokyo, Japan

Alan Andersen (25), Department of Economic Statistics, University of Sidney, New South Wales, Australia 2006

Clive W. J. Granger (25), Department of Economics, University of California, San Diego, La Jolla, California, 92093

Henry L. Gray (39), Department of Statistics, Southern Methodist University, Dallas, Texas 75275

A. Glen Houston (39), NASA Johnson Space Center, Houston, Texas 77058

Richard H. Jones (139), Department of Biometrics, University of Colorado Medical Center, Denver, Colorado 80262

James H. Justice (163), Division of Mathematical Sciences, The University of Tulsa, Tulsa, Oklahoma 74104

Stanley J. Laster (225), Division of Earth Sciences, The University of Tulsa, Tulsa, Oklahoma 74104

Donald D. McIntire (245), W. L. Clayton Research Center, 3333 North Central Expressway, Richardson, Texas 75080

Fred W. Morgan (39), Department of Mathematical Sciences, Clemson University, Clemson, South Carolina 29631

Alan V. Oppenheim (261), Department of Electrical Engineering and Computer Science, Massachusetts Institute of Technology, Cambridge, Massachusetts 02139

Enders A. Robinson (287), 2630 Southwick Street, Houston, Texas 77080

*Jose M. Tribolet** (261), Department of Electrical Engineering and Computer Science, Massachusetts Institute of Technology, Cambridge, Massachusetts 02139

Geoffrey S. Watson (325), Department of Statistics, Princeton University, Princeton, New Jersey 08540

*Present address: Bell Laboratories, Murray Hill, New Jersey 07974

PREFACE

The field of time series analysis or signal processing, as some scientists and engineers prefer to call it, is a child of many parents, as this ambiguity of name already suggests. This is a historical and a present fact. A quick search through the current journals of a technical library is likely to turn up time series papers in publications from the fields of acoustics, earthquake engineering, econometrics, electrical engineering, geology, geophysics, guidance and control engineering, mathematics, mechanical engineering, neurophysiology, psychology, and statistics. In such circumstances, it is easy for a time series analyst to be unaware of important developments in the subject occurring in a field remote from his or her own because there are few opportunities for people working in apparently dissimilar fields to get to know one another.

With this situation in mind, the University of Tulsa is sponsoring a series of Applied Time Series Symposia in which distinguished time series researchers from different disciplines give talks about portions of their research they feel might be of general interest. The first such symposium took place in Tulsa, Oklahoma, on May 14–15, 1976. It featured speakers from econometrics, electrical engineering, geophysics, mathematics, and statistics, and attracted 165 participants from these fields and several others.

The speakers were H. Akaike, R. F. Engle, C. W. J. Granger, H. L. Gray, A. G. Houston, R. H. Jones, J. H. Justice, R. T. Lacoss, S. J. Laster, A. V. Oppenheim, E. Parzen, E. A. Robinson, S. Treitel, and G. S. Watson.

The session chairmen were J. B. Bednar, S. E. Elliot, M. R. Foster, and P. M. Robinson.

The present volume contains papers based on talks presented at the symposium. Because of the novelty of their methods, H. L. Gray, A. G. Houston, and F. W. Morgan were invited to prepare a greatly expanded version of their talk for this publication, which they kindly agreed to do. Professor Gray was also kind enough to recommend inclusion of the paper of D. D. McIntire, which illustrates an important facet of these methods. Three papers, those of H. L. Gray, A. G. Houston, and F. W. Morgan, R. H. Jones, and S. J. Laster, include programs. The TIMSAC program packages of Akaike, Arahata, and Ozaki, discussed in H. Akaike's paper, are available on magnetic tape from the Division of Mathematical Sciences of the University of Tulsa.

The Applied Time Series Symposium was the joint effort of several disciplines and several colleges of the University of Tulsa. Its director was D. F. Findley and its associate director was S. J. Laster, both of whom were given substantial assistance by many colleagues and staff members among whom must be mentioned T. W. Cairns, W. A. Coberly, E. T. Guerrero, N. J. Hyne, D. Murray, and A. R. Soltow. On behalf of all of us, I would like to thank the speakers, session chairmen, and participants for their contributions to a very stimulating conference.

It is my privilege to extend special thanks to the staff of Academic Press for their continued interest in this book, to J. R. Barton for her exceptionally competent work in preparing the manuscript for this book, and to Amoco Research Center and Cities Service Research Laboratory for technical assistance.

David F. Findley

APPLIED TIME SERIES ANALYSIS

TIME SERIES ANALYSIS AND CONTROL THROUGH PARAMETRIC MODELS

Hirotugu Akaike
The Institute of Statistical Mathematics
Minato-ku, Tokyo 106, Japan

Abstract.

The use of parametric models for the analysis and control of time series is discussed using some numerical examples.

1. Introduction.

One of the basic problems of time series analysis is the estimation of power spectra. The procedure of spectrum estimation through the windowing of the sample autocovariance function became practical at the beginning of the 1950s. The procedure was extended to the multivariate case and the statistical estimation of frequency response functions has been practiced within industry since the 1960s. The main difficulty in applying these procedures is the choice of the window to be used for the estimation. Another drawback of these frequency domain estimates is that the results must be transformed into time domain quantities to produce the final outputs of the analysis appropriate, for example, for the prediction or control of time series.

The use of parametric models, such as autoregressive (AR) or autoregressive moving average (AR-MA) model, was well known but the popularity of the frequency domain approach distracted our attention from the potential of the time domain approach. Nevertheless, the advent of computer control of production processes made time domain modelling a necessity and the high resolution in spectrum estimation obtainable through the time domain modelling attracted the attention of scientists, especially of geophysicists.

ISBN 0-12-257250-5

With the introduction of a criterion of fit (Akaike (1970 a,b)), the fitting of an AR-model to time series became quite practical. Especially its application to multivariate time series (Akaike (1971)) was found very useful in developing computer control of stochastic systems (Otomo, Nakagawa, and Akaike (1972)). The use of the AR-MA model as an ultimate model of stationary time series was hindered by both the numerical difficulty of the necessary maximum likelihood computation and the difficulty in choosing the orders of the AR and MA parts. By a theoretical analysis (Akaike (1974a)), it was made clear that a canonical correlation analysis procedure would produce a useful initial guess for the model. Also, a criterion developed for the evaluation of the goodness of fit of a statistical model fitted by the method of maximum likelihood provided a solution to the order determination problem. Based on these results, a computer program which automatically fit an AR-MA model to the data was developed and is now in operation (Akaike, Arahata, and Ozaki (1975)). It is the purpose of this paper to demonstrate the practical use of these recently developed procedures with some numerical examples.

2. Basic Parametric Models.

The basic parametric model to be used in this paper is the linear constant system which generates the time series under consideration from a white noise input. Specifically, the AR model is given by

$$Y_n + B_1 Y_{n-1} + \cdots + B_M Y_{n-M} = X_n ,$$

and the AR-MA model is given by

$$Y_n + B_1 Y_{n-1} + \cdots + B_M Y_{n-M} = X_n + A_1 X_{n-1} + \cdots + A_L X_{n-L},$$

where Y_n is the time series under consideration and X_n is a white noise which is independent of the past observations Y_{n-m} $(m = 1, 2, \ldots)$. Obviously, an AR model is a special case of an AR-MA model, but the extreme simplicity of the computations used for its fitting justifies its independent treatment. The numerical procedure for the fitting of an AR model is described in detail in R. H. Jones' paper in these proceedings.

There is a technical difficulty in fitting an AR-MA model. This is the problem of identifiability or the uniqueness of the model. In the univariate case, by choosing the lowest possible values of the AR order M and the MA order L, the parameters are specified uniquely. In the multivariate case, where the coefficients A_ℓ and B_m are matrices, the solution is not so simple. To avoid this difficulty, we use a canonical Markovian representation,

$$Z_{n+1} = FZ_n + GX_n$$

$$Y_n = HZ_n \ ,$$

where Z_n is the state vector of the system at time n and is specified by a basis of the linear space spanned by the components of the predictors $Y_{n+k}|n$ (k = 0, 1, ...), where $Y_{n+k}|n$ is the best prediction of Y_{n+k} based on the observations Y_n, Y_{n-1}, ... Under the assumption that the process Y_n is Gaussian, the state vector Z_n represents the minimal sufficient information from the present and past needed to predict the future behavior of the process. The assumption of finitness of the dimension of Z_n is fundamental to produce a finite parameter model. This subject is discussed in detail by Akaike (1974 a,b , 1976).

3. AR-MA Model Fitting Procedure.

The numerical procedure for the fitting of an AR-MA model is discussed in detail in Akaike (1976) and the listings of the necessary computer programs are given in Akaike, Arahata, and Ozaki (1975). Here, we will give only a conceptual description of the organization of the procedure. The procedure is as follows:

(a) Fit an AR model $Y_n + C_1 Y_{n-1} + \cdots + C_M Y_{n-M} = X_n$, where the order M is chosen as the one which gives the minimum of the criterion AIC which is to be defined in the next section.

(b) Do the canonical correlation analysis between $(Y_n', Y_{n-1}', \ldots, Y_{n-M}')$ and $(Y_n', Y_{n+1}', \ldots)$, where Y_n' denotes the transpose of Y_n. By this analysis, determine Z_n as the vector of predictors of those variables which are needed to describe those canonical variables within the future set $(Y_n', Y_{n+1}', \ldots)$ that are associated with positive canonical correlation coefficients. Initial estimates of the matrices F and G are obtained by this analysis and H takes a prescribed form.

(c) Do the maximum likelihood computation for the Markovian representation.

(d) Try several possible alternative structures of the state vector Z_n and choose a final estimate by using the criterion AIC.

(e) If necessary, transform the Markovian representation into an AR-MA representation. Comment: For a univariate time series Y_n, direct maximum likelihood computation of an AR-MA model is possible.

4. Criterion for Model Selection.

One serious problem in parametric model fitting of time series is the determination of the orders L and M. This is a common problem when a statistical model contains many parameters. A typical example is the fitting of a polynomial to a set of observational data. If the degree of the polynomial is kept too low, the polynomial would not reproduce the detail of the "true" polynomial. If the degree is set too high, the resulting polynomial would be too sensitive to the stochastic variation of the data.

As a general criterion of fit for a statistical model determined by the method of maximum likelihood, the present author proposed an information criterion defined by

$$\text{AIC} = (-2) \log_e (\text{Maximum likelihood})$$
$$+ 2(\text{number of independently adjusted parameters}) .$$

This criterion AIC is an estimate of minus twice the expected log likelihood. The minused expected log likelihood is, except for an additive constant, identical to the (generalized) entropy of the true stochastic structure with respect to the structure assumed by the model and provides a natural measure of deviation of the true structure from the assumed structure.

Typically, the value of AIC decreases quickly as the number of parameters being adjusted is increased and then increases almost linearly when too many redundant parameters are included. The set of the parameters which produces the minimum of AIC is chosen as our final estimate. This criterion AIC is closely connected with the criterion FPE, which was formerly proposed for the fitting of AR models (Akaike (1970)), and satisfies an approximate equality,

$$\text{AIC} = N \log_e (\text{FPE}) ,$$

where N denotes the data length used for the fitting. The use of AIC in general statistical model fitting is discussed by Akaike (1973, 1974c, 1977a). In the maximum likelihood computation programs by Akaike, Arahata, and Ozaki (1975), an approximate likelihood function of a Gaussian process is used for the fitting of an AR-MA model. The nature of this approximation is discussed by Akaike (1976).

5. Practical Applications.

To show the practical utility of the parametric modelling of time series, some typical numerical examples are illustrated in Figures 1 through 8. It is hoped that these results will help make the parametric approach acceptable to the reader as a standard procedure for spectrum estimation. Also, it is hoped that the tremendous potential of the parametric approach for the identification of system dynamics to be used for the implementation of computer control of a stochastic system will also be recognized through the discussion of some of the results of its application.

Before going into the discussion of the numerical results, the relation between the time domain parametric representation of an AR-MA model and its spectrum is summarized here. For the AR-MA model

$$Y_n + B_1 Y_{n-1} + \cdots + B_M Y_{n-M} = X_n + A_1 X_{n-1} + \cdots + A_L X_{n-L} ,$$

its spectral density p(f) for the univariate case and the spectral density matrix P(f) for the multivariate case are given, respectively, as follows:

$$p(f) = \frac{\left|1 + \sum_{\ell=1}^{L} A_\ell \exp(-i2\pi f\ell)\right|^2}{\left|1 + \sum_{m=1}^{M} B_m \exp(-i2\pi fm)\right|^2} \sigma^2 \qquad (-1/2 \leq f \leq 1/2)$$

and

$$P(f) = B(f)^{-1} A(f) C A(f)^* B(f)^{*-1} \qquad (-1/2 \leq f \leq 1/2) ,$$

where

$$A(f) = I + \Sigma A_\ell \exp(-i2\pi f\ell)$$

and

$$B(f) = I + \Sigma B_m \exp(-i2\pi fm) ,$$

where I denotes an identity matrix, and σ^2 and C denote the variance EX_n^2 and the variance matrix EX_nX_n', respectively. A^* denotes the conjugate transpose of A. Thus, once a set of estimates of the coefficients and the variance (matrix) of the innovation X_n is given, the corresponding estimate of the spectrum is automatically obtained by one of the above formulas. The estimate which is obtained by minimizing AIC will be called the minimum AIC estimate (MAICE).

Figure 1 shows the results of spectrum analysis of the Chandler wobble of the axis of the earth. These results were

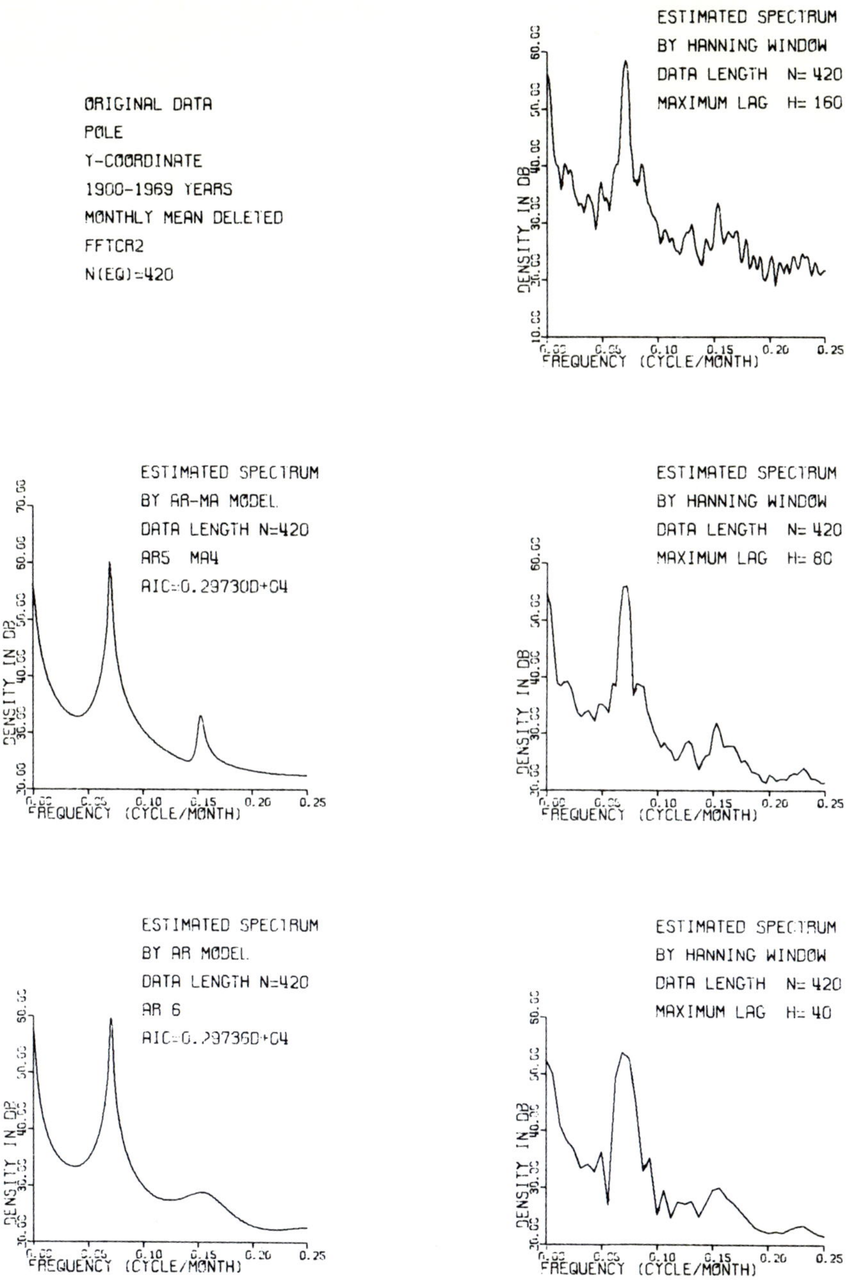

Fig. 1. Comparison of various estimates of the power spectrum of the Chandler wobble.

obtained by the analysis of the montly records of the y-coordinate of the pole for the period 1900-1969. The annual component has been deleted by subtracting the average of each month. Although the original data length was 840 (= 12 x 70), the original set of data was Fourier transformed and the frequency components within the upper half frequency band were put equal to zero. The result was then transformed back into the time domain and resampled so that every second value of the new time series was used for the analysis. This gives a set of data with length N = 420. The results illustrated in Figure 1 clearly indicate the existence of a sharp peak in the spectrum at a frequency of about 0.07 cycles per month, i.e., with a period of approximately 1.185 years. The analysis of this sharp peak has been a subject of intensive discussion within geophysical circles.

Figure 1 also demonstrates the difficulty with applying the classical windowing procedure of spectrum estimation to this case. To get a feeling of the statistical variability of the results, several Monte Carlo experiments were performed with the AR-MA model given in Figure 1. One of the experimental results is illustrated in Figure 2. By looking at the windowed estimates, we can see that the reproduction of the spectrum by the experiment was reasonably successful. The orders chosen by minimizing AIC were identical to the true orders in this case. To further confirm the results, the result of another experiment is illustrated in Figure 3. By observing these figures, it is almost clear that the AR-MA model fitting procedure is producing quite satisfactory results compared with the conventional windowing and AR model fitting procedures. The geophysical meaning of the spectrum is discussed elsewhere (Ooe, Akaike, and Kaneko (1976)).

To show that a finite parameter AR-MA model can sometimes really be useful for the representation of spectral characteristics of some physical processes, an example of its application to a very long record of a physical process is illustrated in Figures 4a and 4b. The data length N was 15,000. The results show that the model with the AR order 5 and the MA order 2 gives an AIC value lower than that of the AR model of order 24 which is the MAICE among the tested AR

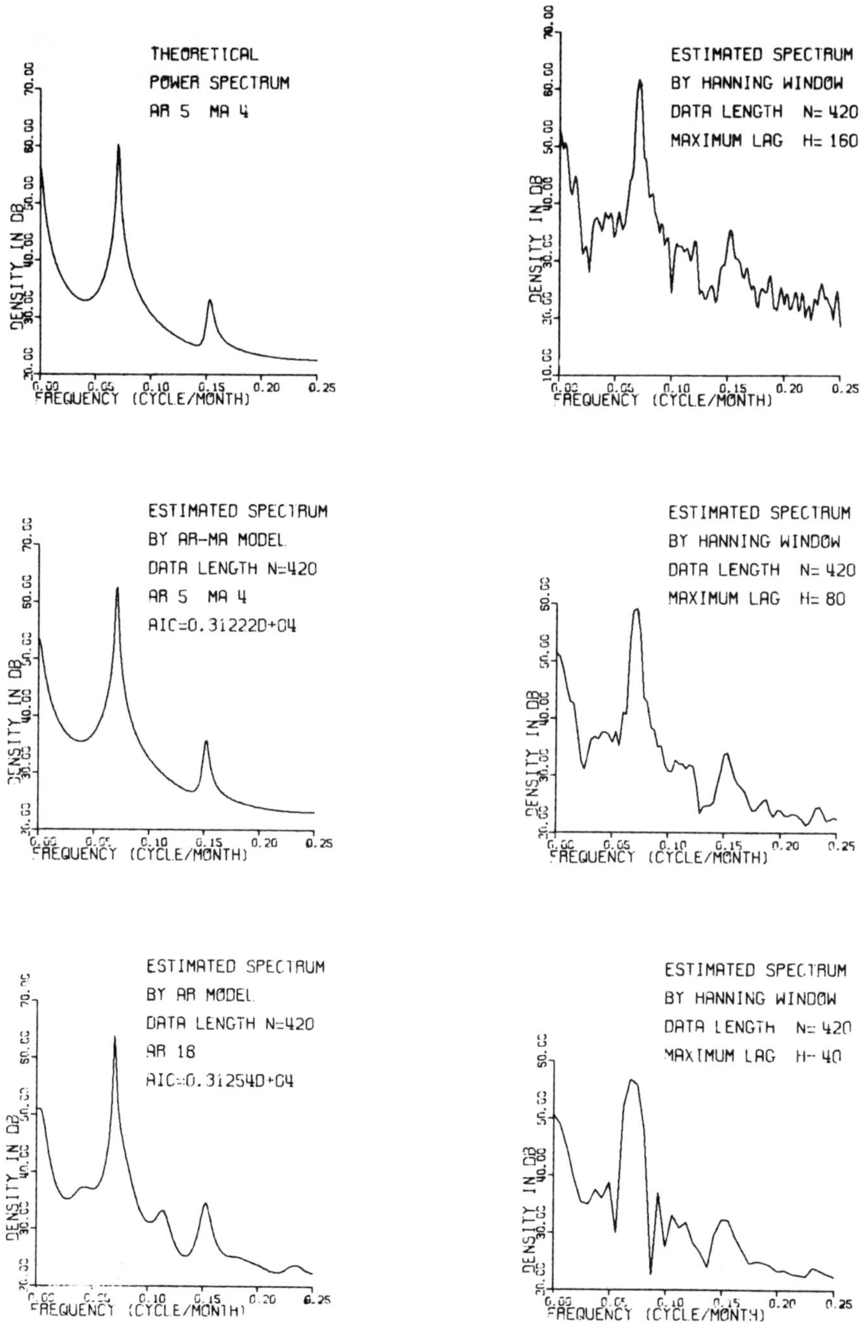

Fig. 2. The results of a Monte Carlo experiment of the Chandler wobble spectrum (1).

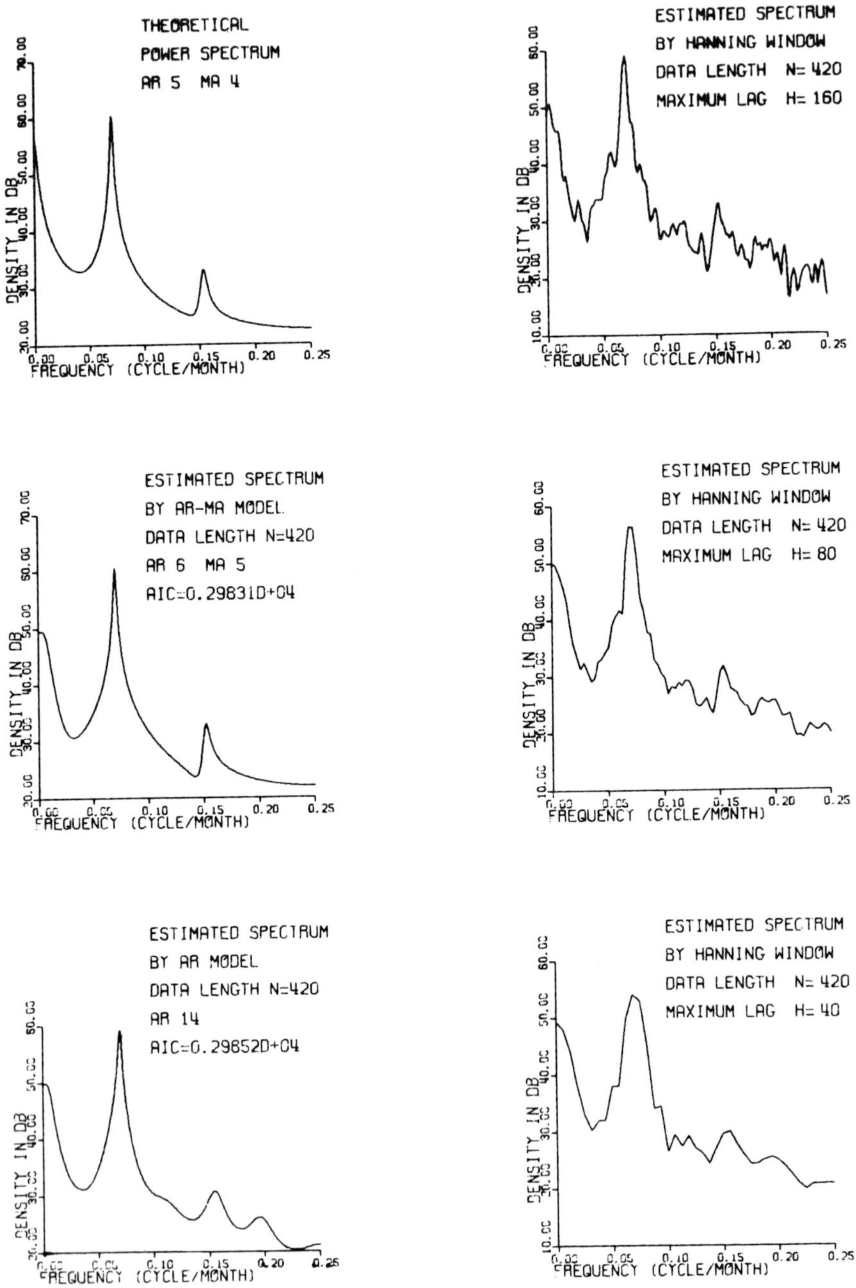

Fig. 3. The results of a Monte Carlo experiment of the Chandler wobble spectrum (2).

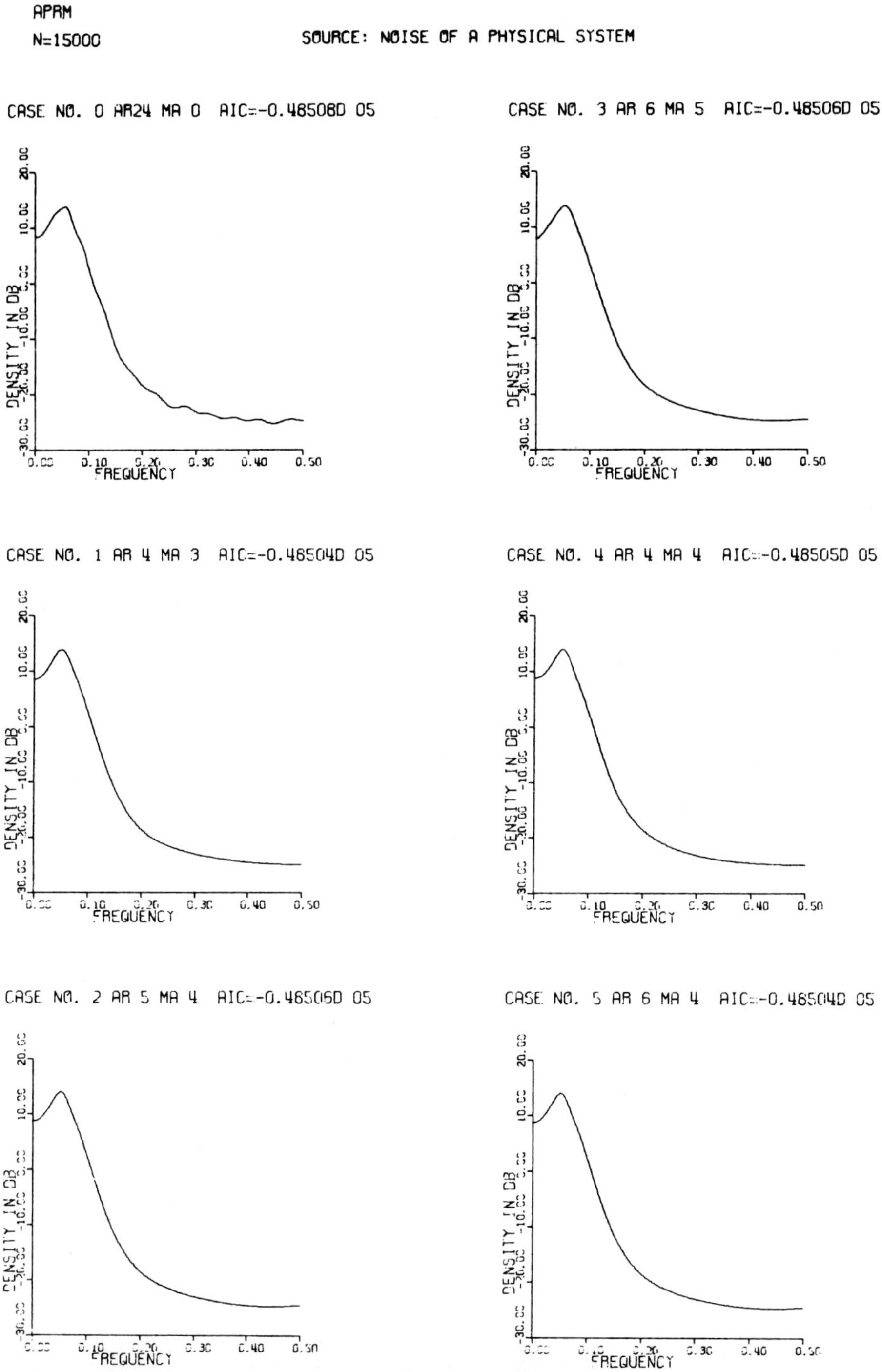

Fig. 4a. An example with a large number of observations from a physical process.

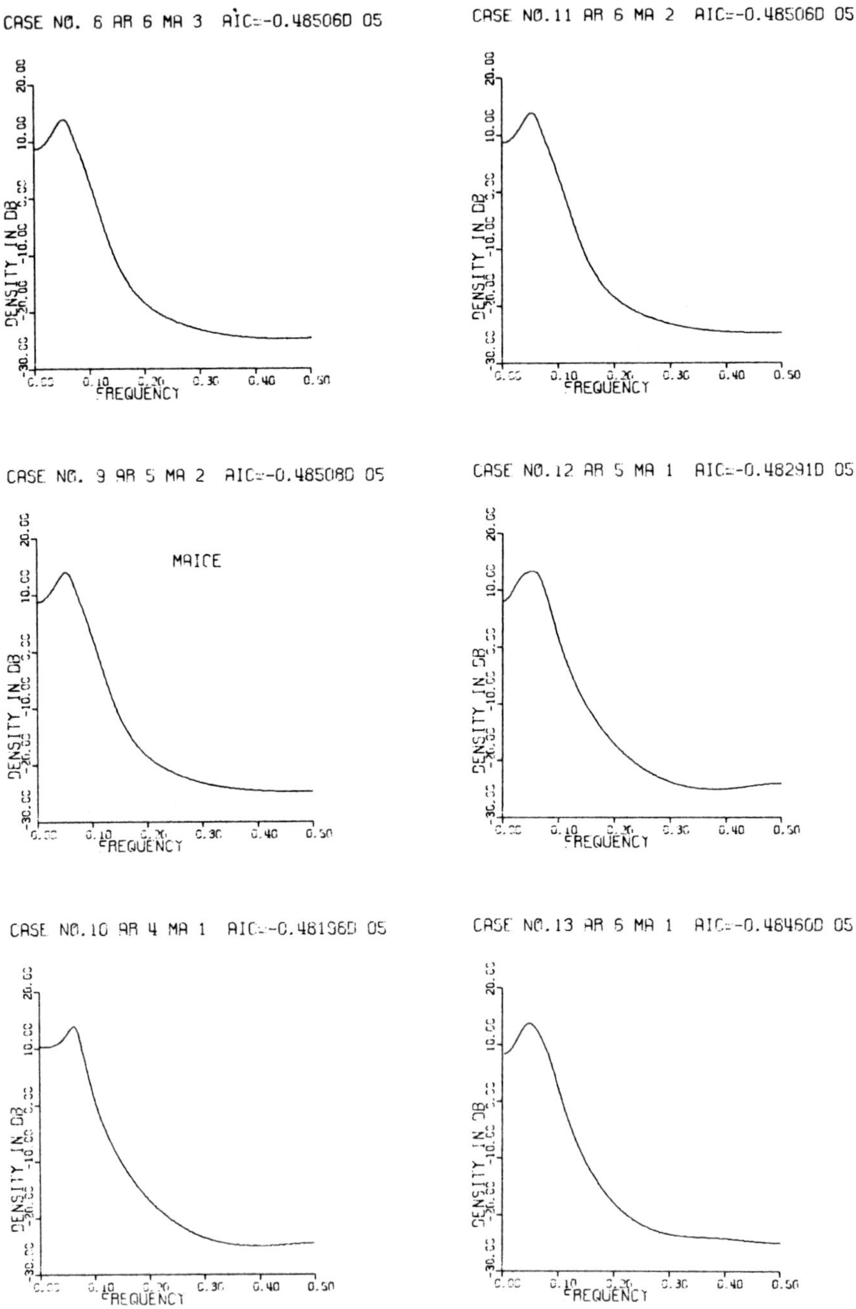

Fig. 4b. Continuation of Fig. 4a.

models. It seems that this estimate by the AR-MA model is quite satisfactory. Some examples of application to the analysis of other physical processes are discussed in Akaike (1977b).

To give an answer, although not quite a definitive one, to the long-disputed problem of the determination of the spectral characteristics of the Beveridge wheat price index series, the fitting of an AR-MA model was tried on the data reported in the book by Anderson (1971). It turned out that the maximum likelihood computation using the data was rather difficult due to the sharp dip of the power spectrum at the zero frequency. Accordingly, the sample covariance sequences were transformed numerically to those corresponding to the time series Z_n generated from the original Y_n by the relation $Z_n = 0.95\ Z_{n-1} + Y_n$. The AR-MA model for Z_n with the minimum value of AIC turned out to be with the AR order 2 and the MA order 1. This gives an AR-MA model of Y_n with AR order 3 and MA order 2 and the corresponding spectrum is illustrated in Figure 5 together with the windowed and AR estimates. The AR estimate does not reproduce the dip of the spectrum at the zero frequency. The most remarkable point of the AR-MA estimate is that it is very smooth and does not suggest the existence of any significant line-like peak. This is quite contrary to the observation made by Granger and Hughes (1971). To check the reliability of the AR-MA estimate, Monte Carlo experiments were performed assuming the estimate to be correct. One of the results of four experiments is illustrated in Figure 5 along with the original estimates. The reproduction of the behaviors of the windowed and AR estimates seems very good and suggests the validity of the AR-MA estimate.

It seems that the enormous potential of the application of multivariate AR models for the analysis of feedback systems and the implementation of computer control of complex industrial processes is not fully recognized yet. The earliest example of such an application is discussed in Ootomo, Nakagawa, and Akaike (1972). The listings of the necessary computer programs are contained in a book by Akaike and Nakagawa (1972) and in the program package called

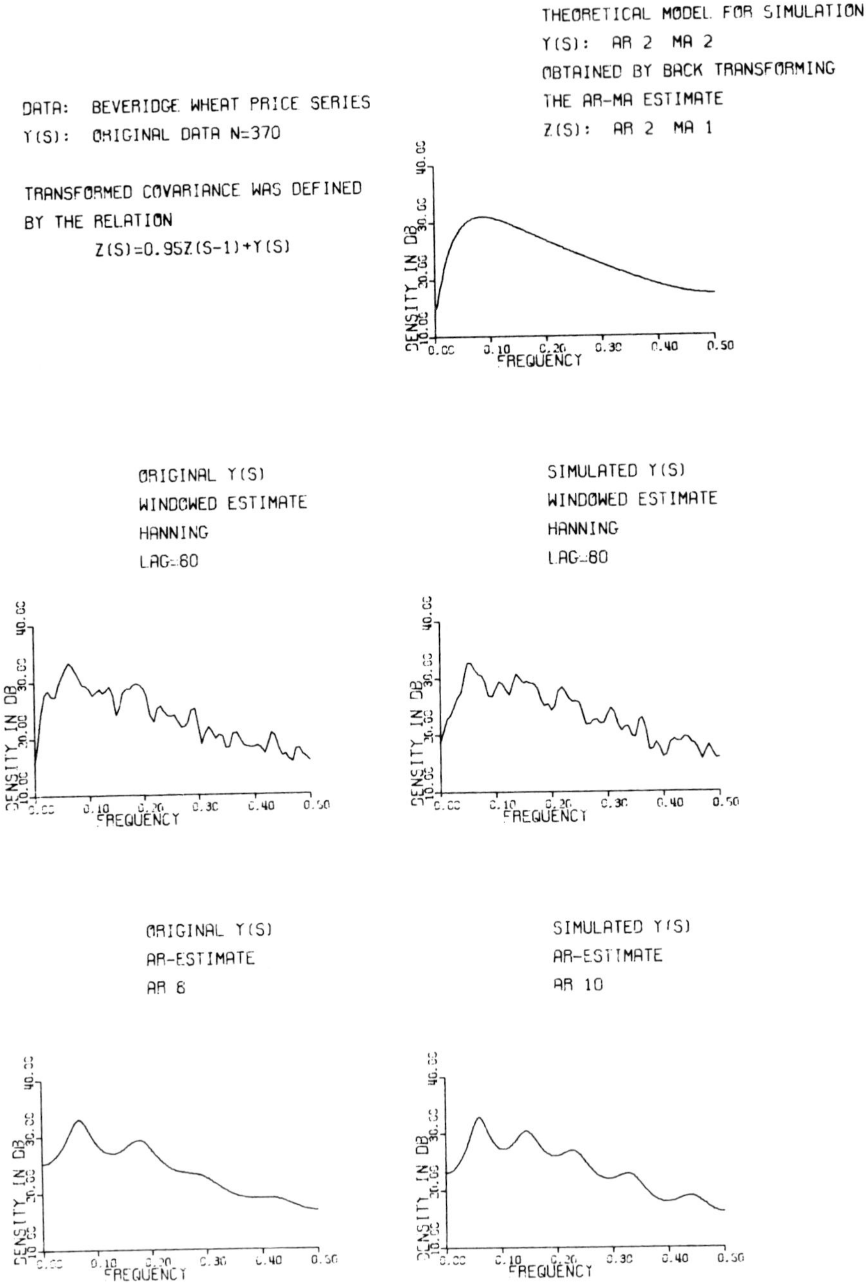

Fig. 5. Power spectrum analysis of the Beveridge wheat price index series.

TIMSAC (Time Series Analysis and Control Program Package).

Just to give a feeling of how the procedure works, an example is illustrated in Figures 6 and 7. The results are based on the recorded operation of an electric power plant simulator. In Figure 6, we have a part of the Monte Carlo reproduction of the behavior of the simulator based on the model obtained by fitting a five-dimensional AR model to an operation record with data length N = 1160. Using the AR model, we can design a control system by following the procedure described in Ootomo, Nakagawa, and Akaike (1972). TIMSAC provides the necessary computing programs. The result of a simulation of implementing a computer control is illustrated in Figure 7. Significant reduction of the main steam pressure (MSP) and the main steam temperature (MST) fluctuation is visible. It has been confirmed that similar results were in fact obtained by a real implementation of the computer control of the simulator. This clearly suggests the feasibility of this type of control of a real power plant. An intensive study of this subject is being conducted by H. Nakamura of Kyushu Electric Company, Fukuoka, Japan. Another interesting example of an application to the control of ships is discussed by Otsu, Horigome, and Kitagawa (1976).

The difficulty in the analysis of multivariate time series is often due to the existence of feedback within the variables. In a lucky situation where the covariance matrix of the white noise, or the innovation, X_n is diagonal, we can decompose the power spectrum of each component variable of Y_n into the components respectively related to the components of the innovation X_n. The results provide a clue to the understanding of the generating mechanism of the statistical fluctuations of Y_n. A very successful application of this technique to the analysis of a nuclear power plant is reported by Fukunishi (1976).

The same technique has been applied to the analysis of the relation between the distribution of the atmosphere and the Chandler wobble by Ooe (1976). The results obtained by Ooe are in good agreement with the expectations based on the understanding of the physical structure of the process.

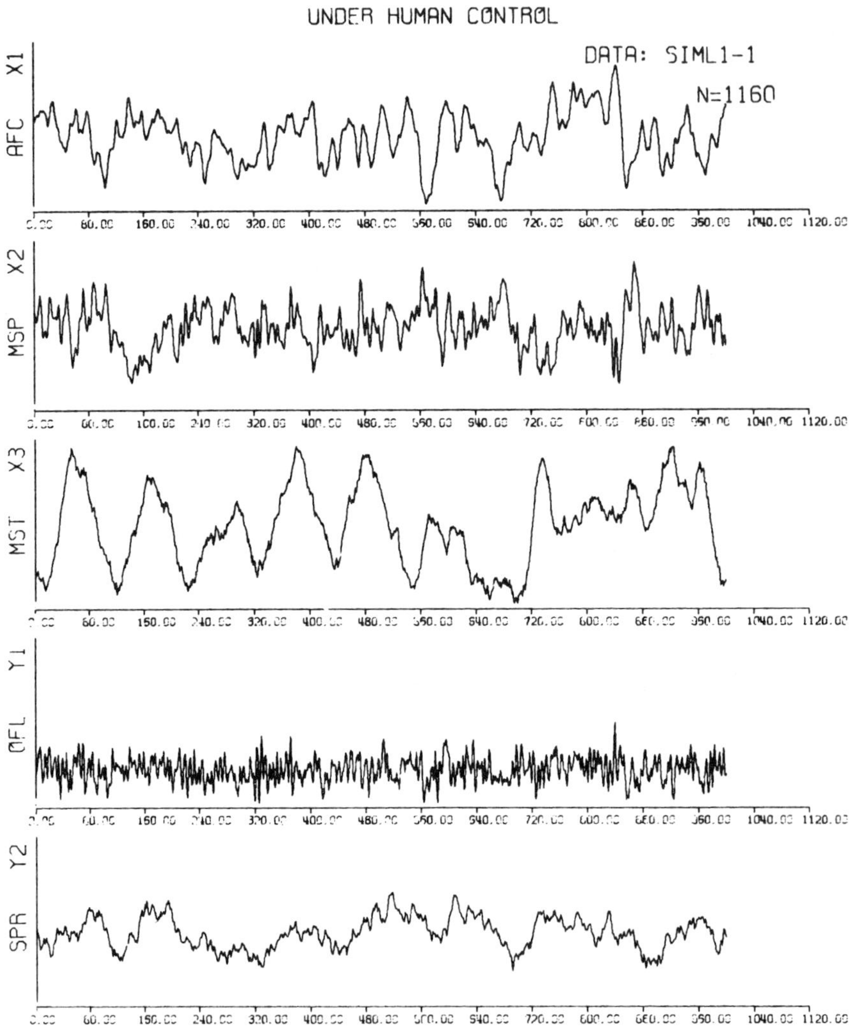

Fig. 6. Simulated behavior of an electric power plant simulator under human control.

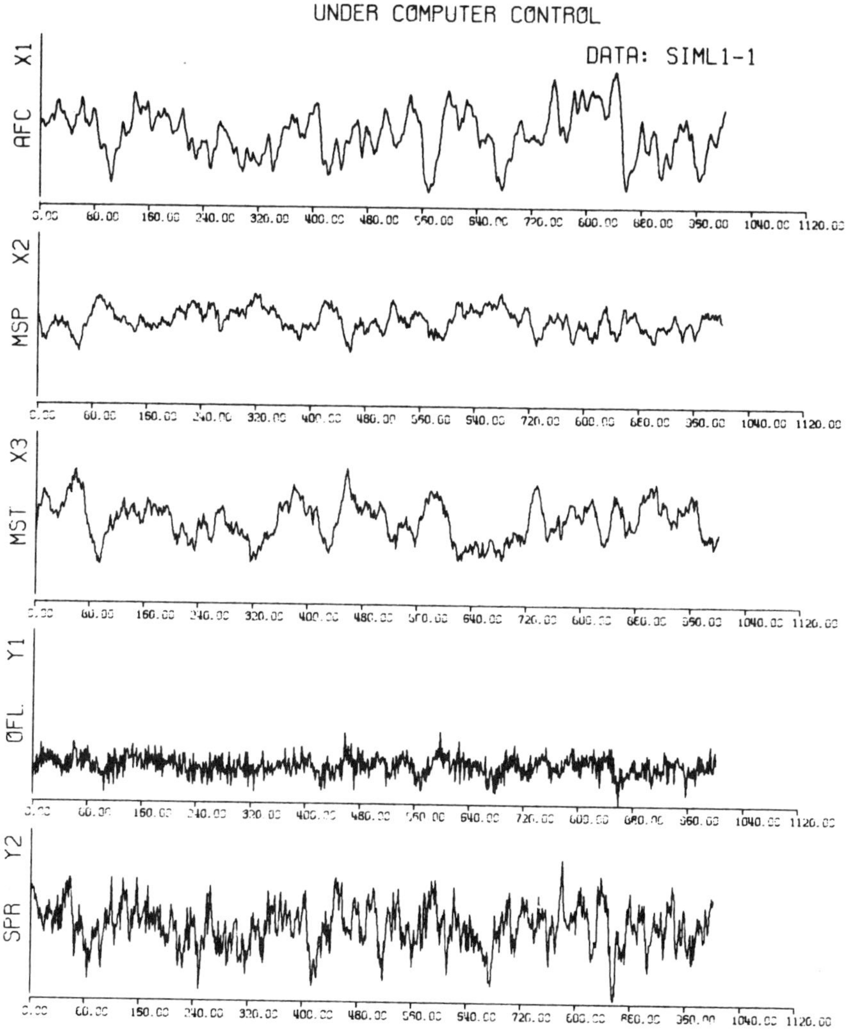

Fig. 7. Simulated behavior of an electric power plant simulator under computer control.

6. Further Topics on AIC.

The statistic AIC provides much insight for the formulation of parametric models. Its applications to other than time series situations are discussed by Akaike (1977). One important area of its application in time series analysis is the analysis of nonstationarity. A practical procedure of fitting of locally stationary AR models is discussed by Ozaki and Tong (1975) and the listing of a computer program is given in Akaike, Arahata, and Ozaki (1976). This program automatically resets the estimate of the power spectrum when a significant change is detected, otherwise it continues to fit an AR model to the span of data which begins with the time of the previous resetting.

Although this is not directly connected with the ordinary time series analysis, AIC has an interesting and perhaps very important application in the field of statistical regularization of ill-posed problems. This problem is closely connected with the subject of signal recovery from a noisy observation. The possibility of application of MAICE to this problem has been intensively investigated by K. Tanabe of the Institute of Statistical Mathematics, and a part of the results is reported in Tanabe (1975).

The statistic AIC has a clearly defined meaning as an estimate of minus twice the expected log likelihood. Yet, the procedure of choosing a model with the minimum value of AIC does not have a clearly defined optimal property. Especially when it is applied to the determination of the order of an AR model, the order chosen by the minimum AIC procedure does not produce a consistent estimate. Results of some Monte Carlo experiments are reported by Jones (1975). In spite of this inconsistency, the corresponding estimate of the spectrum is consistent. This aspect is discussed by Shibata (1976).

A modification of AIC for the fitting of AR or AR-MA models to univariate time series is given by

$$\mathrm{BIC} = (N - k) \log_e \left(\frac{S(k)}{N - k} \right) + k \log_e \left(\frac{S(0) - S(k)}{k} \right),$$

where S(k) denotes N (= data length) times the maximum likelihood estimate, under the Gaussian assumption, of the variance of the innovation X_n and k is the number of the coefficients (= L + M). By minimizing BIC, we get consistent estimates of the orders. When the minimum BIC procedure is applied to a situation where in fact an infinite order model is required, the procedure tends to produce an oversmoothed spectrum estimate compared with the minimum AIC procedure. A reasonable interpretation is that the minimum AIC produces an upper limit of the order of a model to be fitted to a set of data and that the minimum BIC procedure produces more consistent results in repeated applications. The estimate which is obtained by minimizing BIC will be called the minimum BIC estimate (MBICE). Table 1 shows the results of a Monte Carlo experiment for the comparison of MAICE and MBICE. The basic model was the third-order AR process defined by

$$Y_n - 0.56163\, Y_{n-1} + 0.57986\, Y_{n-2} - 0.92705\, Y_{n-3} = X_n .$$

The frequency distributions of the orders determined by the two procedures illustrated in Table 1 clearly show the difference between the two. Table 2 shows the frequency distributions when the procedures are applied to the fitting of AR models to data generated by an AR-MA model with the AR order 5 and the MA order 4 which requires an infinite order AR model for its representation. Figure 8 shows the results obtained by fitting AR and AR-MA models to the record of the x-coordinate of the pole of the earth. In the AR-MA spectrum of the MBICE, the sharp peak at the frequency of about 0.16 cycle per month is ignored. Comparing the present results with those of Figure 1, it is almost clear that this second peak should not be ignored. If the results obtained by various procedures are in good agreement, we can be confident in using the MBICE. Otherwise, we must be careful not to ignore some of the possibly meaningful details of the spectrum. The derivation of the statistic BIC is discussed in Akaike (1977a).

TABLE 1

Order Determination by MAICE and MBICE

(frequencies of orders selected in 100 trials)

Order	2	3*	4	5	6	7	8	9	10	11
MAICE	0	63	22	4	6	1	1	1	2	0
MBICE	0	95	5	0	0	0	0	0	0	0

*True Order of the AR Model.

TABLE 2

Order Determination by MAICE and MBICE

(frequencies of orders selected in 100 trials)

Order	5	6	7	8	9	10	11	12	13	14	Above 14*
MAICE	0	27	7	0	2	17	16	8	3	5	15
MBICE	0	98	2	0	0	0	0	0	0	0	0

*The true order of the model is infinite.

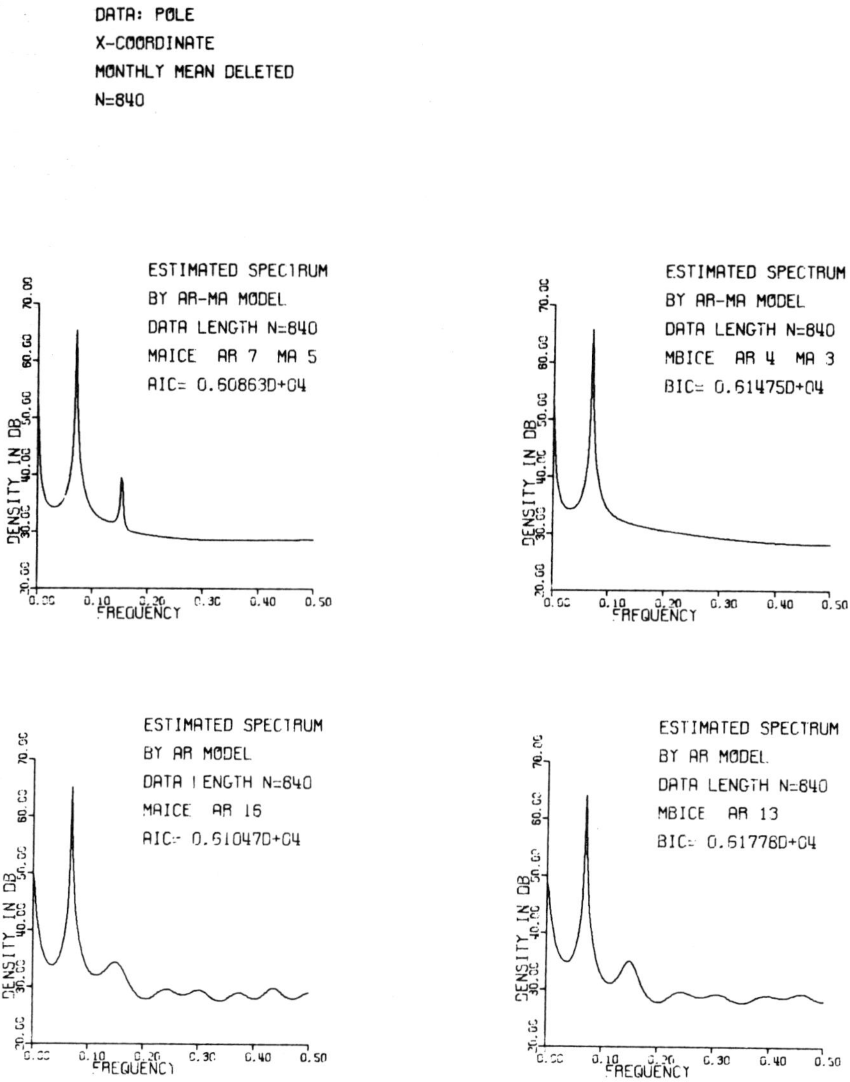

Fig. 8. Comparison of MAICE and MBICE.

Acknowledgment.

The preparation of this report was facilitated by the Vinton Hayes Serior Fellowship at the Division of Engineering and Applied Physics, Harvard University.

References.

Akaike, H. (1970a) On a semiautomatic power spectrum estimation procedure, *Proc. 3rd Hawaii Int. Conf. System Sciences*, 974-977.

Akaike, H. (1970b) Statistical predictor identification, *Ann. Inst. Statist. Math. 22*, 203-217.

Akaike, H. (1971) Autoregressive model fitting for control, *Ann. Inst. Statist. Math. 23*, 163-180.

Akaike, H. (1973) Information theory and an extension of the maximum likelihood principle, *2nd International Symposium on Information Theory*, B. N. Petrov and F. Csaki (eds.), Akademiai Kiado, Budapest, 267-281.

Akaike, H. (1974a) Markovian representation of stochastic processes and its application to the analysis of autoregressive moving average processes, *Ann. Inst. Statist. Math. 26*, 363-387.

Akaike, H. (1974b) Stochastic theory of minimal realization, *IEEE Trans. Automatic Control AC-19*, 667-674.

Akaike, H. (1974c) A new look at the statistical model identification, *IEEE Trans. Automatic Control AC-19*, 716-723.

Akaike, H. (1976) Canonical correlation analysis of time series and the use of an information criterion, *System Identification: Advances and Case Studies*, R. K. Mehra and D. G. Lainiotis (eds.), Academic Press, New York.

Akaike, H. (1977a) On entropy maximization principle, Proceedings of the Symposium on Applications of Statistics, Wright State University, Dayton, Ohio, June 14-18, 1976, to be published.

Akaike, H. (1977b) Spectrum estimation through parametric model fitting, Proceedings of the IUTAM Symposium on Stochastic Problems in Dynamics, Southampton, July 19-23, 1976, to be published.

Akaike, H., Arahata, E., and Ozaki, T. (1975) TIMSAC-74, a time series analysis and control program package (1), *Computer Science Monographs*, No. 5, The Institute of Statistical Mathematics.

Akaike, H., Arahata, E., and Ozaki, T. (1976) TIMSAC-74, a time series analysis and control program package (2), *Computer Science Monographs*, No. 6, The Institute of Statistical Mathematics.

Akaike, H., and Nakagawa, T. (1972) *Statistical Analysis and Control of Dynamic Systems*, Saien su-sha, Tokyo (In Japanese, with the list of TIMSAC program package in a Fortran IV type language.).

Anderson, T. W. (1971) *The Statistical Analysis of Time Series*, Wiley, New York.

Fukunishi, K. (1976) Diagnostic analyses of a nuclear power plant using multivariate autoregressive processes, *Nuclear Science and Engineering*, to appear.

Granger, C. W. J., and Hughes, A. O. (1971) A new look at some old data: the Beveridge wheat price series, *J. Royal Statist. Soc., Ser. A 134*, 413-428.

Jones, R. H. (1975) Fitting autoregression, *J. Amer. Statist. Assoc. 70*, 590-592.

Ooe, M. (1976) Role of the atmosphere as an excitation source of the Chandler wobble, submitted to *Publ. Astron. Soc. Japan.*

Otomo, T., Nakagawa, T., and Akaike, H. (1972) Statistical approach to computer control of cement rotary kilns, *Automatica 8*, 35-48.

Otsu, K., Horigome, M., and Kitagawa, G. (1976) On the prediction and stochastic control of ship's motion, *Ship Operation Automation II*, M. Pitkin, J. J. Roche, and T. J. Williams (eds.), North-Holland, Amsterdam.

Ozaki, T., and Tong, H. (1975) On the fitting of non-stationary autoregressive models in time series analysis, *Proc. 8th Hawaii Int. Conf. System Sciences*, 225-226.

Shibata, R. (1976) Selection of the order of an autoregressive model by Akaike's information criterion, *Biometrika 63*, 117-126.

Tanabe, K. (1976) Statistical approach to ill-posed problems, *Suri Kagaku (Mathematical Sciences)*, No. 153, 60-64 (In Japanese).

NON-LINEAR TIME SERIES MODELLING

C. W. J. Granger and A. Andersen

1. Non-linear Models.

Single-series time-series models have been analyzed and used for a very long time and it is hardly necessary to defend further consideration of them. Their use in providing forecasts for comparison with alternative techniques is justification enough if any is needed. The traditional search has been for a class of models that can be analyzed with reasonable ease yet are sufficiently general to be able to well approximate most series that arise in the real world. The class of models considered by Box and Jenkins (1970) known as the integrated mixed autoregressive, moving average or ARIMA models, have this property and have been used with considerable success in many scientific fields, including economic forecasting. It is natural to ask if there exists a more general class of models that will provide a yet better fit to reality. The obvious directions to go are to consider either models with time-varying parameters or non-linear models. This investigation is more concerned with the second of these possibilities.

When considering single-series non-linear models, it seems obvious to start with forms such as

$$x_t = h(x_{t-1}) + \varepsilon_t \tag{1.1}$$

where h( ) is some function, and ε_t is a white noise series. Wiener (1958), for example, suggests that the function h( ) should be a polynomial. However, even simple models such as

$$x_t = ax_{t-1}^2 + \varepsilon_t \tag{1.2}$$

are almost invariably highly explosive unless a is very small and the distribution of ε_t is truncated and very limited in extent. As actual economic series are not generally explosive,

ISBN 0-12-257250-5

such models seem to be of limited value. If in (1.1) the function h is chosen so that

$$|h(x)| < |x|$$

then stationary series can occur, but this seems to be a restriction that is not easily explained or readily interpreted in economic terms. For example, if h(x) = sine x, then a stationary model is achieved but it has very little intuitive appeal. Thus, models of form (1.1) and the obvious generalizations do not appear to be a profitable class to consider.

Nelson and Van Ness (1973) have considered a somewhat related class of models of the form

$$x_t = \alpha y_{t-1} + \beta y_{t-1}^2 + \gamma y_{t-2} + \delta y_{t-2}^2 + \dots + \varepsilon_t \qquad (1.3)$$

where y_t is a "driving" or "causal" stationary series generated, say, by

$$y_t = \sum_{j=1}^{p} a_j y_{t-j} + \eta_t \ . \qquad (1.4)$$

Thus, y_t is generated by a linear model and x_t by a non-linear function of past y's. However, the method of analysis proposed by Nelson and Van Ness specifically excludes the use of non-linear terms of past x's on the right-hand side of (1.3) or of linear terms of past x's on the right-hand side of (1.4). These limitations again seem to be unnecessarily restrictive in the eyes of a time-series analyst, who generally expects to start explaining x_t from past x_t, so this class of models does not seem to be the natural extension of the Box-Jenkins class.

A further related class of models can be called generalized moving average models, an example being

$$x_t = \varepsilon_t + \alpha\varepsilon_{t-1} + \beta\varepsilon_{t-1}^2 + \gamma\varepsilon_{t-2}\varepsilon_{t-1} + \dots \qquad (1.5)$$

where ε_t is a white noise series. Although similar in form to the models just considered, these are very different when analysis is attempted, since the y_t's are taken to be observed in (1.3) but the ε_t's in (1.5) are not directly observed. To estimate ε_t even with the α, β, γ, etc., given, one would have to invert (1.5), that is solve ε_t in terms of past and present

x_t. Although not necessarily impossible, such invertibility will be difficult to establish. We delay an exact definition of the invertibility of a non-linear model until a later paper. An appropriate definition seems to be: if a model is completely known and estimates $\hat{\varepsilon}_t$ of the residuals ε_t are obtained by using a set of assumed starting values and solving the iterative equation, then the model is said to be invertible if

$$\text{plim } \hat{\varepsilon}_t = \varepsilon_t$$

whatever (sensible) set of starting values are used. If models such as (1.5) are invertible, then the series generated may have interesting properties. For example, suppose x_t is formed by

$$x_t = \varepsilon_t + \beta\varepsilon_{t-1}^2$$

where ε_t is a Gaussian zero-mean white noise, then

$$E(x_t) = 0$$

and

$$\text{corr}(x_t, x_{t-\tau}) = 0 \qquad \text{all } \tau \neq 0,$$

so that x_t will appear to be white noise if just covariances are considered even though it is _potentially_ forecastable from its own past, although in a non-linear fashion. However, it may be shown that this model is never invertible, for any non-zero value of β, and so in practice it cannot be used for forecasting.

One further class of non-linear models that have been considered, by Granger and Newbold (1976a) for example, involves taking an instantaneous transform $y_t = g(x_t)$ of the original data x_t and then building a linear model on the y_t's. Again, this is a very restrictive class of non-linear models and there is considerable difficulty in the proper choice of the function g( ). These models are further considered in the next section.

The other class of models considered in this paper in any detail is one that we believe might be sufficiently non-linear to pick up part of any real non-linearity in the system but are still in a form for which analysis is possible. These bilinear models are linear both in the x_t's and in the innovation series ε_t, an example being

$$x_t = \beta x_{t-1}\varepsilon_{t-1} + \varepsilon_t .$$

There has been some interest in such models, and their continuous time counterparts, by control engineers in recent years; a good survey of the results achieved in the field is given by Bruni, Di Pillo, and Koch (1974). However, we know of no previous work discussing the statistical or time series aspects of these models.

2. Instantaneous Data Transformations.

It has become common practice amongst econometricians to build models not on their raw data, X_t, but rather on instantaneously transformed data $Y_t = T(X_t)$. By far the most usual transformation is the logarithmic although there has recently been considerable interest in using the transformation suggested by Box and Cox,

$$Y_t = \frac{(X_t + m)^{\theta} - 1}{\theta} . \tag{2.1}$$

Here m is a constant chosen to ensure that $X_t + m$ is positive and θ is the main transformation parameter. The main advantage of the Box-Cox transformations is that they include the linear ($\theta = 1$) and logarithmic ($\theta = 0$) as special cases. The usual stated reason for making such transformations is to produce models with homogeneous error terms. In the single series forecasting context, the question that has to be asked is - are superior forecasts produced for the original series X_t from model building on the Y_t's. It has been suggested that superior forecasts do frequently occur.

For whatever reason, instantaneous transformations are often used in practice, and it is important to study the time-series properties of the resulting series and also the question of how to forecast X_{n+h} given that one only has available a model for $Y_t = T(X_t)$, where T( ) is the transformation used. These questions have been considered in some detail by Granger and Newbold (1976a) and so only the main results will be presented here, largely without proof.

Let X_t be a stationary, Gaussian series with mean μ, variance σ^2 and autocorrelation sequence

$$\operatorname{corr}(X_t, X_{t-\tau}) = \rho_\tau .$$

Set

$$Z_t = \frac{X_t - \mu}{\sigma}$$

so that Z_t, $Z_{t-\tau}$ will be jointly distributed as bivariate Normal with zero means, unit variance and correlation ρ_τ. Consider an instantaneous transformation of the form

$$Y_t = T(Z_t)$$

where T( ) can be expanded in terms of Hermite polynomials in the form:

$$T(Z) = \sum_{j=0}^{m} \alpha_j H_j(Z) \tag{2.2}$$

and where m can be infinite. The jth Hermite polynomial $H_j(Z)$ is a polynomial in Z of order j with, for example, $H_0(Z) = 1$, $H_1(Z) = Z$, $H_2(Z) = Z^2 - 1$, $H_3(Z) = Z^3 - 3Z$, and so forth. If X and Y are Normally distributed random variables with zero means, unit variances and correlation ρ, these polynomials have the important orthogonal properties

$$\left.\begin{aligned} E[H_n(X)H_k(X)] &= 0 \qquad n \neq k \\ &= n! \qquad n = k \end{aligned}\right\} \tag{2.3}$$

and

$$\left.\begin{aligned} E[H_n(X)H_k(Y)] &= 0 \qquad n \neq k \\ &= \rho^n n! \qquad n = k \end{aligned}\right\} \tag{2.4}$$

Using these properties, it is easy to show that $E(Y_t) = \alpha_0$ and

$$\operatorname{cov}(Y_t, Y_{t-\tau}) = \sum_{j=1}^{m} \alpha_j^2 j! \rho_\tau^j .$$

Thus, the linear properties of the transformed series can be determined. It follows, for example, that if

$$Y_t = a + bX_t + cX_t^2 \tag{2.5}$$

then

$$\mathrm{cov}(Y_t, Y_{t-\tau}) = (b + 2c\mu)^2\sigma^2\rho_\tau + 2c^2\sigma^2\rho_\tau^2$$

and if

$$Y_t = \exp(X_t)$$

then

$$\mathrm{cov}(Y_t, Y_{t-\tau}) = \exp(2\mu + \sigma^2)(\exp(\sigma^2\rho_\tau) - 1) .$$

Further, if X_t is MA(q), Y_t has autocovariances appropriate for MA(q), but if X_t is AR(p) and Y_t is given by (2.5), then its autocovariances correspond to a mixed ARMA (1/2 p(p + 3), 1/2 p(p + 1)) process.

It may also be shown that if X_t is an integrated ARIMA (p, 1, q) process then Y_t will also be such that its first differences can be modelled as an ARMA process.

If X_t is Gaussian, or is at least assumed to have this property, and a model has been formed for it, then standing at time n an optimum forecast, $f_{n,h}$ of X_{n+h} can be easily formed by classical methods as have been discussed in Chapters 4 and 5 of Granger and Newbold (1976b), and it will be linear in X_{n-j}, $j \geq 0$. Suppose, however, that X_{n+h} is not the quantity for which a forecast is required, but rather

$$Y_{n+h} = T\left(\frac{X_{n+h} - \mu}{\sigma}\right)$$

needs to be forecast. An example would be an economist who has built a model for log price but actually wants to forecast-price. The optimum forecast of Y_{n+h} under a least-squares criterion is given by

$$g_{n,h}^{(1)} = E[Y_{n+h} | I_n]$$

where I_n is the information set X_{n-j}, $j \geq 0$. As Y_{n+h} is a function of a Gaussian variable, it can be shown that this optimum, generally non-linear, forecast is given by

$$g_{n,h}^{(1)} = \sum_{j=0}^{m} \alpha_j A^j H_j(P)$$

where

$$A = (1 - S^2(h)/\sigma^2)^{1/2} \quad ; \quad P = (f_{n,h} - \mu)/(\sigma^2 - S^2(h))^{1/2} \quad ,$$

$f_{n,h}$ is the optimum forecast of X_{n+h} and $S^2(h)$ is the variance of $(X_{n+h} - f_{n,h})$.

Define as a measure of forecastability of a series Y_t, having finite variance, using a given information set, the quantity

$$R_{h,y}^2 = \frac{\operatorname{var}(f_{n,h}^{(y)})}{\operatorname{var}(Y_{n+h})}$$

where $f_{n,h}^{(y)}$ is the optimum forecast of Y_{n+h}, possibly a non-linear forecast, based on the information set and using a least-squares criterion. Clearly, the nearer this quantity is to unity, the more forecastable is the series. It may be shown that any non-linear instantaneous transformation of a Gaussian process X_t is always less forecastable than X_t, provided X_t is not simply a white noise series.

As an example, suppose that X_t has $\mu = 0$ and $\sigma = 1$ and is Gaussian, but that a forecast is required of

$$Y_{n+h} = \exp(X_{n+h}) \quad .$$

This would correspond to the example given earlier where X_t is log price and Y_t is price. Let, as before, $f_{n,h}^{(x)}$ be the optimum forecast of X_{n+h} using $I_n: X_{n-j}$, $j \geq 0$ and $S^2(h)$ be the variance of the h-step forecast error of X_{n+h}. The optimum forecast of Y_{n+h} is then given by

$$g_{n,h}^{(1)} = \exp(f_{n,h}^{(x)} + 1/2\ S^2(h)) \quad .$$

The usual naive forecast is

$$g_{n,h}^{(2)} = \exp(f_{n,h}^{(x)})$$

which is seen to be biased and, naturally, to lead to a higher expected squared error. In this case, it is fairly easily shown that

$$R^2_{h,y} = e^{-S^2(h)}$$

and of course

$$R^2_{h,x} = 1 - S^2(h) \ .$$

At first sight, these results seem both very general and potentially very useful. Provided an instantaneous transformation can be found that turns the original series X_t into a Gaussian series Y_t, the usual model building techniques can be applied to Y_t, forecasts derived for Y_t and thus optimum forecasts for X_t. The forecasts for X_t so derived might well be expected to be better than those from a linear model fitted directly to the X_t series. The main problems with this approach are the choice of an appropriate transformation and the heavy reliance placed on the normality of the transformed series. Some unpublished results by H. Nelson (1976) at the University of California, San Diego, have suggested that the Box-Cox transformation, with the optimal value of θ, by no means achieves normality and does not generally produce superior forecasts of the original series compared to using a linear model.

It is possible that a different class of transformation would produce better results, but there is no inherent reason why the type of non-linearity that occurs in the real world can be captured by using linear models of transformed series. We believe that the bilinear class of models, to be discussed in the remainder of this paper, has greater potential.

3. Introduction to Bilinear Models.

As was mentioned in the first section, the class of bilinear models has some features which are interesting in themselves and which also may be useful to the time series analyst. The general Bilinear Autoregressive Moving Average model of order (p,q,Q,P), BARMA(p,q,Q,P), is of the form

$$X_t = \sum_{j=1}^{p} \alpha_j X_{t-j} + \sum_{i=0}^{q} \theta_i \varepsilon_{t-i} + \sum_{k=0}^{Q} \sum_{\ell=1}^{P} \beta_{k\ell} \varepsilon_{t-k} X_{t-\ell} , \tag{3.1}$$

where $\theta_0 = 1$.

It may be noted that the model (3.1) is a simplification of the general non-linear model, with the property of linearity in the X variables and the ε variables separately, but not in both. At this point, it is convenient to give names to several subclasses of the general model.

Firstly, if $p = 0$, the model is said to be homogeneous in the output, which if $q = 0$, one has homogeneity in the input. The models to be considered in this paper are homogeneous in both variables. That is, (3.1) reduces to

$$X_t = \sum_{k=0}^{Q} \sum_{\ell=1}^{P} \beta_{k\ell} \varepsilon_{t-k} X_{t-\ell} + \varepsilon_t . \tag{3.2}$$

Let B be the $((Q + 1) \times p)$ matrix of the β coefficients. If $\beta_{ij} = 0$ for all $j < i$, the model is called superdiagonal. Here, the multiplicative terms with non-zero coefficients are such that the input factor ε_{t-i} occurs after the output factor X_{t-j}, so that these two terms are independent. Correspondingly, the model is subdiagonal if $\beta_{ij} = 0$ for all $j \geq i$ and is said to be diagonal if $\beta_{ij} = 0$ for all $i \neq j$. The superdiagonal models are the easiest to analyze but are not necessarily the most realistic. The restriction $\beta_{0j} = 0$ for all j gives models of particular interest, as they can always be solved in the form

$$X_t = \varepsilon_t + H(\varepsilon_{t-1}, \varepsilon_{t-2}, \ldots)$$

where H( ) is some non-linear function. This means that if the model is invertible, forecasting is particularly easy.

Before proceeding to analyze some special cases, it is interesting to ask if bilinear models can arise from the linear models with which econometricians and time series analysts are more familiar. To a certain extent they can, as the following examples show:

(a) Suppose that a true rate of return for a series X_t is generated by an MA(1) model, so that

$$\frac{X_t - X_{t-1}}{X_{t-1}} = \varepsilon_t + b\varepsilon_{t-1}$$

then, this may be written

$$X_t = X_{t-1} + \varepsilon_t X_{t-1} + b\varepsilon_{t-1} X_{t-1}$$

which is already in a bilinear model form. If now one cannot observe X_t directly but only with additive noise, so that the available data are

$$Y_t = X_t + \eta_t ,$$

then a non-homogeneous bilinear model for Y_t becomes appropriate and analysis of the rate of return series for Y_t may be less appropriate.

(b) If a pair of series is related by the model

$$X_t = a + bY_t + cX_{t-1} + \varepsilon_t$$

$$\log Y_t = \alpha \log Y_{t-1} + \log \eta_t$$

so that one equation is linear, and the second is log-linear, then if one only has data for X_t, this will appear to be generated by

$$(X_t - a - cX_{t-1} - \varepsilon_t) = (X_{t-1} - a - cX_{t-2} - \varepsilon_{t-1})^{\alpha} b^{1-\alpha} \eta_t .$$

Approximating the term on the right-hand side that is raised to the power α by the first few terms (linear and cross-product) of a Taylor expansion will again produce a model of a bilinear type, or at least one that will analyze like a bilinear model rather than a linear one.

(c) If a series X_t is generated by

$$X_t = \text{expectation of } X_t \text{ made at time } t\text{-}1 + \varepsilon_t$$

and if these expectations are a function of the most recent data available at time t-1, that is X_{t-1} and ε_{t-1}, the previous expectation error, of the form

$$\text{expectation} = g(X_{t-1}, \varepsilon_{t-1}) ,$$

there is no overpowering reason to suppose that this function will be linear. One way of picking up at least part of this non-linearity is to use the approximation

$$g(X, \varepsilon) = AX + CX\varepsilon + D\varepsilon$$

which then gives a bilinear model for the series X_t.

A number of examples have been given by Mohler (1973) of how bilinear models, often multivariate, can arise in a variety of fields, but particularly in a control theory context. The continuous time version of bilinear models arise from consideration of the stochastic difference equations, see for example Brockett (1975).

However, these examples of how bilinear models can arise are not necessarily completely convincing to time-series analysts. Our major reasons for considering these models are largely pragmatic; they do offer a class of models that are potentially capable of analysis, that may pick up part of any non-linearity in the data and thus could suggest improved methods of forecasting and also indicate model mis-specification if a linear model has been used. As will be shown in the following sections, bilinear models have interesting properties which can match known properties of real data. This is true both of the distributional properties of the data generated by these models and also the covariance properties. The present techniques of time series analysis concentrate on interpretation of the autocovariance sequence, which we shall term COVA. It is shown in the following sections that such analysis can be misleading, for example, a series that is white noise according to COVA may well be forecastable from its own past, although obviously not in a linear fashion. The fact that such possibilities exist is clearly sufficient reason for bilinear models to be worth considering.

4. Particular Case 1: A Diagonal Model.

The model discussed in this section is the diagonal BARMA(0,0,1,1) model, with $\beta_{01} = 0$ and $\{\varepsilon_t\}$ a sequence of independent, identically distributed Normal $(0, \sigma^2)$ random variables. The model chosen has a particularly simple form and some interesting properties. The assumption of normality in the inputs is vital for some results, and may be ignored for others, although the assumption of independence for the ε_t sequence is always required. It will be stated whenever the normality assumption may be ignored. The distribution of ε_t will always be taken to be symmetric.

The model is written

$$X_t = \beta_{11} X_{t-1} \varepsilon_{t-1} + \varepsilon_t \; . \tag{4.1}$$

In what follows it is convenient to use the notation $\lambda = \beta_{11}\sigma$.

We have given the details with proofs of the properties of this model elsewhere (Andersen and Granger (1976)) and so here only list the most interesting of these properties.

(1) If $|\lambda| < 1$, the process is stationary.

(2) The autocorrelation structure of the process is the same as for a moving average of order one, so that $\rho_k \equiv \text{corr}(X_t, X_{t-k}) = 0$, $k > 1$.

(3) ρ_1 increases with $\lambda > 0$ up to a maximum value of 0.155 at $\lambda = 0.6$ but then decreases.

(4) The model is certainly invertible up to $|\lambda| < 0.6$ and is clearly not invertible beyond $|\lambda| > 0.707$, but the issue is less clear-cut for intermediate values of λ.

(5) The distribution of X_t is generally skewed and is non-normal. Its third moment increases with $\lambda > 0$ taking a maximum at $\lambda = 0.8$ and decreasing thereafter. The fourth moment does not exist for $\lambda > 0.75$ and for any non-zero λ some higher moments do not exist.

The complicated fashion in which the properties of this bilinear model varies with λ contrasts sharply with the simple properties of MA(1) and AR(1) models.

If data are actually generated by such a model, they will be identified by ordinary linear, COVA analysis as MA(1) but if it is then forecast using this MA(1) model rather than by the true model, considerable forecasting loss results. For example, if $\text{var}(\varepsilon_t) = 1$ and $\lambda = \beta = 0.55$, the error variance from the MA(1) model is nearly double that from using the true model. However, it should be asked how one could identify if one's data actually do come from an MA(1) model or the suggested bilinear model. One way of doing this is to examine the autocorrelations of X_t^2, as using the results of the previous section it can be shown that if X_t is MA(1), then so will be X_t^2, but it can also be shown that if X_t is bilinear of form (4.1), then the autocorrelations of X_t^2 are the same as for an ARMA(1, 1) model. Thus, identification is possible, in principle.

5. Particular Case 2: White Noise Models.

Now consider the model

$$X_t = \alpha X_{t-2}\varepsilon_{t-1} + \varepsilon_t \tag{5.1}$$

where ε_t is Gaussian, zero-mean white noise. This is a particular example of a superdiagonal model. The process is stationary if $|\alpha\sigma| < 1$ where $\sigma^2 = \text{var}(\varepsilon_t)$ and has the autocorrelation structure of a white noise process, so that

$$\text{corr}(X_t\ , X_{t-k}) = 0\ , \qquad \text{all } k \neq 0\ .$$

It is thus possible that a COVA analysis has suggested that some series, such as the residuals of a Box-Jenkins model-fitting procedure, are white noise, but that in reality they are generated by a model such as (5.1). This has very clear consequences for forecasting. The "white noise" property suggests that the series cannot be forecast from its own past, whereas if actually given by (5.1) it can be so forecast, although in a non-linear fashion. Identification between the white noise and bilinear models can again be achieved by looking at the autocorrelations of X_t^2, which will be as for a white noise process if X_t is really white noise but as for an ARMA(2, 1) model if X_t is generated by (5.1).

We have found by experience with both real and simulated data that there is no particular difficulty in estimating the parameter of models such as (5.1), using a non-linear, search procedure.

As an example of this work, a bilinear model similar to (5.1) was fitted to the residuals obtained from a linear

Box-Jenkins model. The complete details of the identification and estimation are given in Andersen and Granger (1976). The final result was that the variance of the residuals was decreased. More importantly though, the variance of the one step ahead forecast errors, in a period not used for estimation, was reduced by eight percent. This suggests a useful improvement over the usual naive forecasts.

Bibliography

Andersen, A. P. and C. W. J. Granger, (1976), "Introduction to Bilinear Time Series Models," Department of Economics Discussion Paper 76-5, University of California, San Diego.

Box, G. E. P. and G. M. Jenkins, (1970), *Time Series Analysis, Forecasting and Control*, San Francisco: Holden Day.

Brockett, R. W., (1975), "Parametrically Stochastic Linear Differential Equation," Mimeo: Harvard University, Department of Control Engineering.

Bruni, C., G. DiPillo, and G. Koch, (1974), "Bilinear Systems: An Appealing Class of 'Nearly Linear' Systems in Theory and Applications," *IEEE*, *4*, August, 1974.

Granger, C. W. J. and P. Newbold, (1976a), "Forecasting Transformed Series," *J. R. Statist. Soc. B*, *38*.

Granger, C. W. J. and P. Newbold, (1976b), *Forecasting Economic Time Series*, to be published by Academic Press.

Johnson, N. L., E. Nixon, and D. E. Amos, "Table of Percentage Points of Pearson Curves, for given $\sqrt{\beta_1}$ and $\sqrt{\beta_2}$, Expressed in Standard Measure," *Biometrika*, *50*, 3 and 4.

Mohler, R. R., (1973), *Bilinear Control Processes (with Applications to Engineering, Ecology and Medicine)*, Academic Press: New York.

Nelson, J. Z. and J. W. Van Ness, (1973), "Formulation of a Nonlinear Predictor," *Technometrics*, 15, 1-12.

Nelson, H., (1976), unpublished Ph.D. thesis, University of California, San Diego.

Wiener, N., (1958), *Nonlinear Problems in Random Theory*, Technology Press of Massachusetts Institute of Technology, Clapman & Hall.

ON G-SPECTRAL ESTIMATION*

H. L. Gray and A. G. Houston
Southern Methodist University

F. W. Morgan
Clemson University

I. Introduction.

In an unpublished manuscript (Gray, 1974), an estimator referred to as the G-spectral estimator was introduced by H. L. Gray and a few of its properties were developed. Later that estimator was introduced to the published literature in an article by Gray and Foster (1975) although again it was not studied to any extent. The purpose of this paper is to establish in some detail the properties of that estimator and to demonstrate it as a viable new tool for spectral estimation.

Initially, the so-called G_n- and e_n-transformations are defined and some simple examples are given to demonstrate their ability to increase the rate of convergence of certain types of sequences. Following this, theorems are given which establish sufficient conditions for a sequence to be extrapolated from a finite number of terms to its limit via these transformations. The above-described results are kept short so as not to be a deterrent, but are included since an understanding of them will arm the reader with excellent insight regarding the behavior of the G-spectral estimator given later. Adequate references are included for those who are interested in pursuing such ideas in greater depth. Section II is then concluded with Theorem 3 which is one of the more important theorems in the paper. However, there is a danger in interpreting that result too restrictively (see concluding

* This work was supported by ONR Contract #N00014-75-C-0439.

ISBN 0-12-257250-5

remarks) as Examples 1 and 2 are intended to imply. Section III introduces the G-spectral estimator and shows it to be a consistent estimator of the spectrum when the autocorrelation is eventually a linear combination of exponentials with real or complex arguments. In Section IV, it is shown that we can restrict ourselves to a calculation algorithm suggested by Pye and Atchison and retain a certain amount of desirable flexibility. The results of that section are essentially summarized by Theorem 11 which is important for selecting the appropriate G-spectral estimator in practice. For data which is of sufficient length to allow adequate analysis in a single pass by the computer program included in the Appendix 2, Theorem 11 can be ignored. However, the shorter the record, the more important an adequate understanding of Theorem 11 becomes. The remaining part of the paper is dedicated to exemplifying the method suggested by the theoretical structure established to this point. The examples given are far from exhaustive and in fact somewhat incomplete. However, with the program included in Appendix 2, they furnish an adequate basis for the reader to investigate the method in whatever setting seems appropriate. The examples included are outlined below. We use the notation AR(p) for pth order autoregressive, MA(q) for qth order moving average, and ARMA(p,q) for the mixed process.

Outline of Examples.

A. Autocorrelation known over finite record.

Example 4, an AR(1) model ... pg. 61
Example 5, an MA(1) model ... pg. 65
Example 6, an ARMA(2,1) model ... pg. 66
Example 7, an ARMA(1,2) model ... pg. 73

B. Comparison with other methods when the autocorrelation is known over a finite record.

Example 8, an AR(1) model ... pg. 78
Example 9, an MA(1) model ... pg. 78
Example 10, an ARMA(1,1) model ... pg. 80
Example 11, an ARMA(1,2) model ... pg. 82
Example 12, an ARMA(3,2) model ... pg. 82

C. Estimated autocorrelation and comparison with other methods.

Example 13, an AR(1) model ... pg. 85
Example 14, an MA(1) model ... pg. 88
Example 15, an ARMA(1,1) model ... pg. 91
Example 16, an ARMA(1,2) model ... pg. 97
Example 17, an ARMA(3,2) model ... pg. 102
Example 18, a signal plus noise model ... pg. 102
Example 19, a real data example ... pg. 106

II. The G_n- and e_n-Transformations.

Without some knowledge of the so-called G_n-transformation (Gray et al., 1971) and e_n-transformation (Shanks, 1955), a clear understanding of G-spectral estimation seems unlikely and hence we devote this section to those topics.

DEFINITION 1. Let

$$H_{k+1}(z(j);\ a_{ij}) = \begin{vmatrix} z(0) & z(1) & \cdots & z(k) \\ a_{00} & a_{01} & \cdots & a_{0k} \\ \vdots & & & \\ a_{k-1,0} & \cdots & & a_{k-1,k} \end{vmatrix}, \qquad (1)$$

and

$$F(t) = \int_a^t f(x)\,dx. \qquad (2)$$

Then we define the G_n-transformation of $F(t)$ by

$$G_n[F(t);\ h] = \frac{\begin{vmatrix} F(t) & F(t+h) & \cdots & F(t+nh) \\ f(t) & f(t+h) & \cdots & f(t+nh) \\ \vdots & & & \\ f(t+(n-1)h) & \cdots & & f(t+(2n-1)h) \end{vmatrix}}{\begin{vmatrix} 1 & 1 & \cdots & 1 \\ f(t) & f(t+h) & \cdots & f(t+nh) \\ \vdots & & & \\ f(t+(n-1)h) & \cdots & & f(t+(2n-1)h) \end{vmatrix}}$$

$$= \frac{H_{n+1}(F(t+jh);\ f(t+(i+j)h))}{H_{n+1}(1\ ;\ f(t+(i+j)h))}. \qquad (3)$$

If $H_{n+1}(1;\ f(t_0+(i+j)h) = 0$, we define $G_n[F(t_0);\ h]$ as the limit as $t \to t_0$ of $G_n[F(t);\ h]$, if this exists. If $G_n[F(t);\ h] \equiv 0/0$ over some interval I, we define $G_n[F(t);\ h] = G_{n-1}[F(t);\ h]$, the latter being referred to as the *drop back rule* .

DEFINITION 2. Let $k \geq 1$ and $A_k = \sum_{i=0}^{k} a_i$. Then for a fixed positive integer ℓ, the e_n-transformation of A_k is defined by

$$e_n[A_k;\ \ell] = \frac{\begin{vmatrix} A_k & A_{k+\ell} & \cdots & A_{k+n\ell} \\ a_k & a_{k+\ell} & \cdots & a_{k+n\ell} \\ \vdots & & & \\ a_{k+(n-1)\ell} & a_{k+n\ell} & \cdots & a_{k+(2n-1)\ell} \end{vmatrix}}{\begin{vmatrix} 1 & 1 & \cdots & 1 \\ a_k & a_{k+\ell} & \cdots & a_{k+n\ell} \\ \vdots & & & \\ a_{k+(n-1)\ell} & a_{k+n\ell} & \cdots & a_{k+(2n-1)\ell} \end{vmatrix}}$$

$$= \frac{H_{n+1}(A_{k+j\ell}\ ;\ a_{k+(i+j)\ell})}{H_{n+1}(1\ ;\ a_{k+(i+j)\ell})} . \tag{4}$$

If $e_n[A_k;\ \ell] \equiv 0/0$, we define $e_n[A_k;\ \ell] = e_{n-1}[A_k;\ \ell]$. Again this is referred to as the *drop back rule* . For the motivation of (3) and (4), see [23, 24]. It should be remarked that the definition given by Shanks in [53] is actually the case $\ell = 1$ of (4) and that the transformation for general values of ℓ has only previously been studied for $n = 1$. However, for our purposes, it will be useful to allow different values of ℓ. Now (3) and (4) were originally developed for the purpose of increasing the rate of convergence of improper integrals and infinite series, respectively. We demonstrate this by the following well worn but useful examples.

Example 1 (Shanks, 1955). Consider the Leibnitz expansion for π. That is,

$$\pi = 4 - \frac{4}{3} + \frac{4}{5} - \frac{4}{7} + \ldots .$$

From a practical point of view, the above series appears rather worthless since it takes more than 4×10^7 terms to achieve eight figure accuracy. However, it is possible to increase the rate of convergence of this series dramatically as demonstrated by the following table, where e_1^k is to be understood as the kth application of $e_1(A_0; 1)$. It should be noted that only the first 10 terms in the above series are used while the element in the last row and column is correct for π rounded to eight figures.

TABLE 0

Accelerating the Convergence of the Leibnitz Expansion

$\pi = 3.14159265 \ldots$

n	A_n	$e_1(A_n)$	$e_1^2(A_n)$	$e_1^3(A_n)$	$e_1^4(A_n)$
0	4.0000000				
1	2.6666667	3.1666667			
2	3.4666667	3.1333333	3.1421053		
3	2.8952381	3.1452381	3.1414502	3.1415993	
4	3.3396825	3.1396825	3.1416433	3.1415909	3.1415928
5	2.9760462	3.1427129	3.1415713	3.1415933	3.1415927
6	3.2837385	3.1408814	3.1416029	3.1415925	
7	3.0170718	3.1420718	3.1415873		
8	3.2523659	3.1412548			
9	3.0418396				

<u>Example 2</u> (Gray and Atchison, 1967). Let

$$F(t) = \int_0^t \frac{\sin x}{x}\,dx \to \frac{\pi}{2} \quad \text{as } t \to \infty .$$

Then

$$G[F(t);\pi] = \frac{t+\pi}{2t+\pi}\int_0^{t+\pi} \frac{\sin x}{x}\,dx + \frac{t}{2t+\pi}\int_0^{t} \frac{\sin x}{x}\,dx.$$

Thus

$$G[F(9\pi);\pi] = \frac{10}{19}\,\mathrm{Si}(10\pi) + \frac{9}{19}\,\mathrm{Si}(9\pi) \approx 1.5707886 \ldots .$$

Note that $\pi/2 = 1.5707963\ldots$ while $F(10{,}000) = 1.5709\ldots$ and hence the G-transform significantly increases the rate of convergence of F(t).

The above examples suggest that

$$\lim_{t\to\infty} G_n[F(t);\ h] = \lim_{t\to\infty} F(t) = F(\infty) \tag{5}$$

and

$$\lim_{k\to\infty} e_n[A_k;\ \ell] = \lim_{k\to\infty} A_k = A_\infty \ . \tag{6}$$

Conditions under which (5) and (6) are true have been studied in several places, see [22, 38]. In particular when n = 1 it has been shown that (5) holds if

$$\lim_{t\to\infty} \frac{f(t+h)}{f(t)} \neq 1 \tag{7}$$

and (6) holds if

$$\lim_{k\to\infty} \frac{a_{k+\ell}}{a_k} \neq 1 \ . \tag{8}$$

Moreover, if in addition the limit in (7) is not zero, then the convergence of G[F(t); h] is more rapid than that of F(t) and similarly for $e_1[A_k;\ \ell]$. The examples above demonstrate this more rapid convergence. In general, the effectiveness of the transformations tend to improve with n although this is not always true. However, it can be shown under rather general conditions that for $t \geq a$ (see [61]),

$$\lim_{n\to\infty} \lim_{h\to 0} G_n[F(t);\ h] = F(\infty) \ . \tag{9}$$

Thus, one can consider (3) as simply a truncated expansion of $F(\infty)$. The properties above combined with the development of very efficient computational algorithms referred to as the G-algorithm (Pye and Atchison, 1972) and the ε-algorithm (Wynn, 1956) make the G_n- and e_n-transformations attractive numerical tools. Although the ability of the e_n- and G_n-transformations to increase the rate of convergence of certain types of sequences has by far been the primary stimulus for their study, it is the following theorems which attract our interest in them as tools for spectral estimation.

DEFINITION 1. *A function f will be said to be an element of L(n,d) over an interval I (denoted f ε L(n,d)) if there exists a smallest integer n > 0 and a set of c_i's such that f is a solution of the differential equation*

$$y^{(n)}(x) - c_1 y^{(n-1)}(x) \dots - c_n y(x) = 0, \quad x \in I \,. \tag{10}$$

DEFINITION 2. *A function f will be said to be an element of L(n,Δ) over a set of integers I if there exists a smallest integer n > 0 and a set of c_i's such that f is a solution of the difference equation*

$$y_k + c_1 y_{k-1} + \dots + c_n y_{k-n} = 0, \quad k \in I \,. \tag{11}$$

THEOREM 1. *Let f ε L(n,d) over [α,∞) and suppose the improper Riemann integral over [a,∞) exists, a ≤ α. Then*

$$G_m[F(t);\, h] \equiv F(\infty) = \int_a^\infty f(x)dx \tag{12}$$

for every h > 0 and t ≥ α, and m ≥ n.

Proof. The theorem is a trivial extension of the one given in [24], i.e., here we require f ε L(n,d) over [α,∞) while there the condition is f ε L(n,d) over (a,∞); that is, f need only be in L(n,d) for t sufficiently large. As mentioned, this requires only a trivial change in the proof.

THEOREM 2. *Suppose a_i ε L(n,Δ) for i ≥ h+n and*

$$\lim_{k\to\infty} \sum_{i=0}^{k} a_i = \lim_{k\to\infty} A_k = A_\infty \neq \mp \infty \,.$$

Then

$$e_m(A_k;\, \ell) \equiv A_\infty \tag{13}$$

for every m ≥ n, and k ≥ h.

Proof. See Appendix 1.

Henceforth, whenever infinite series or improper integrals are displayed, it will be tacitly assumed they converge.

The e_n- and G_n-transformations above are defined over intervals which are infinite in only one direction. Since both transformations are non-linear, we use the following device to extend our definition.

Let $g(x) = f(x) + f(-x)$ and let

$$b_i = a_i + a_{-i},\ i = 1, 2, \ldots .$$

Then we define

$$G_n[\int_{-t}^{t} f(x)dx;\ h] = G_n[\int_{0}^{t} g(x)dx;\ h] \tag{14}$$

and

$$e_n[\sum_{-m}^{m} a_i;\ \ell] = e_n[a_0 + \sum_{i=1}^{m} b_i;\ \ell] . \tag{15}$$

It should be emphasized that in general

$$G_n[\int_{0}^{t} g(x)dx;\ h] \neq G_n[\int_{0}^{t} f(x)dx] + G_n[\int_{0}^{t} f(-x)dx]$$

and similarly for e_n. We now have the following corollaries to Theorems 1 and 2.

COROLLARY 1. *If* $f(x)$ *and* $f(-x) \in L(n,d)$ *for* $x \in [t,\infty)$, *then there exists an* $s \leq 2n$ *such that*

$$G_m[\int_{0}^{t} g(x)dx;\ h] \equiv \int_{-\infty}^{\infty} f(x)dx \tag{16}$$

for all $t > 0$, $h > 0$, *and* $m \geq s$.

The result follows by noting that $f(x)$, $f(-x) \in L(n,d)$ on $[t,\infty)$ implies $g(x) \in L(s,d)$ on (t,∞) for some $s \leq 2n$. It should be noted that f need not satisfy the same differential equation on (t,∞) as on $(-\infty,-t)$. An obvious example of this is $f(x) = e^{-|x|}$, where in this case $s = n = 1$.

COROLLARY 2. *If* a_i *and* $a_{-i} \in L(n,\Delta)$ *when* $i \geq h+n$, *then there exists an* $s \leq 2n$ *such that*

$$e_m[a_0 + \sum_{i=1}^{m} b_i;\ \ell] \equiv \sum_{-\infty}^{\infty} a_i$$

for all $m \geq s$ *and* $k \geq h$.

The following simple examples demonstrate the use of

these theorems in evaluating Fourier transforms. The transforms selected are trivial, the only purpose of the examples being to demonstrate the above theorems. That is, to demonstrate that for $f \in L(n,\Delta)$ or $L(n,d)$, the leakage problem can be eliminated.

Example 3. Use Corollary 1 to find the Fourier transform of (i) $e^{-\alpha|x|}$ and (ii) $e^{i2\pi\alpha x}$, $\alpha > 0$.

(i) From Corollary 1, we know that for $t \geq 0$ and $h > 0$,

$$G_2[2\int_0^t e^{-\alpha x} \cos 2\pi\omega x dx;\ h] \equiv \int_{-\infty}^{\infty} e^{-i2\pi\omega x}\, e^{-\alpha|x|} dx \triangleq s(\omega)\ .$$

Thus, taking $t = 0$, we have

$$s(\omega) = 2\,\frac{\begin{vmatrix} 0 & \int_0^h e^{-\alpha x}\cos 2\pi\omega x dx & \int_0^{2h} e^{-\alpha x}\cos 2\pi\omega x dx \\ 1 & e^{-\alpha h}\cos 2\pi\omega h & e^{-2\alpha h}\cos 4\pi\omega h \\ e^{-\alpha h}\cos 2\pi\omega h & e^{-2\alpha h}\cos 4\pi\omega h & e^{-3\alpha h}\cos 6\pi\omega h \end{vmatrix}}{\begin{vmatrix} 1 & 1 & 1 \\ 1 & e^{-\alpha h}\cos 2\pi\omega h & e^{-2\alpha h}\cos 4\pi\omega h \\ e^{-\alpha h}\cos 2\pi\omega h & e^{-2\alpha h}\cos 4\pi\omega h & e^{-3\alpha h}\cos 6\pi\omega h \end{vmatrix}}$$

$$= \frac{2\alpha}{\alpha^2 + (2\pi\omega)^2}\ .$$

If $\omega = j/h$, $j = 0, 1, 2, \ldots$, the right side above takes the form 0/0 and the drop back rule must be used to obtain the final result. See [24].

(ii) Again, Corollary 1 yields for $t \geq 0$ and $h > 0$.

$$G_2[\int_{-t}^{t} e^{-i2\pi(\omega - \alpha)x} dx;\ h] \equiv \int_{-\infty}^{\infty} e^{-i2\pi(\omega - \alpha)x} dx\ .$$

But taking $t = 0$ gives

$$G_2[\int_{-t}^{t} e^{-i2\pi(\omega - \alpha)x} dx;\ h] = \begin{cases} 0 & \text{if } \omega \neq \alpha \\ \infty & \text{if } \omega = \alpha\ . \end{cases}$$

That is, $s(\omega) \sim \delta(\omega - \alpha)$. Thus, although we have assumed that the integrals involved exist as improper Riemann integrals, this example demonstrates that the assumption is not necessary. See [24].

The following theorem is of primary importance and will be basic to the remaining part of the paper.

THEOREM 3. *Let $\{x(t)\}$ be a stationary autoregressive moving average process of order p,q, i.e., ARMA(p,q) with autocorrelation $\rho(\tau)$. Then for every ω and $h > 0$ there exists a $k \leq 2p$ such that*

$$G_m[2 \int_0^\tau \rho(t) \cos 2\pi\omega t \, dt; h] = 2 \int_0^\infty \rho(t) \cos 2\pi\omega t \, dt \tag{17}$$

for all $m \geq k$ and $\tau \geq 0$, when $x(t)$ is a continuous index process. If the process is a discrete index process, then there exists a $k \leq 2p$ such that for $\tau \geq \max\{1,q-p+1\}$ and $m \geq k$.

$$e_m[1 + 2 \sum_{m=1}^{\tau} \rho(m) \cos 2\pi\omega m; \ell] = 1 + 2 \sum_{m=1}^{\infty} \rho(m) \cos 2\pi\omega m . \tag{18}$$

Proof. For the case $p > 0$, we use the well known fact (cf. J. L. Doob, *Stochastic Processes*, Wiley, 1953, for example) that the autocorrelations of an ARMA(p,q) process satisfy

$$\rho^{(p)}(\tau) - c_1\rho^{(p-1)}(\tau) - \dots - c_p\rho(\tau) = 0$$

for $\tau > 0$ in the case of continuous τ; or, in the discrete time case, with $k \geq q + 1$:

$$\rho(k) - c_1\rho(k-1) - \dots - c_p\rho(k-p) = 0 .$$

The coefficients c_k, $k = 1, \dots, p$ are those of the AR part of the equation defining the process.

In the discrete case, we observe that if $k \geq q + 1$,

$$\rho(k)e^{\pm i2\pi\omega k} - (c_1e^{\pm i2\pi\omega})\rho(k-1)e^{\pm i2\pi\omega(k-1)} - \dots - (c_pe^{\pm i2\pi\omega p})\rho(k-p)e^{\pm i2\pi\omega(k-p)} = e^{\pm i2\pi\omega k}(\rho(k) - c_1\rho(k-1) - \dots - c_p\rho(k-p)) = 0 .$$

This shows that $\rho(k)e^{i2\pi\omega k}$ and $\rho(k)e^{-i2\pi\omega k}$ belong to $L(p,\Delta)$ for $k \geq q+1$. Applying Corollary 2, we now obtain (18). When $p = 0$, (18) follows from the definition of the transform.

A similar proof, which we omit, applies in the case of continuous τ, where p is required to be positive.

As is clear from the examples, if p is known, m in (17) or (18) (whichever is appropriate) is also known. However, in the work that follows, p will not be assumed known, since as will be seen it is easily estimated when needed.

III. The G-Spectral Estimator.

The previous examples and theorems suggest the following definition.

DEFINITION 3. *Let $\{x(t)\}$ be a continuous or discrete index set covariance stationary stochastic process with autocorrelation function $\rho(\tau)$, and let $\hat{\rho}_T(\tau)$ be an estimator for $\rho(\tau)$ based on a sample record from $-T$ to T. Further let*

$$\hat{S}_T(\omega;\, T_0) = \int_{-T_0}^{T_0} e^{-i2\pi\omega\tau}\, \hat{\rho}_T(\tau)\,d\tau\ , \tag{19}$$

where $0 < T_0 + (2k-1)h < T$ *and*

$$\hat{S}_T(\omega;\, N_0) = \sum_{m=-N_0}^{N_0} e^{-i2\pi\omega m}\, \hat{\rho}_T(m) \tag{20}$$

where $0 < N_0 + (2k-1)h < T$ *and* $|\omega| \leq 1/2$.

We define the G-spectral estimator, denoted $G_T(\omega;\cdot,\cdot,\cdot)$, *by*

$$G_T(\omega;k,T_0,h) = G_k[\hat{S}_T(\omega;T_0);h] = G_k[\int_{-T_0}^{T_0} e^{-i2\pi\omega\tau}\, \hat{\rho}_T(\tau)\,d\tau;h\ ]$$

if the index set is continuous and by

$$G_T(\omega;k,N_0,h) = e_k[\hat{S}_T(\omega;N_0);h] = e_k[\sum_{-N_0}^{N_0} e^{-i2\pi\omega m}\,\hat{\rho}_T(m);h\,] \tag{21}$$

<u>if the index set is discrete</u>.

Although it is not necessary for our results to hold, it is convenient in the future to assume real processes. Moreover, we will also hereafter take $\hat{\rho}_T(\tau)$ as the sample autocorrelation function, i.e.,

$$\hat{\rho}_T(\tau) = \frac{\int_0^{T-|\tau|} (x(t)-\bar{x})(x(t+|\tau|)-\bar{x})\,dt}{\int_0^T [x(t)-\bar{x}]^2 dt} \tag{22}$$

for continuous data and integrals replaced by sums for discrete data. With these assumptions, we obtain

$$G_T(\omega;k,T_0,h) = 2G_k[\int_0^{T_0} \hat{\rho}_T(\tau)\cos 2\pi\omega\tau\, d\tau;\ h\,] \tag{23}$$

and

$$G_T(\omega;k,N_0,h) = e_k[\,1 + 2\sum_{m=1}^{N_0} \hat{\rho}_T(m)\cos 2\pi\omega m;\ h\,], \tag{24}$$

which we will take hereafter as the definition of the G-spectral estimator. We have at once the following theorems.

<u>THEOREM 4</u>. <u>Let</u> $\{X(t);t=0,\mp 1,\mp 2,\ldots\}$ <u>be a stochastic process such that</u> $\rho(\tau)\ \varepsilon\ L(n,\Delta)$, $\tau \geq N_0$. <u>Then there exists an integer</u> m <u>such that</u> $n \leq m \leq 2n$ <u>and</u>

$$\lim_{T\to\infty} G_T(\omega;k,N_0,h) = 1 + 2\sum_{\tau=1}^{\infty} \rho(\tau)\cos 2\pi\omega\tau, \text{ a.e.} \tag{25}$$

<u>for all</u> $k \geq m$.

Proof. Since $\rho(\tau)\ \varepsilon\ L(n,\Delta)$, $\rho(\tau)\cos 2\pi\omega\tau\ \varepsilon\ L(m,\Delta)$ for some m such that $n \leq m \leq 2n$. Moreover, it is well known that $\hat{\rho}_T(\tau)$ converges a.e. to $\rho(\tau)$. Then, employing the drop back rule for $k > m$, we have

$$\lim_{T\to\infty} G_T(\omega;k,N_0,h) = \lim_{T\to\infty} e_m[1 + 2 \sum_{m=1}^{N_0} \hat{\rho}_T(m) \cos 2\pi\omega m;h]$$

$$= e_m[1 + 2 \sum_{m=1}^{N_0} \rho(m) \cos 2\pi\omega m;h]$$

$$= 1 + 2 \sum_{m=1}^{\infty} \rho(m) \cos 2\pi\omega m \qquad \text{a.e.}$$

by Theorem 2.

In the above theorem, it should be noted that one will normally take $k = m$ since if $k > m$, $G_T(\omega;k,N_0,h) \equiv 0/0$ and the drop back rule then forces one back to $k = m$. It should also be noted that m is a function of ω. For most values of ω we have $m = 2n$. However, at $\omega = 0$ and $\omega = 1/2$, we see $m = n$, while if $\omega = \text{Tan}^{-1}(\beta_j/\alpha_j)$, $\alpha_j + i\beta_j$ a root of the associated characteristic equation, we have $m = 2n - 1$ (we are tacitly assuming no repeated roots). We will later show how this information is utilized in properly selecting k. A similar result holds for the continuous case (Theorem 5 below). It was first shown in [21] and comments corresponding to the above apply equally well to it.

THEOREM 5. *Let* $\{X(t); -\infty < t < \infty\}$ *be a stochastic process with* $\rho(\tau) \in L(n,d)$ *when* $\tau \geq T_0$. *Then there exists an integer* $m \leq 2n$ *(possibly less than n) such that*

$$\lim_{T\to\infty} G_T(\omega;k,T_0,h) = 2 \int_0^{\infty} \rho(\tau) \cos 2\pi\omega\tau d\tau , \qquad (26)$$

for every $k \geq m$.

Proof. See [26].

IV. Calculation of the G-Spectral Estimator.

Efficient methods for calculating the G_n- and e_n-transformations have, as mentioned before, been developed. Some discussion on this is necessary to fully describe the

G-spectral estimator. We shall proceed with such a discussion after Theorem 6 below. Let us first, however, make the following observations. When strictly numerical data is available, the G_n-transformation requires numerical approximation of the integrals in the top row. For the remaining part of this paper, we will use the trapezoid rule approximation unless stated to the contrary. The next theorem then shows that the G-spectral estimator may be calculated for both discrete and continuous processes by the same algorithm regardless of whether the process under investigation is continuous or discrete.

THEOREM 6. *Let* $\tilde{S}(\omega;M)$ *be the trapezoidal rule approximation of*

$$S(\omega;M) = 2\int_0^M \rho(\tau)\cos 2\pi\omega\tau d\tau$$

based on (ω.l.o.g.) *a unit sampling interval, where* M *is an integer. Then for* $|\omega| \leq 1/2$, *and* $n \geq 1$,

$$G_n[\tilde{S}(\omega;M);h] \equiv e_n[1 + 2\sum_{k=1}^{M} \rho(k)\cos 2\pi\omega k;h] , \tag{27}$$

h *an integer equal to or greater than one* (*normally* h = 1).

Proof. The proof follows by substituting the trapezoid approximation in the G transform and applying simple elementary row operations.

Due to Theorem 6, we will now only discuss the algorithm of Pye and Atchison for purposes of spectral estimation. In this way, we leave open the option of approximating the integrals involved by quadrature methods other than the trapezoid rule when the process sampled is continuous in time. We again stress that here, however, we will always assume the trapezoid approximation. For convenience, let us also abbreviate our notation somewhat as well as introduce some new relationships for future use. That is, we let

$$H_{n+1}^{(m)}[F_j;f_{i+j}] = H_{n+1}[F(mh+jh);f(mh+(i+j)h)] .$$

Then

$$G_n[F(mh);h] = H_{n+1}^{(m)}[F_j;f_{i+j}]/H_{n+1}^{(m)}[1;f_{i+j}]$$

and we shorten this to

$$G_n^{(m)} = G_n[F(mh);h] \ .$$

Further, define

$$G_0^{(m)} = F(mh)$$

$$f_m \equiv f(mh)$$

$$H_n^{(m)}[f_j] = \begin{vmatrix} f_m & f_{m+1} & \cdots & f_{m+n-1} \\ f_{m+1} & f_{m+2} & \cdots & f_{m+n} \\ \vdots & & & \\ f_{m+n-1} & f_{m+n} & \cdots & f_{m+2n-2} \end{vmatrix} ,$$

and

$$H_0^{(m)}[f_j] \equiv 1 \ .$$

Finally, let

$$r_n^{(m)}(f) = \frac{H_n^{(m)}[f_j]}{H_n^{(m)}[1;f_{i+j}]} \tag{28}$$

and

$$s_n^{(m)}(f) = \frac{H_{n+1}^{(m)}[1;f_{i+j}]}{H_n^{(m)}[f_j]} \ .$$

When the argument on the left side in (28) is clear, we will suppress it in our notation. We then have the following theorem which was first shown in [51].

THEOREM 7. *If* $H_n^{(m+1)}[f_j]$, $H_{n+1}^{(m)}[f_j]$, $H_{n+1}^{(m)}[1;f_{i+j}]$, $H_{n+1}^{(m+1)}[1;f_{i+j}]$ *and* $H_{n+2}^{(m)}[1;f_{i+j}]$ *are not zero, then*

$$G_{n+1}^{(m)} = G_n^{(m)} + \frac{G_n^{(m+1)} - G_n^{(m)}}{1 - R_{n+1}^{(m)}} , \tag{29}$$

where

$$R_{n+1}^{(m)} = \frac{r_{n+1}^{(m+1)}}{r_{n+1}^{(m)}} . \tag{30}$$

In order to calculate the G-transform recursively, it is clear from (29) that a recursive calculation of $R_{n+1}^{(m)}$ also is needed. This is accomplished by the following relations which also were established in [51].

$$r_{n+1}^{(m)} = \begin{cases} r_n^{(m+1)} \left[\dfrac{s_n^{(m+1)}}{s_n^{(m)}} - 1 \right] & , s_n^{(m)} \neq 0 \\ -r_n^{(m+1)} & , s_n^{(m)} = 0 \end{cases}$$

$$s_n^{(m)} = \begin{cases} s_{n-1}^{(m+1)} \left[\dfrac{r_n^{(m+1)}}{r_n^{(m)}} - 1 \right] & , r_n^{(m)} \neq 0 \\ -s_{n-1}^{(m+1)} & , r_n^{(m)} = 0 \end{cases} \tag{31}$$

for $n = 1, 2, \ldots$ and $m = 0, 1, 2, \ldots$, and

$$s_0^{(m)} = 1, \; m = 0, 1, 2, \ldots$$

$$r_1^{(m)} = f_m, \; m = 0, 1, 2, \ldots .$$

The algorithm which we shall use is to calculate the G_n-transform via (29) and (31) with the integrals replaced with their trapezoid approximations unless stated to the contrary. We repeat, in this way the algorithm will calculate the correct transform even if the process is strictly discrete. Thus, in calculating the G-spectral estimator, we take

$$f(t) = 2\,\hat{\rho}(t)\cos 2\pi\omega t$$

and proceed as described. This is demonstrated schematically by Figure 0.

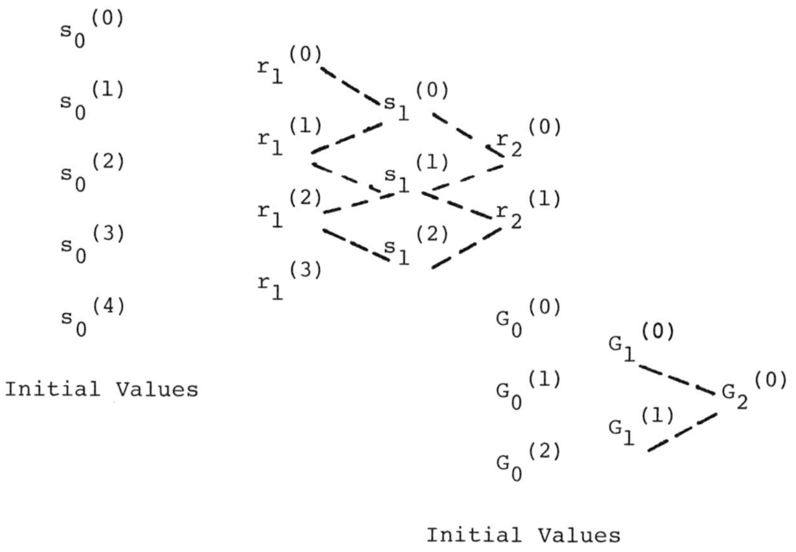

Fig. 0. The calculation of the r-, s-, and G-matrices. The dashed lines show the quantities needed to calculate the indicated element on the far right.

In the definition of $R_{n+1}^{(m)}$, we have assumed that $r_n^{(m)} \not\equiv 0$. If $r_n^{(m)} \equiv 0$, we simply take $G_{n+1}^{(m)} = G_n^{(m)}$ which is consistent with the drop back rule. Also, we have assumed $R_{n+1}^{(m)} \not\equiv 1$. This assumption is of no practical consequence since if $R_{n+1}^{(m)} \equiv 1$, we simply use a different h so that $R_{n+1}^{(m)} \not\equiv 1$, i.e., we take h in (3) greater than one if necessary.

V. Selecting the Order of the G-Spectral Estimator.

Theoretically, neither p nor q need be known to take advantage of the exactness theorems in the previous sections. That is, if $p > q$, it is clear q need not be known since we will normally take $\tau > 0$ in practice. But p need not be known either, since we can select the order of the G-transform too large and the drop back rule will lead to the correct spectrum. Moreover, if $q \geq p$, it will be observed that the correct spectrum will still be obtained by taking n sufficiently large (see Theorem 12). Thus, one could adopt the procedure of simply calculating the G-spectral estimator recursively until G_n remained essentially constant with n. This will be referred to as the "stabling off" procedure. The authors primarily used this approach but find it requires special attention at isolated frequencies. That is, the procedure is only recommended as an initial approach since it can lead to unnecessary instabilities due to division by very small numbers. Fortunately, the algorithm for calculating the G-spectral estimator furnishes us with an excellent method for supplementing the above-suggested method. The validity of this statement is seen by the next few theorems. Due to Theorem 6 and the fact that a sampled ARMA(p,q) continuous process can be modeled as an ARMA(p,p-1) discrete process (see [49]), for spectral purposes we restrict ourselves to discrete sampled data.

THEOREM 8. *Let $n > 0$ and suppose $s_n^{(m)}$ and $r_n^{(m)}$ are defined and $s_n^{(m)} \neq 0$. Then $f_m \in L(n,\Delta)$ for $m \geq m_0 + n$ if and only if $s_n^{(m)}$ is constant, as a function of m, for $m \geq m_0$.*

Proof. See Appendix 1.

We should remark that the restriction that $s_{(n)}^{(m)}$ be defined and non-zero does not limit the generality of the above theorem since $s_n^{(m)}$ is zero only when the associated characteristic equation has a unit root, i.e., the process is nonstationary. This property can in fact be used as a check for stationarity. Of course, if $s_n^{(m)}$ and $r_n^{(m)}$ are not defined, it simply means f_m satisfies a lower order difference equation. The following corollary is immediate.

COROLLARY 3. *Let* $n > 0$ *and suppose* $s_n^{(m)}$ *and* $r_n^{(m)}$ *are defined and* $s_n^{(m)} \neq 0$. *Then* $f_m \in L(n,\Delta)$ *for* $m \geq m_0 + n$ *if and only if* $r_{n+1}^{(m)} = 0$ *for* $m \geq m_0 + n$.

THEOREM 9. *If* $\{X(t); t=0,\mp 1,\mp 2, \ldots\}$ *is a stationary* ARMA(p,q) *process, then for some constant* c *and integer* m

$$\left.\begin{array}{ll} s_n^{(\tau)}(\rho) & \equiv c \\ r_{n+1}^{(\tau)}(\rho) & \equiv 0 \end{array}\right\} \text{ when } \tau \geq m-n+1 \quad \text{and} \quad \begin{array}{l} s_n^{(m-n)}(\rho) \neq c \\ r_{n+1}^{(m-n)}(\rho) \neq 0 \end{array}$$

if and only if $m = q$ *and* $n = p$. *Moreover,*

$$\left.\begin{array}{ll} s_{n-1}^{(\tau)} & \not\equiv c \\ r_n^{(\tau)} & \not\equiv 0 \end{array}\right\} \text{ when } \tau \geq m-n+2 \,.$$

Proof. If $p = 0$, the result follows by definition of the quantities involved. Therefore, take $p > 0$. Since $X(t)$ is ARMA(p,q), $\rho(k) \in L(n,\Delta)$ for $k \geq m-n+1$ if and only if $n = p$ and $m = q$ and the result follows from Theorem 8 and Corollary 3. It should be noted that Theorem 9 furnishes a way to estimate p and q from the estimated autocorrelations. This will not be pursued here but its ramifications are great in the study of autoregressive processes and will be presented in a future paper (where theorems such as Theorem 9 will be greatly expanded). For our purposes, however, Theorem 9 will suffice. Now, note that Corollary 3 actually only pertains to our problem when $\omega = 0$ or $\omega = 1/2$, in (27). For other values of ω, we have the following result.

THEOREM 10. *Let* $\{X(t); t=0,\mp 1,\mp 2, \ldots\}$ *be a stationary* ARMA(p,q) *process whose characteristic equation has roots* $\alpha_j \mp i\beta_j$. *Then for* $\omega \neq 0, 1/2, \mp\theta_j/2\pi, .5 \pm \theta_j/2\pi$, *where* $\theta_j = \mathrm{Tan}^{-1}\, \beta_j/\alpha_j \neq \theta_i$, *and some constant* c *and integer* m,

$$\left.\begin{array}{l} s_{2n}^{(\tau)}(\rho(k)\cos 2\pi\omega k) \equiv c \\ \\ r_{2n+1}^{(\tau)}(\rho(k)\cos 2\pi\omega k) \equiv 0 \end{array}\right\} \underline{\text{when}}\ \tau \geq m-n+1\ \underline{\text{and}}\ \begin{array}{l} s_{2n}^{(m-n)}(\rho(k)\cos 2\pi\omega k) \neq c \\ \\ r_{2n+1}^{(m-n)}(\rho(k)\cos 2\pi\omega k) \neq 0 \end{array}$$

<u>if and only if</u> $m = q$ <u>and</u> $n = p$. <u>Moreover</u>,

$$\begin{array}{l} s_{2n-1}^{(\tau)}(\rho(k)\cos 2\pi\omega k) \not\equiv c \\ \\ r_{2n}^{(\tau)}(\rho(k)\cos 2\pi\omega k) \not\equiv 0 \end{array} \quad \underline{\text{when}}\ \tau \geq m-n+2 .$$

Proof. The proof follows by noting that $\rho(k)\cos 2\pi\omega k \in L(2p,\Delta)$ for $\tau \geq q-p+1$ if $\omega \neq 0, 1/2, \theta_j/2\pi$ and applying Theorem 8 and Corollary 3.

The above theorems establish the behavior of $s_n^{(\tau)}$ and $r_n^{(\tau)}$ for all $\omega \neq \theta_j/2\pi$. However, the situation there is also clear and we include it in the following theorem which is simply a summary of the above observations.

<u>THEOREM 11</u>. <u>Let</u> $\{X(t);t=0,\mp1,\mp2, \ldots\}$ <u>be a stationary</u> ARMA(p,q) <u>process with spectral density</u> $S(\omega)$. <u>Then if</u> $\tilde{S}(\omega;m)$ <u>is defined by</u> (27) <u>and the calculation procedure described above is used, we have the following</u>:

1. (i) <u>If</u> $\omega \neq 0, 1/2, \mp\theta_j/2\pi, .5 \pm \theta_j/2\pi$ <u>and</u> $m \geq \max\{1,q-p+1\}$

$$G_{2p-1}[\tilde{S}(\omega;m);h] \neq G_{2p}[\tilde{S}(\omega;m);h] = G_{2p+1}[\tilde{S}(\omega;m);h] = \ldots = S(\omega) , \tag{32}$$

(ii) <u>If</u> $\omega = \theta_j/2\pi, .5 - \theta_j/2\pi, \theta_i \neq \theta_j$ <u>and</u> $m \geq \max\{1,q-p+1\}$

$$G_{2p-2}[\tilde{S}(\omega;m);h] \neq G_{2p-1}[\tilde{S}(\omega;m);h] = G_{2p}[\tilde{S}(\omega;m);h] = \ldots = S(\omega), \tag{33}$$

(iii) *If* $\omega = 0$ *or* $1/2$ *and* $m \geq \max\{1, q-p+1\}$

$$G_{p-1}[\tilde{S}(\omega;m);h] \neq G_p[\tilde{S}(\omega;m);h] = G_{p+1}[\tilde{S}(\omega;m);h] = \ldots$$
$$= S(\omega) \ . \qquad (34)$$

2. *Under conditions* (i), (ii), *and* (iii) *respectively with* $m = M$, p *is the smallest positive integer such that*

(i) $r_{2p+1}^{(m)} \equiv 0$ *and* $r_{2p+1}^{(q-p)} \neq 0$, $s_{2p}^{(m)} \equiv c \neq 0$ *and*

$$s_{2p}^{(q-p)} \neq c \qquad (35)$$

(ii) $r_{2p}^{(m)} \equiv 0$ *and* $r_{2p}^{(q-p)} \neq 0$, $s_{2p-1}^{(m)} \equiv c \neq 0$ *and*

$$s_{2p-1}^{(q-p)} \neq c \qquad (36)$$

(iii) $r_p^{(m)} \equiv 0$ *and* $r_p^{(q-p)} \neq 0$, $s_p^{(m)} \equiv c \neq 0$ *and*

$$s_p^{(q-p)} \neq c \ . \qquad (37)$$

3. *The process* $\{x(t)\}$ *is non-stationary if for some* m_0, $s_p^{(m)} \equiv 0$ *when* $m \geq m_0$ *and* $\omega = 0$ or $.5$.

We are now in a position to fully describe the method of G-spectral estimation (GSPEC). Treat the data as if it were sampled from a continuous process, and approximate the integrals involving $\rho_T(\tau) \cos 2\pi\omega\tau$ in the G_n-transform by the trapezoid rule (This will be correct even if the data is inherently discrete, see Theorem 6.) with $\rho_T(\tau)$ replaced by $\hat{\rho}_T(\tau)$. Use the algorithm of Pye and Atchison described above to calculate the $G_i^{(j)}$, $s_i^{(j)}$, and $r_i^{(j)}$ recursively to some preassigned maximum values of i and j. Select the proper $G_i^{(j)}$ as G-spectral estimator from the "stabling off" property. If $G_i^{(j)}$ does not stable off, inspect the $r_i^{(j)}$ and $s_i^{(j)}$ to determine p and q. In addition in a neighborhood of drop back frequencies such as $\omega = 0$, $.5$ and $\theta_j/2\pi$ (i.e., 0, .5, and near peaks), it is recommended that the $r_i^{(j)}$ and $s_i^{(j)}$ be considered for possibly a better choice of j and i in $G_i^{(j)}$. This procedure is implemented in the program given in Appendix 2.

VI. Examples and Simulations.

In this section, examples are given to demonstrate the application of the previous theory. Initially, only data-limited problems are considered and it is demonstrated that in these cases, the leakage problem is indeed eliminated as claimed by the previous theoretical development. In the later examples, data which are both time-limited and in error, i.e., when $\rho_T(\tau)$ is replaced by $\hat{\rho}_T(\tau)$, are considered in detail. Spectral estimates also are obtained by the Tukey, Bartlett, and Parzen windows as well as by the maximum entropy method (MEM) and the autoregressive spectral estimator (ARSPEC). These methods are then compared to the GSPEC method. In these comparisons, the best window estimate obtainable from considering several possible window lengths was used. Hence, the window estimators appear much better than they would be in practice. Thus, those comparisons should be considered as a comparison to the best possible window of this type. The MEM and ARSPEC were implemented via Akaike's (1970a) procedure (FPE) for selecting the order of the prediction error filter (PEF) and hence they should be realistic.

The models selected for comparing the different spectral density estimators are the discrete ARMA(p,q) model, the continuous AR(n) model, and the signal plus noise model given, for a fixed frequency ω, by

$$X(t) = \sin(2\pi\omega t + \phi) + Z(t)$$

where ϕ is uniformly distributed over $(-\pi, \pi)$ and $Z(t)$ is white noise. The manner of generating the data for the discrete ARMA(p,q) process is obvious from the form of the model. The data from the continuous processes are actually discrete and obtained by sampling the processes at equally spaced time intervals. This was accomplished by modeling the AR(n) as an ARMA(n,n-1).

The discrete parameters were, however, obtained from the underlying continuous parameters according to the relationships derived by Phadke and Wu (1974) and hence w.l.o.g. all simulations appear simply as discrete AR or ARMA processes. The white noise input for each of the processes is Gaussian and was generated from the standard normal variates subroutine NORMAL, which is available in the Southern Methodist University Statistical Research Laboratory collection of random number generators known as STATPAK. This subroutine uses the Box-Muller transformation based on the polar method to obtain normal deviates from uniform deviates obtained from subroutine RANF. The computer employed was a CDC Cyber 72. The graphs were obtained from the Zeta Plotter 230.

In order to demonstrate the use of the $s_n^{(\tau)}$ and $r_n^{(\tau)}$ in determining the G-spectral estimator, examples are presented utilizing the true autocorrelation function of the processes considered. In each example, h = 1, unless otherwise indicated, and the $s_n^{(\tau)}$ and $r_n^{(\tau)}$ are presented in number arrays referred to as the s-matrix and r-matrix, respectively. This terminology is used even though the arrays are obviously not mathematical matrices. It should be kept in mind that in all cases the models are not assumed known but their order may be identified and utilized by the GSPEC method.

Example 4. An AR(1) model.

Consider the discrete AR(1) process given by

$$X_t = a_1 X_{t-1} + Z_t$$

with autocorrelation function

$$\rho_k = a_1^{|k|},\ k = 0, \pm 1, \pm 2, \ldots ,$$

and spectral density function

$$S(\omega) = \frac{1 - a_1^2}{1 + a_1^2 - 2a_1 \cos(2\pi\omega)} ,\ -\frac{1}{2} \le \omega \le \frac{1}{2} ,$$

where Z_t is a discrete white noise process with constant variance σ_z^2 and $|a_1| < 1$. The spectral density function has a peak at $\omega = 0$ or peaks at $\omega = -1/2$ and $\omega = 1/2$ depending on whether $a_1 > 0$ or $a_1 < 0$. In either case, the magnitude of the peak is $(1 + |a_1|)/(1 - |a_1|)$.

As an example, let $a_1 = .9$ and $\sigma_z^2 = 1.0$. Then, the peak of the spectral density function occurs at $\omega = 0$ with a magnitude of 19.0. The s-matrix, r-matrix, and G-matrix are displayed in Table 1. Note that $s_n^{(m)}$ becomes constant and $r_{n+1}^{(m)} = 0$ for $n = 1$ and $m \geq 0$. The GSPEC procedure then takes the G-spectral estimator to be $G_1^{(1)} = 19.0$. Of course, here variability is not a problem so that $G_1^{(0)}$ could be used as well.

The values of p and q are then clearly p = 1 and since q - p + 1 = 0, q = 0. Thus, the s- and r-matrices identify the model as AR(1) as well as identifying the element of the G-matrix that is the G-spectral estimator. Of course, for exact data, even though time limited, the G algorithm will automatically produce $G_1^{(0)}$ without any recourse to inspection of the s- and r-matrix. See Table 1. We should emphasize, however, for non-exact data, $G_n^{(0)}$ is not used for the G_n-spectral estimator but $G_n^{(m)}$ where $m \geq 1$ instead.

Now, consider the s-, r-, and G-matrices evaluated at $\omega = .3$ and exhibited in Table 2. Note that $s_n^{(m)}$ becomes constant and $r_{n+1}^{(m)} = 0$ at $n = 2$ and $m \geq 0$. Hence, the G-spectral estimator is given by $G_2^{(1)} = .080$, which is the true value of the spectral density function at $\omega = .3$. Again, for exact data, we could take $G_2^{(0)}$. This phenomenon will be observed on the remaining illustrations when $\rho(\tau)$ is known over a finite interval and we shall feel no further need to explain it. Note also that at $\omega = .3$, the order of the G-spectral estimator (i.e., column of s-matrix) is twice the order at frequency 0, whereas the row behavior is the same. This, of course, reinforces the modeling of the data as an AR(1) process, as stated earlier.

TABLE 1

Spectral Density Evaluated at Frequency 0 Based on the True acf of the Process $X_t = .9X_{t-1} + Z_t$

	s-matrix		r-matrix	
m	$s_0^{(m)}$	$s_1^{(m)}$	$r_1^{(m)}$	$r_2^{(m)}$
-3*	1.000	.111	1.458	0.000
-2*	1.000	.111	1.620	0.000
-1*	1.000	.111	1.800	-3.800
0	1.000	-.100	2.000	0.000
1	1.000	-.100	1.800	0.000
2	1.000	-.100	1.620	
3	1.000		1.458	
4	1.000			

	G-matrix		
m	$G_0^{(m)}$	$G_1^{(m)}$	$G_2^{(m)}$
0	0.000	19.000	19.000
1	1.900	19.000	
2	3.610		

True spectral density = 19.000

G-spectral estimator = $G_1^{(1)}$ = 19.000

*The values of r's and s's for negative m follow the same rule as (31). The reasons for calculating them are discussed on page 113.

TABLE 2

Spectral Density Evaluated at Frequency .3 Based on the True acf of the Process $X_t = .9X_{t-1} + Z_t$

	s-matrix			r-matrix		
m	$s_0^{(m)}$	$s_1^{(m)}$	$s_2^{(m)}$	$r_1^{(m)}$	$r_2^{(m)}$	$r_3^{(m)}$
-3	1.000	-2.111	2.921	1.179	.953	.045
-2	1.000	- .576	2.888	-1.311	-3.885	.297
-1	1.000	-4.596	2.293	- .566	-1.444	.037
0	1.000	-1.278	2.366	2.000	1.146	0.000
1	1.000	1.356	2.366	- .566	3.147	0.000
2	1.000	-1.900	2.366	-1.311	- .772	
3	1.000	- .656		1.179	2.012	
4	1.000	-3.912		.405		
5	1.000			-1.181		
6	1.000					

	G-matrix			
m	$G_0^{(m)}$	$G_1^{(m)}$	$G_2^{(m)}$	$G_3^{(m)}$
0	0.000	.565	.080	.080
1	.722	1.410	.080	
2	- .211	- .246		
3	- .277			

True spectral density = .080

G-spectral estimator = $G_2^{(1)}$ = .080

Example 5. An MA(1) model.

Consider the discrete MA(1) process given by

$$X_t = Z_t - b_1 Z_{t-1}$$

with autocorrelation function

$$\rho_k = \begin{cases} 1 & , k = 0 \\ \dfrac{-b_1}{1 + b_1^2} & , k = \pm 1 \\ 0 & , k = \pm 2, \pm 3, \ldots \end{cases}$$

and spectral density function

$$S(\omega) = \frac{1 + b_1^2 - 2b_1 \cos(2\pi\omega)}{1 + b_1^2}, \quad -\frac{1}{2} \le \omega \le \frac{1}{2},$$

where Z_t is discrete white noise with constant variance σ_z^2. There are no restrictions on b_1 for this process to be stationary. If $b_1 < 0$, the spectral density function is dominated by low frequencies, i.e., $S(\omega)$ is monotone decreasing over the interval [0, 1/2]. Conversely, if $b_1 > 0$, the spectral density function is dominated by high frequencies. In either case, the magnitude of the peak is $1 + [2|b_1|/(1 + b_1^2)]$.

Let $b_1 = .9$ and $\sigma_z^2 = 1.0$. Then the peak of the spectral density function occurs at $\omega = 1/2$ with a value of 1.99. For $\omega = 1/2$, the s- and r-matrices and the G-matrix are displayed in Table 3. Note that $r_1^{(m)} = 0$ for $m \ge 2$. This, of course, indicates that the G-spectral estimator is given by $G_0^{(2)} = 1.99$, which is the true spectral density at this frequency. Thus, p = 0, q = 2, as would be seen by looking at the autocorrelations alone. However, this example simply exemplifies that for this simple case, the GSPEC procedure will automatically produce the natural estimate.

This example also illustrates that no matter what value of m ($m \ge 0$) is used, the transform eventually yields the true spectral density. This is not a property restricted to this example alone and in fact is true in general. For an ARMA(p,q) process with $p > 0$, the result can be seen by the previous theorems and our adopted convention when $r_n^{(\tau)} = 0$. Theorem 12 in Appendix 1 establishes the result when $p = 0$ as well, although it may be of limited value for real data.

The s-, r-, and G-matrices evaluated at $\omega = .2$ for the above example are given in Table 4. They illustrate that the correct order transform does not change with frequency for purely moving average processes. For example, at frequency 1/2, $r_1^{(m)} = 0$ for $m \geq 2$. Hence, the G-spectral estimator is $G_0^{(2)} = .69$, which is the true spectral density at $\omega = .2$.

Example 6. An ARMA(2,1) model.

Consider the discrete ARMA(2,1) process given by

$$X_t - a_1X_{t-1} - a_2X_{t-2} = Z_t - b_1Z_{t-1}$$

with autocovariance function

$$\gamma_0 = \sigma_z^2 \cdot \frac{a_1b_1 + a_1a_2b_1 + (1 - a_2)(1 + a_1b_1 + b_1^2)}{1 - a_2 - a_1^2 - a_1^2a_2 - a_2^2(1 - a_2)}$$

$$\gamma_1 = \frac{a_1}{1 - a_2}\gamma_0 + \frac{b_1}{1 - a_2}\sigma_z^2$$

$$\gamma_k = a_1\gamma_{k-1} + a_2\gamma_{k-2}, \quad k \geq 2$$

and spectral density function

$$S(\omega) = \frac{\sigma_z^2}{\gamma_0}\left[\frac{1 + b_1^2 + 2b_1\cos(2\pi\omega)}{1 + a_1^2 + a_2^2 - 2a_1(1 - a_2)\cos(2\pi\omega) - 2a_2\cos(4\pi\omega)}\right],$$

$$-\frac{1}{2} \leq \omega \leq \frac{1}{2},$$

TABLE 3

Spectral Density Evaluated at Frequency .5 Based on the True acf of the Process $X_t = Z_t - .9Z_{t-1}$

	s-matrix			r-matrix		
m	$s_0^{(m)}$	$s_1^{(m)}$	$s_2^{(m)}$	$r_1^{(m)}$	$r_2^{(m)}$	$r_3^{(m)}$
-3	1.00	-1.00	1.00	0.00	0.00	.99
-2	1.00	-1.00	.50	0.00	-2.00	- .98
-1	1.00	1.01	.67	.99	-2.99	.49
0	1.00	- .50	1.00	2.00	.98	0.00
1	1.00	-1.00	1.00	.99	0.00	0.00
2	1.00	-1.00	1.00	0.00	0.00	
3	1.00	-1.00		0.00	0.00	
4	1.00	-1.00		0.00		
5	1.00			0.00		
6	1.00					

	G-matrix			
m	$G_0^{(m)}$	$G_1^{(m)}$	$G_2^{(m)}$	$G_3^{(m)}$
0	0.00	2.98	1.99	1.99
1	1.49	1.99	1.99	
2	1.99	1.99		
3	1.99			

True spectral density = 1.99

G-spectral estimator = $G_0^{(2)}$ = 1.99

TABLE 4

Spectral Density Evaluated at Frequency .2 Based on the True acf of the Process $X_t = Z_t - .9Z_{t-1}$

	s-matrix			r-matrix		
m	$s_0^{(m)}$	$s_1^{(m)}$	$s_2^{(m)}$	$r_1^{(m)}$	$r_2^{(m)}$	$r_3^{(m)}$
-3	1.00	-1.00	1.00	0.00	0.00	- .31
-2	1.00	-1.00	1.15	0.00	-2.00	- .04
-1	1.00	-7.51	1.18	- .31	-1.69	- .01
0	1.00	-1.15	1.00	2.00	.04	0.00
1	1.00	-1.00	1.00	- .31	0.00	0.00
2	1.00	-1.00	1.00	0.00	0.00	
3	1.00	-1.00		0.00	0.00	
4	1.00	-1.00		0.00		
5	1.00			0.00		
6	1.00					

	G-matrix			
m	$G_0^{(m)}$	$G_1^{(m)}$	$G_2^{(m)}$	$G_3^{(m)}$
0	0.00	.73	.69	.69
1	.85	.69	.69	
2	.69	.69		
3	.69			

True spectral density = .69

G-spectral estimator = $G_0^{(2)}$ = .69

where Z_t is discrete white noise with constant variance σ_z^2. The stationarity restrictions on the autoregressive parameters are

$$a_2 + a_1 < 1$$

$$a_2 - a_1 < 1$$

$$-1 < a_2 < 1$$

with no restrictions on b_1. Of course, the autocorrelation function is given by

$$\rho_k = \frac{\gamma_k}{\gamma_0}, \; k \geq 0 .$$

The class of spectral density functions for this process contains densities with peaks and troughs, i.e., relative maxima and minima at frequencies other than 0 and 1/2, as well as those which are monotone, depending on the values of the parameters. The peaks and troughs will occur when the characteristic equation $1 - a_1B - a_2B^2 = 0$ has complex roots.

As an example, let $a_1 = .6$, $a_2 = -.8$, $b_1 = -.6$, and $\sigma_z^2 = 1.0$. The spectral density function will have a peak at the frequency $\omega = .19556^+$. The s-, r-, and G-matrices with $\omega = 0$ are exhibited in Table 5. Note that $s_n^{(m)}$ becomes constant and $r_{n+1}^{(m)} = 0$ when $n = 2$ and $m \geq 0$. Hence, the G-spectral estimator is $G_2^{(0)} = .32$, which is the true spectral density at this frequency. Thus, values of p and q are given by $p = 2$ and $q = 1$, since $q - p + 1 = 0$.

The s-, r-, and G-matrices evaluated at $\omega = .234$ are displayed in Table 6. Note that $s_n^{(m)}$ becomes constant and $r_{n+1}^{(m)} = 0$ when $n = 4$ and $m \geq 0$. Hence, the G-spectral estimator is given by $G_4^{(0)} = 1.25$, which is the true value of the spectral density function for $\omega = .234$. As in the previous examples, this reinforces the model identified at frequency 0, which, in this example, is the ARMA(2,1) model.

In order to demonstrate the tendency of the transform to "nearly drop back" near a peak frequency, the s-, r-, and G-matrices evaluated at $\omega = .195$ are given in Table 7. Note

TABLE 5

Spectral Density Evaluated at Frequency 0 Based on the True acf of the Process $X_t - .6X_{t-1} + .8X_{t-2} = Z_t + .6Z_{t-1}$

	s-matrix			r-matrix		
m	$s_0^{(m)}$	$s_1^{(m)}$	$s_2^{(m)}$	$r_1^{(m)}$	$r_2^{(m)}$	$r_3^{(m)}$
-3	1.00	- .14	1.50	-1.31	-12.82	.18
-2	1.00	-1.70	1.32	-1.13	- 1.50	.42
-1	1.00	1.54	1.12	.79	- 2.79	.16
0	1.00	- .61	1.20	2.00	2.37	0.00
1	1.00	-2.43	1.20	.79	1.20	0.00
2	1.00	.16	1.20	-1.13	10.26	
3	1.00	-1.09		-1.31	- 1.03	
4	1.00	8.47		.12		
5	1.00			1.21		
6	1.00					

	G-matrix			
m	$G_0^{(m)}$	$G_1^{(m)}$	$G_2^{(m)}$	$G_3^{(m)}$
0	0.00	2.30	.32	.32
1	1.39	1.32	.32	
2	1.22	8.87		
3	.01			

True spectral density = .32

G-spectral estimator = $G_2^{(0)}$ = .32

TABLE 6

Spectral Density Evaluated at Frequency .234 Based on the True acf of the Process $X_t - .6X_{t-1} + .8X_{t-2} = Z_t + .6Z_{t-1}$

m	s-matrix $s_0^{(m)}$	$s_1^{(m)}$	$s_2^{(m)}$	$s_3^{(m)}$	$s_4^{(m)}$	r-matrix $r_1^{(m)}$	$r_2^{(m)}$	$r_3^{(m)}$	$r_4^{(m)}$	$r_5^{(m)}$
-2	1.00	- .93	- .78	- .33	.49	1.10	-2.15	3.23	1.43	- .42
-1	1.00	24.29	.43	- .95	.19	.08	-2.08	.75	.69	.06
0	1.00	- .96	.15	- .79	.22	2.00	-1.15	-4.14	.52	0.00
1	1.00	12.97	.67	-1.34	.22	.08	-1.16	.76	.44	0.00
2	1.00	- .65	12.49	-2.21	.22	1.10	.04	.68	.39	
3	1.00	- .72	.50	-1.83		.39	- .70	-2.31	.35	
4	1.00	3.96	1.96	2.50		.11	- .79	- .15		
5	1.00	- 1.88	- 6.65			.54	.03	- .09		
6	1.00	- 1.75	- 2.86			- .47	.16			
7	1.00	- 2.54				.35	.34			
8	1.00	- .95				- .55				
9	1.00					- .03				
10	1.00									

G-matrix

m	$G_0^{(m)}$	$G_1^{(m)}$	$G_2^{(m)}$	$G_3^{(m)}$	$G_4^{(m)}$	$G_5^{(m)}$
0	0.00	1.08	8.87	3.67	1.25	1.25
1	1.04	.99	2.72	3.28	1.25	
2	1.63	2.78	2.78	3.08		
3	2.38	2.72	4.13			
4	2.63	2.54				
5	2.95					

True spectral density = 1.25

G-spectral estimator = $G_4^{(0)}$ = 1.25

TABLE 7

Spectral Density Evaluated at Frequency .195 Based on the True acf of the Process $X_t - .6X_{t-1} + .8X_{t-2} = Z_t + .6X_{t-1}$

	s-matrix					r-matrix				
m	$s_0^{(m)}$	$s_1^{(m)}$	$s_2^{(m)}$	$s_3^{(m)}$	$s_4^{(m)}$	$r_1^{(m)}$	$r_2^{(m)}$	$r_3^{(m)}$	$r_4^{(m)}$	$r_5^{(m)}$
-2	1.0000	- .6927	-1.1770	- .2064	.3849	.8686	-2.7687	3.2281	1.3096	.0172
-1	1.0000	6.4939	.4991	- .3492	.3355	.2669	-2.2669	1.8929	- .1337	- .0002
0	1.0000	- .8666	- .4838	- .3349	.0355	2.0000	- .9612	3.2596	.0002	0.0000
1	1.0000	2.2545	1.6049	- .3349	.0355	.2669	- .7549	2.5793	.0002	0.0000
2	1.0000	.2949	.7538	- .3350	.0355	.8686	-4.8639	1.4332	.0002	
3	1.0000	- .9804	- .2074	- .3350		1.1247	-1.1239	3.7478	.0002	
4	1.0000	48.9392	.4872	- .3351		.0221	-1.1191	1.1707		
5	1.0000	- .7485	- .9728			1.1023	- .3907	1.5739		
6	1.0000	.3065	-2.7756			.2772	.8494			
7	1.0000	1.0253				.3621	-1.4499			
8	1.0000	-1.0015				.7334				
9	1.0000					- .0011				
10	1.0000									

	G-matrix					
m	$G_0^{(m)}$	$G_1^{(m)}$	$G_2^{(m)}$	$G_3^{(m)}$	$G_4^{(m)}$	$G_5^{(m)}$
0	0.0000	1.3080	- .6790	9.0529	9.0460	9.0460
1	1.1334	.8816	1.3520	9.0522	9.0460	
2	1.7012	-1.6787	4.7736	9.0515		
3	2.6978	3.2827	-2.1354			
4	3.2712	3.2597				
5	3.8334					

True spectral density = 9.0460

G-spectral estimator = $G_4^{(0)}$ = 9.0460

that $s_n^{(m)}$ becomes constant and $r_{n+1}^{(m)}$ is exactly zero when $n = 4$ and $m \geq 0$. This yields the G-spectral estimator $G_4^{(0)} = 9.0460$, which is the true spectral density at this frequency. However, note also that $s_n^{(m)}$ is nearly constant and $r_{n+1}^{(m)} \approx 0$ for $n = 3$ and $m \geq 0$. In this case, the G-spectral estimator is $G_3^{(0)} = 9.0529$, which is very nearly exact. If $\omega = .195$ were the peak frequency, $G_3^{(0)}$ would be exact. This illustrates the behavior of the transform near a peak frequency. In fact, this behavior also occurs near the frequencies 0 and 1/2. Therefore, the s- and r-matrices may be used to indicate the "drop back" frequencies.

Note also that a "stabling off" procedure with $m = 0$ or $m = 1$ would probably select $G_3^{(m)}$ as the G-spectral estimator. Morgan (1975) noticed this and suggested that the same order transform used at a "drop back" frequency should also be used within ±.03 of the "drop back" frequency. The s- and r-matrices in the above example suggest that this is not an unreasonable guideline to follow.

Example 7. An ARMA(1,2) model.

Consider the discrete ARMA(1,2) process given by

$$X_t - a_1 X_{t-1} = Z_t - b_1 Z_{t-1} - b_2 Z_{t-2}$$

with autocorrelation function

$$\gamma_0 = \sigma_z^2 \cdot \frac{1 - a_1 b_1 + b_2^2 - (a_1 - b_1)(b_1 + 2a_1 b_2)}{1 - a_1^2}$$

$$\gamma_1 = a_1 \gamma_0 - \sigma_z^2 (b_1 + a_1 b_2 - b_1 b_2)$$

$$\gamma_2 = a_1 \gamma_1 - b_2 \sigma_z^2$$

$$\gamma_k = a_1 \gamma_{k-1} \, , \quad k \geq 3$$

and spectral density function

$$S(\omega) = \frac{\sigma_z^2}{\gamma_0}\left[\frac{1 + b_1^2 + b_2^2 - 2b_1(1-b_2)\cos(2\pi\omega) - 2b_2\cos(4\pi\omega)}{1 + a_1^2 - 2a_1\cos(2\pi\omega)}\right],$$

$$-\frac{1}{2} \le \omega \le \frac{1}{2},$$

where Z_t is discrete white noise with constant variance σ_z^2. The stationarity requirement is that $-1 \le a_1 \le 1$. The spectral density function will have a peak or a trough at certain frequencies when the characteristic equation $1 - b_1B - b_2B^2 = 0$ has complex roots.

As an example, let $a_1 = .6$, $b_1 = -.75$, $b_2 = 2.5$, and $\sigma_z^2 = 1.0$. The s-, r-, and G-matrices are presented only at frequency 0 to demonstrate the selection of p and q. This is displayed in Table 8. Note that $s_n^{(m)}$ becomes constant and $r_{n+1}^{(m)} = 0$ first at $n = 1$ and $m \ge 2$. Therefore, the G-spectral estimator is $G_1^{(2)} = .48$, and clearly, $p = 1$ and $q = 2$. Hence, the r- and s-matrices identify the ARMA(1,2) model and the G-spectral estimator yields the true spectral density.

For other frequencies, the G-spectral estimator would be given by $G_2^{(2)}$ for the same reason given in previous examples, i.e., the order of the transform is twice that at $\omega = 0$, while m remains the same. Note also in the above example that, for fixed m, the transform eventually becomes exact. For $m = 1$, $G_2^{(1)} = .48$ and for $m = 0$, $G_3^{(0)} = .48$. This again suggests a "stabling off" procedure for selecting the G-spectral estimator.

In each of the examples thus far, the true autocorrelation function over a limited set of points was used to demonstrate the exactness of the G-spectral estimator for this class of processes as well as to point out what to look for when the procedure is applied to the sample autocorrelation function. Morgan (1975) demonstrated that the transform yields more stable estimates for $m \ge 1$ when applied to the sample autocorrelation function. Therefore, in practice, even if the

TABLE 8

Spectral Density Function Evaluated at Frequency 0 Based on the True acf of the Process $X_t - .6X_{t-1} = Z_t + .75Z_{t-1} - 2.5Z_{t-2}$

	s-matrix			r-matrix		
m	$s_0^{(m)}$	$s_1^{(m)}$	$s_2^{(m)}$	$r_1^{(m)}$	$r_2^{(m)}$	$r_3^{(m)}$
-2	1.00	-2.20	3.55	- .40	-1.17	1.80
-1	1.00	3.17	.97	.48	-2.48	.01
0	1.00	- .76	.99	2.00	.68	- .18
1	1.00	-1.83	.40	.48	.31	0.00
2	1.00	- .40	.40	- .40	0.00	
3	1.00	- .40		- .24	0.00	
4	1.00	- .40		- .14		
5	1.00			- .09		
6	1.00					

	G-matrix			
m	$G_0^{(m)}$	$G_1^{(m)}$	$G_2^{(m)}$	$G_3^{(m)}$
0	0.00	1.63	.95	.48
1	1.24	1.26	.48	
2	1.28	.48		
3	.96			

True spectral density = .48

G-spectral estimator = $G_1^{(2)}$ = .48

s- and r-matrices indicate the estimator to be evaluated at m = 0, it is suggested that the G-spectral estimator be selected at m = 1. Of course, the s- and r-matrices may indicate $m \geq 1$ and in this case the estimator corresponding to that particular value of m should be used.

Another important observation that should be made here is that when the G-spectral estimator is given by $G_n^{(m)}$, it is calculated using only the first 2n + m - 1 lags of the autocorrelation function. This will be important when applying the G-spectral estimator to real data, since the best estimates of the sample autocorrelation function will occur at the shorter lags.

In practice, it is not necessary to examine the s- and r-matrices at each and every frequency. In fact, it is suggested that they only be examined when there is reason to suspect the estimate obtained by using the stabling off property of the G-matrix. Thus, in the neighborhood of indicated peaks and near 0 and .5, the "stabling off" procedure should be considered.

The above guidelines were followed in obtaining the G-spectral estimator for the simulated data as well as real data in later examples.

Before considering data which are not error free, we should mention that our previous r- and s-matrices may be misleading on real data. That is, our definition of $r_n^{(m+1)}$ when $s_n^{(m)} = 0$ and $s_n^{(m)}$ when $r_n^{(m)} = 0$ is in some instances artificial and will not always reflect the behavior of these quantities when $s_n^{(m)}$ and $r_n^{(m)}$ are small. However, it is not difficult to ascertain the behavior of these quantities in that case. The table below shows the behavior of columns p and p+1 in the s-matrix and p+1 and p+2 in the r-matrix when $\rho(\tau)$ is replaced by the sample estimate $\hat{\rho}(\tau)$. The ε's indicate numbers near zero.

m		s-matrix $s_p^{(m)}$	s-matrix $s_{p+1}^{(m)}$		r-matrix $r_{p+1}^{(m)}$	r-matrix $r_{p+2}^{(m)}$
⋮		⋮			⋮	
q-p	⋯	s_p^{q-p}	$(C_1 + \varepsilon_{11})\left[\frac{\varepsilon_1}{r_{p+1}^{q-p}} - 1\right]$		r_{p+1}^{q-p}	$\varepsilon_1[\cdots]$
q-p+1	⋯	$C_1 + \varepsilon_{11}$	$(C_1 + \varepsilon_{12})\left[\frac{\varepsilon_2}{\varepsilon_1} - 1\right]$		ε_1	$\varepsilon_2[\cdots]$
q-p+2	⋯	$C_1 + \varepsilon_{12}$	$(C_1 + \varepsilon_{13})\left[\frac{\varepsilon_3}{\varepsilon_2} - 1\right]$		ε_1	⋮
q-p+3	⋯	$C_1 + \varepsilon_{13}$	$(C_1 + \varepsilon_{14})\left[\frac{\varepsilon_4}{\varepsilon_3} - 1\right]$		ε_3	
⋮						

From the above, it should now be clear that when $\rho(\tau)$ is estimated, one would expect column p to be identifiable by <u>two</u> properties, its stability from some row on down followed by a highly variable column. On the other hand, the zero behavior in the r column should continue to some extent. The reader will find it useful to examine Table 18 now where the reality of these remarks is clear.

VII. <u>Comparisons based on the True Autocorrelation Function.</u>

In order to compare the effects of leakage in other spectral estimators with the G-spectral estimator, which is leakage free for ARMA processes, the true autocorrelation function for several of these processes is used to obtain the spectral density. In the examples which follow, the Parzen window has been selected to represent the window methods and was based on the number of lags that gave the best representation of the true spectral density. The window methods of Bartlett and Tukey would yield similar estimates of the spectral density, compared to the Parzen window, if their respective bandwidths were selected equal to the bandwidth of the Parzen window (see Jenkins and Watts (1968, pp. 272-273)). In each example, the autoregressive spectral estimator was calculated using up to 20 autoregressive coefficients. Akaike's minimum FPE criterion then selected the one to use as the spectral density estimator. The G was determined by the GSPEC procedure using $T_0 = mh$ and $h = 1$. The maximum entropy method is of course here the same as the AR estimate.

In the following, the Parzen window spectral estimator will be denoted by PARZEN, the autoregressive spectral estimator will be denoted by ARSPEC, and the G-spectral estimator will be denoted simply by G. They will be called "estimates" of the spectral density function with the understanding that this is a degenerate case, since the true autocorrelation function is used over a finite data window. Also, in each example, the true spectral density function coincides with the G-spectral estimator as proved by earlier theorems. The spectral density obtained by each of the above estimators for each example was calculated over the interval [0, .5] at increments of .02 and plotted. The number in parentheses in the legend of each plot is the number of lags, excluding lag 0, of the autocorrelation function used to calculate the estimator. For ARSPEC, this is also the number of autoregressive coefficients used in obtaining the autoregressive spectral estimator.

Example 8. An AR(1) model.

The spectral estimates, based on the true autocorrelation function of the discrete AR(1) process

$$X_t = .9X_{t-1} + Z_t , \tag{38}$$

as calculated by PARZEN, ARSPEC, and G, are given in Figure 1. PARZEN was determined using 19 lags of the true acf. Akaike's FPE criterion selected a first-order model with minimum FPE = .66084 for ARSPEC. The G-spectral estimator was given by $G_1^{(0)}$ at frequencies 0 and .5 and $G_2^{(0)}$ elsewhere. That is, the G was based on one lag at frequencies 0 and .5 and three lags elsewhere. Since this is a true finite autoregressive process, ARSPEC yields the true spectral density function and coincides with the G. Note that PARZEN underestimates the peak and overestimates the trough. This is typical of the window methods and is caused by the leakage problem.

Example 9. An MA(1) model.

The spectral estimators, PARZEN, ARSPEC, and G, based on the true acf of the discrete MA(1) process

$$X_t = Z_t - .9Z_{t-1} , \tag{39}$$

are displayed in Figure 2. PARZEN was based on 11 lags of the

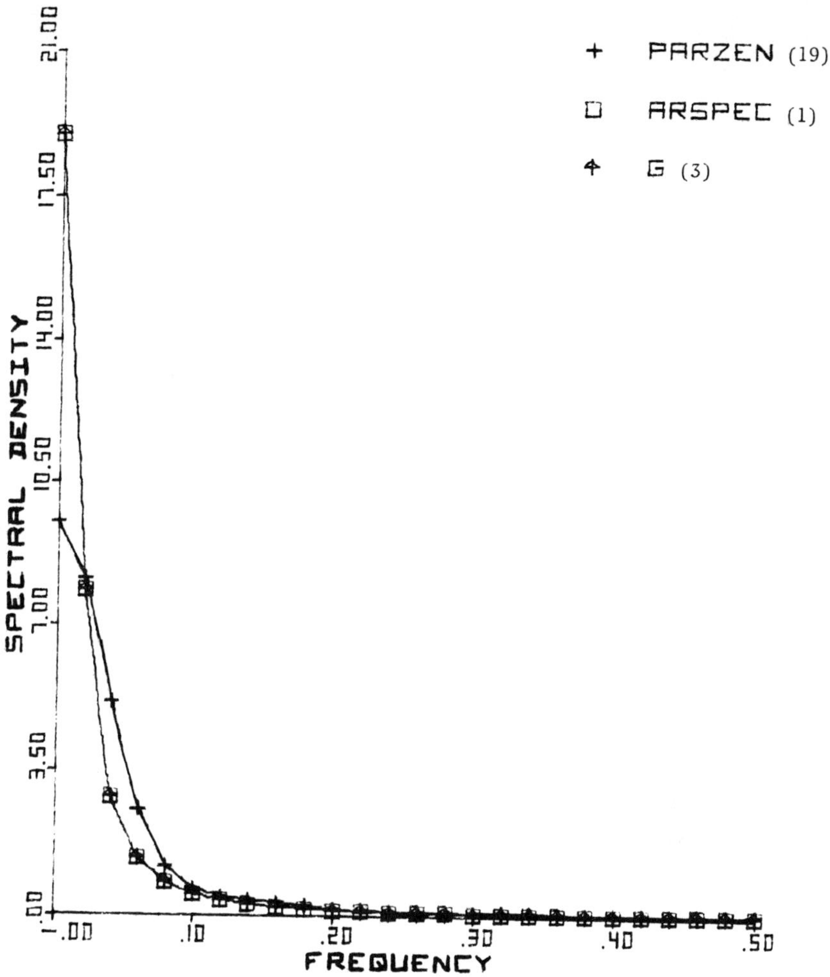

Fig. 1. Spectral density estimators based on the true acf of the AR(1) process $X_t = .9X_{t-1} + Z_t$.

true acf, while the G was based on one lag and given by $G_0^{(2)}$ at all frequencies. Note again that PARZEN overestimates the trough and underestimates the peak. However, in this case, since there is not much power at the peak, the bias incurred is very small. The autoregressive process that corresponds to this purely moving average process is, in fact, infinite. The FPE criterion, however, selected a fifth-order model for ARSPEC with minimum FPE = 1.42285. This yielded spurious peaks as seen in Figure 2.

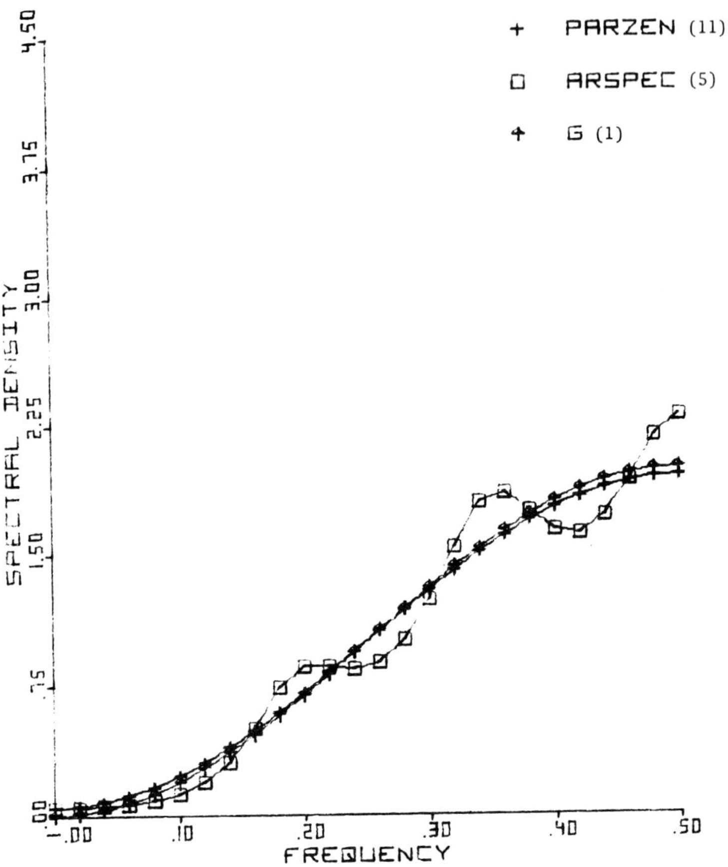

Fig. 2. Spectral density estimators based on the true acf of the MA(1) process $X_t = Z_t - .9Z_{t-1}$.

<u>Example 10</u>. An ARMA(1,1) model.

Consider the discrete ARMA(1,1) process given by

$$X_t = .25X_{t-1} + Z_t - .95Z_{t-1} \; . \tag{40}$$

The spectral density of this process, based on the true acf, is given in Figure 3 for the three estimators. Note that PARZEN essentially coincides with the G in this example. However, PARZEN was determined using 28 lags of the true acf, while the G was based on only two lags at frequencies 0 and .5

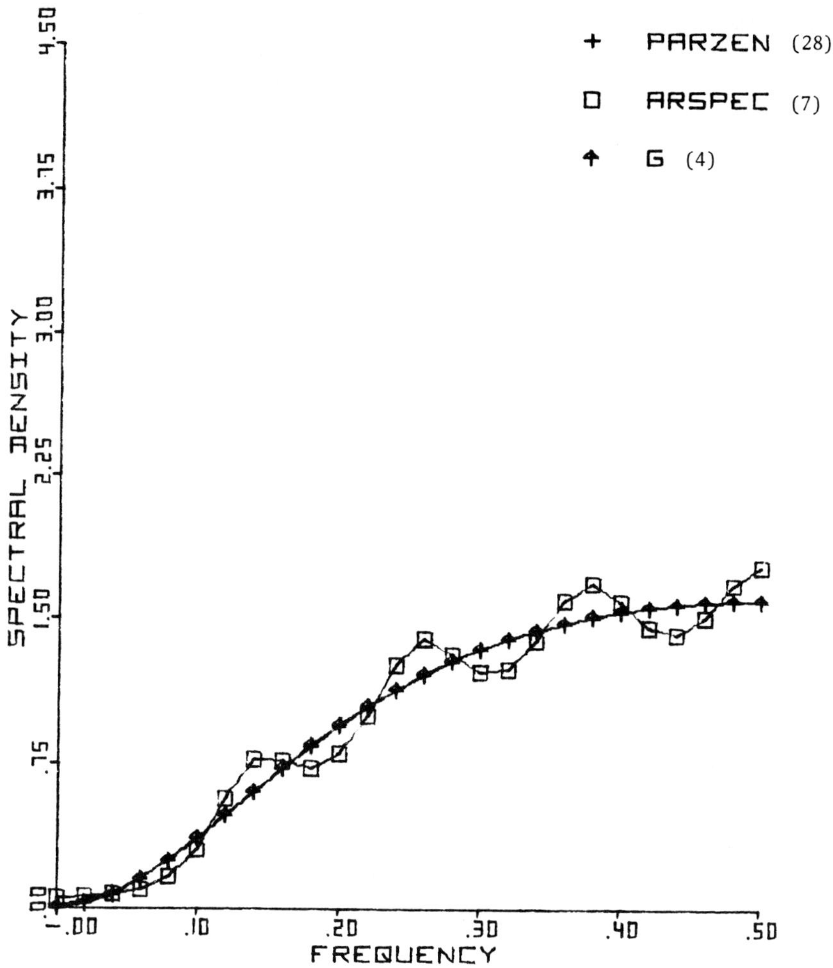

Fig. 3. Spectral density estimators based on the true acf of the ARMA(1,1) process $X_t = .25X_{t-1}$.

and four lags elsewhere. At frequencies 0 and .5, the G-spectral estimator was given by $G_1^{(1)}$. At other frequencies, it was given by $G_2^{(1)}$. ARSPEC was calculated using seven autoregressive coefficients with minimum FPE = 1.31970. Since the autoregressive process corresponding to this mixed process is infinite and the autoregressive coefficients damp out slowly, ARSPEC yielded spurious peaks.

Example 11. An ARMA(1,2) model.

Consider the discrete ARMA(1,2) process given by

$$X_t = -.5X_{t-1} + Z_t - 1.75Z_{t-1} + .8Z_{t-2} . \tag{41}$$

The spectral density estimates based on the true acf of this process are exhibited in Figure 4. PARZEN has very good resolution in this example, but was based on 28 lags of the true acf. ARSPEC also does a good job in this case. The FPE criterion selected a 13th-order autoregressive process with minimum FPE = .90135. The G-spectral estimator was given by $G_1^{(2)}$ at frequencies 0 and .5 and by $G_2^{(2)}$ at other frequencies. That is, the G was based on at most five lags of the true acf.

Example 12. An ARMA(3,2) model.

Consider the discrete ARMA(3,2) process given by

$$X_t = 1.5X_{t-1} - 1.21X_{t-2} + .455X_{t-3} + Z_t -$$

$$1.75Z_{t-1} + .8Z_{t-2} . \tag{42}$$

The estimators, based on the true acf of this process, of the spectral density are exhibited in Figure 5. PARZEN was calculated using 46 lags of the true acf and yielded very good estimates with slight underestimates near and at the peak. ARSPEC was determined using 11 autoregressive coefficients with minimum FPE = 1.06216. ARSPEC also did a good job with a slight overestimate at the peak. The G-spectral estimator was given by $G_3^{(0)}$ at frequencies 0 and .5 and by $G_6^{(0)}$ at the other frequencies. Hence, the G is based on at most 11 lags of the true acf.

In the above we have demonstrated the leakage effect. Possibly a more appropriate approach would have been to limit all of the estimates to the smallest length data window to achieve exactness. This, of course, would seriously degrade the performance of the windows and to a lesser extent ARSPEC. Thus, possibly the previous examples best demonstrate how

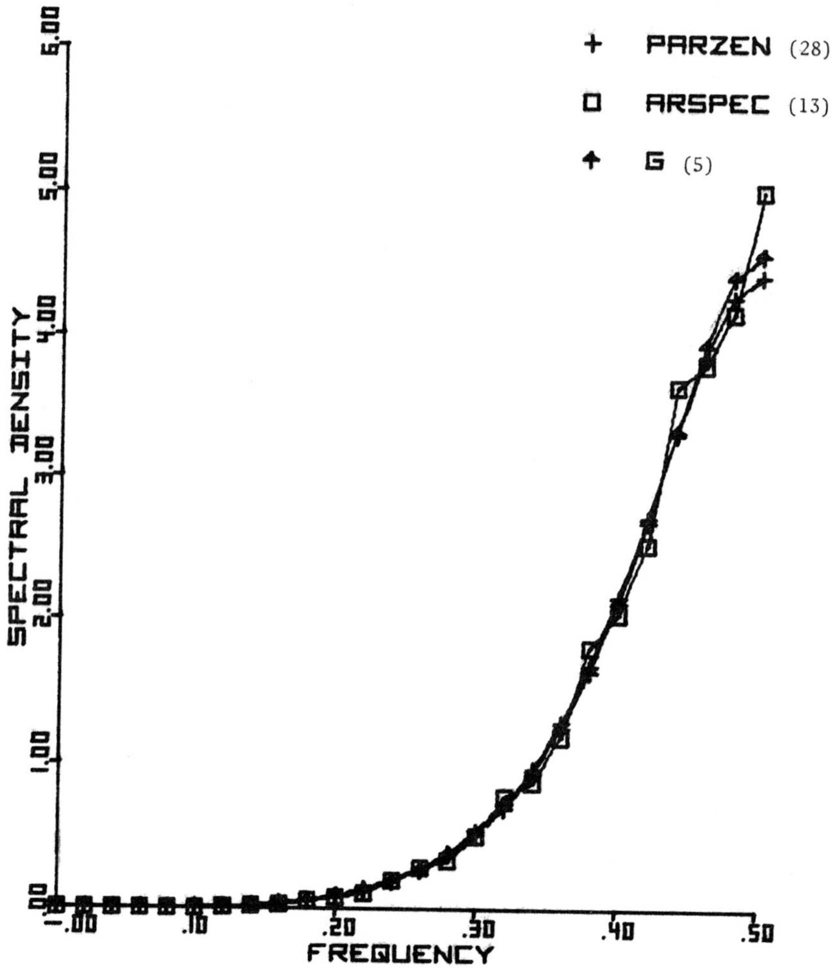

Fig. 4. Spectral density estimators based on the true acf of the ARMA(1,2) process $X_t + .5X_{t-1} = Z_t - 1.75Z_{t-1} + .8Z_{t-2}$.

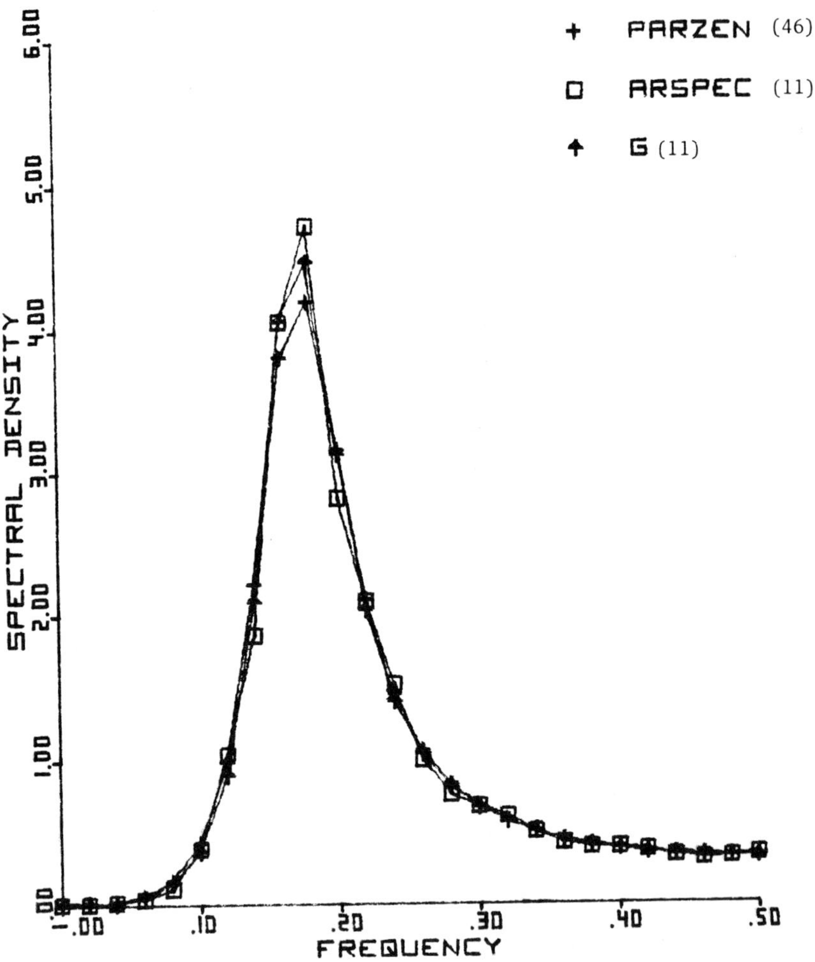

Fig. 5. Spectral density estimators based on the true acf of the ARMA(3,2) process $X_t = 1.5X_{t-1} - 1.21X_{t-2} + .455X_{t-3} + Z_t - 1.75Z_{t-1} + .8Z_{t-2}$.

much larger a data window would be required by the window and ARSPEC method to reduce the leakage effect to something comparable to the G.

VIII. Comparisons based on the Sample Autocorrelation Function.

In order to further compare the spectral density estimators, 30 realizations of each process discussed above were generated and a spectral analysis performed using each of the estimators. The sample mean, sample variance, and sample mean squared error (MSE) were then calculated at several frequencies for each estimator. One realization was selected at random for each process and the spectral estimates were plotted over the interval [0, .5] using .02 as the increment for the Parzen window spectral estimator (PARZEN), the autoregressive spectral estimator (ARSPEC), the maximum entropy spectral estimator (MEM), and the G-spectral estimator (G). In each of the tables and graphs, the number in parentheses next to the name of the estimator is the number or range of lags (autoregressive coefficients for ARSPEC and MEM) used in calculating the estimator. Also, the true spectral density function is plotted on each graph and denoted by TRUE. The true value of the spectral density at a particular frequency ω is denoted by $S(\omega)$ in the tables.

The window spectral estimators of Bartlett, denoted by BARTLETT, and Tukey, denoted by TUKEY, also were calculated. These results will appear in the tables containing the average value and mean squared error comparisons. The number of lags used to calculate these estimators was selected so that their bandwidths were equal to the bandwidth of the Parzen window spectral estimator.

Example 13. An AR(1) model.

Thirty realizations of 100 observations each were generated from the discrete AR(1) process given by (38). The summary statistics for each of the spectral density estimators are given in Table 9. Note that the windows have about the same values for the summary statistics. This is

TABLE 9

Summary Statistics for Spectral Estimates of the Discrete AR(1) Process $X_t = .9X_{t-1} + Z_t$.

Statistics Based on 30 Realizations of 100 Observations Each.

Frequency (ω)	$S(\omega)$	Estimator	Sample Mean	Sample Variance	Sample MSE
0.00	19.000	BARTLETT(15)	7.731	2.575	129.560
		TUKEY(13)	7.724	1.631	128.775
		PARZEN(19)	7.964	2.162	123.953
		ARSPEC(1-8)	14.514	65.148	85.275
		MEM(1-10)	15.152	61.820	76.627
		G(2-6)	15.886	48.141	57.839
0.06	1.393	BARTLETT	2.684	.412	2.078
		TUKEY	3.173	.159	3.328
		PARZEN	2.999	.259	2.838
		ARSPEC	1.629	.254	.309
		MEM	1.764	.719	.856
		G	2.146	1.210	2.031
0.24	.112	BARTLETT	.230	.0090	.0230
		TUKEY	.162	.0084	.0109
		PARZEN	.127	.0106	.0109
		ARSPEC	.131	.0015	.0019
		MEM	.130	.0038	.0041
		G	.143	.0060	.0069
0.50	.053	BARTLETT	.104	.0015	.0042
		TUKEY	.070	.0015	.0019
		PARZEN	.099	.0014	.0036
		ARSPEC	.064	.0005	.0006
		MEM	.062	.0011	.0012
		G	.070	.0011	.0014

the result of using equal bandwidths for the three estimators. Note also that they have the highest MSE at the peak when compared to the other estimators.

Since this is a finite autoregressive process, both ARSPEC and MEM inherently assume the correct model. The FPE criterion selected the correct order (first-order) 24 times for ARSPEC and 18 times for MEM. It selected as high as an 8th-order model for ARSPEC and a 10th-order model for MEM. This may or may not contribute to the variability of these estimators at the peak. However, away from the peak, these estimators yielded smaller MSE than the other estimators.

For each of the 30 realizations, the stabling effect of $G_n^{(m)}$ as well as the s- and r-matrices indicated the first-order transform, $G_1^{(1)}$, to be used as the G-spectral estimator at frequency 0.5. Since the estimated power at .5 is small, $G_1^{(1)}$ also was used as the estimator at frequency 0, even though inspection of the s- and r-matrices may or may not have suggested this. The inherent variability of the sample autocorrelation function is exemplified by the s- and r-matrices at peak frequencies. Therefore, the model is usually more easily identified by the s- and r-matrices at those frequencies with less power. Again, however, we should stress, these quantities are only examined at certain frequencies. In fact, the program included at the end of this paper uses only the stability property of $G_n^{(m)}$ to select the G-spectral estimate.

For these realizations, the G-spectral estimator was given by $G_1^{(1)}$ for frequencies 0, .02, .48, and .50. At the other frequencies, $G_2^{(1)}$ was selected as the estimator. However, if any of these estimates were negative, the estimator was given by the same order transform using one more lag, e.g., if $G_2^{(1)}$ was negative, then $G_2^{(2)}$ was selected as the estimator if it was positive. It happened that for some realizations, both $G_2^{(1)}$ and $G_2^{(2)}$ were negative so that $G_2^{(3)}$ was taken as the estimator.

In order to demonstrate that a mild reduction in MSE can usually be obtained by examining the $s_n^{(m)}$ and $r_n^{(m)}$ at drop back or suspicious frequencies, the r- and s-matrices were examined at $\omega = 0$, and .06 as well as the G-matrix for 30 realizations. At frequency 0.00, the "stabling off" procedure alone yielded a sample mean of 14.669, a sample variance of 43.833 and a sample MSE of 62.592. At $\omega = .06$, it yielded a sample mean of 2.513, a sample variance of 1.133 and a sample MSE of 2.388. In Table 9, however, the s- and r-matrices were also examined at 0 and .06 before selecting the proper estimate. In both cases, the latter procedure has slightly smaller MSE. Also the "stabling off" procedure selected estimates based on from 2 to 12 lags of the acf, while the closer examination used only two to six lags.

One realization was randomly selected from the 30 realizations and the spectral estimates obtained from this realization are plotted in Figures 6 and 7. For this particular realization, the minimum FPE criterion selected a first-order model for both ARSPEC and MEM which is correct for this process. Hence, both yielded very good estimates of the spectral density with slight overestimates at the peak. The G had the best point estimate of the power at frequency 0.00 while PARZEN had the worst. However, each spectral estimator did indicate the correct peak frequency for this realization.

Example 14. An MA(1) model.

The summary statistics for each estimator based on 30 realizations of 100 observations each from the discrete MA(1) process in (39) are displayed in Table 10. The window estimators have very good estimates for this process. One reason being, speaking loosely, that the peak frequency does not have very much power so that the bias incurred by the windows is negligible.

The autoregressive process corresponding to this purely moving average process is infinite and its coefficients damp out slowly. Hence, both ARSPEC and MEM inherently model the data poorly, yielding spurious peaks in their spectral estimates. Table 10 indicates this at frequency .42 where both ARSPEC and MEM have a higher sample mean than at frequency .50 where the peak should be. Also, these estimators have the highest MSE at all frequencies except 0, where there is very little power. Note that ARSPEC was based on from 2 to 15 autoregressive coefficients and MEM was based on from 4 to 17 autoregressive coefficients, while the G used only two to eight lags and had the smallest MSE at the frequencies with the higher power, as seen in Table 10.

The spectral estimates for the randomly selected realization are plotted in Figures 8 and 9. Note that the G gave the best representation of the true spectral density. It was calculated by a first- and second-order transform, i.e., two to four lags of the sample acf. Both ARSPEC and MEM

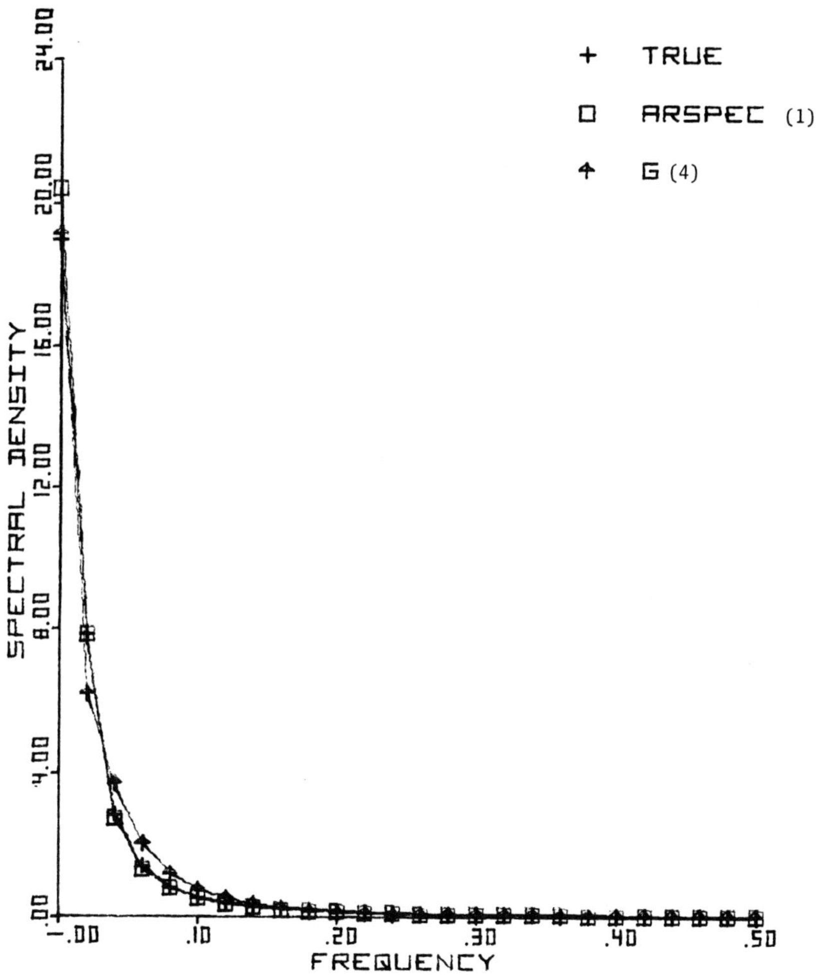

Fig. 6. Spectral estimates based on the sample acf of the AR(1) process $X_t = .9X_{t-1} + Z_t$.

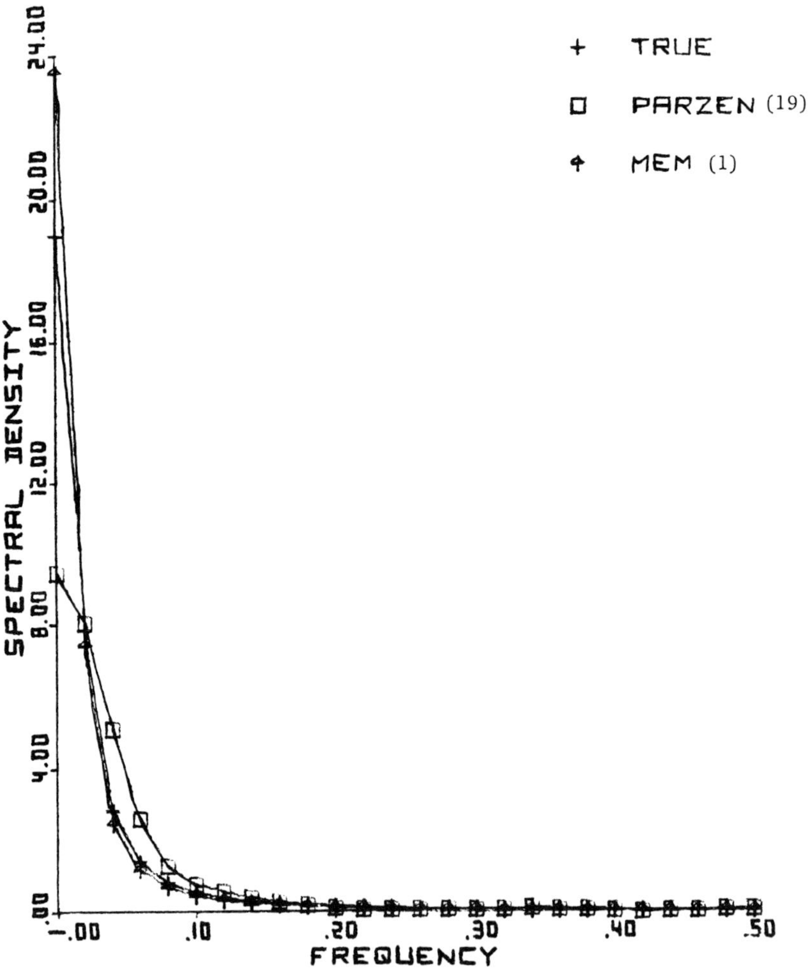

Fig. 7. Spectral estimates based on the sample acf of the AR(1) process $X_t = .9X_{t-1} + Z_t$.

TABLE 10

Summary Statistics for Spectral Estimates of the Discrete MA(1) Process $X_t = Z_t - .9Z_{t-1}$.
Statistics Based on 30 Realizations of 100 Observations Each.

Frequency (ω)	S(ω)	Estimator	Sample Mean	Sample Variance	Sample MSE
0.00	.006	BARTLETT(9)	.120	.0004	.0135
		TUKEY(8)	.045	.0002	.0018
		PARZEN(11)	.053	.0003	.0025
		ARSPEC(2-15)	.039	.0006	.0018
		MEM(4-17)	.025	.0002	.0006
		G(2-8)	.069	.0026	.0066
0.12	.275	BARTLETT	.371	.0087	.0178
		TUKEY	.318	.0079	.0097
		PARZEN	.324	.0081	.0105
		ARSPEC	.316	.0283	.0299
		MEM	.299	.0417	.0423
		G	.294	.0101	.0104
0.36	1.634	BARTLETT	1.542	.138	.147
		TUKEY	1.603	.094	.095
		PARZEN	1.594	.105	.106
		ARSPEC	1.749	1.165	1.178
		MEM	1.677	1.711	1.713
		G	1.627	.003	.003
0.42	1.871	BARTLETT	1.780	.090	.099
		TUKEY	1.815	.073	076
		PARZEN	1.816	.076	.079
		ARSPEC	2.343	1.514	1.736
		MEM	2.085	1.178	1.224
		G	1.849	.070	.071
0.50	1.994	BARTLETT	1.771	.301	.351
		TUKEY	1.866	.337	.354
		PARZEN	1.853	.322	.342
		ARSPEC	2.085	2.253	2.261
		MEM	2.073	2.439	2.445
		G	1.997	.160	.160

yielded spurious peaks as expected and were about the same, since the minimum FPE criterion selected a 6th-order model for each of them. PARZEN was based on 11 lags of the sample acf and also indicated spurious peaks.

Example 15. An ARMA(1,1) model.

Table 11 contains the summary statistics for each of the spectral estimators for the discrete ARMA(1,1) process given by (40). These statistics are based on 30 realizations of 200 observations each. The windows and the G yielded

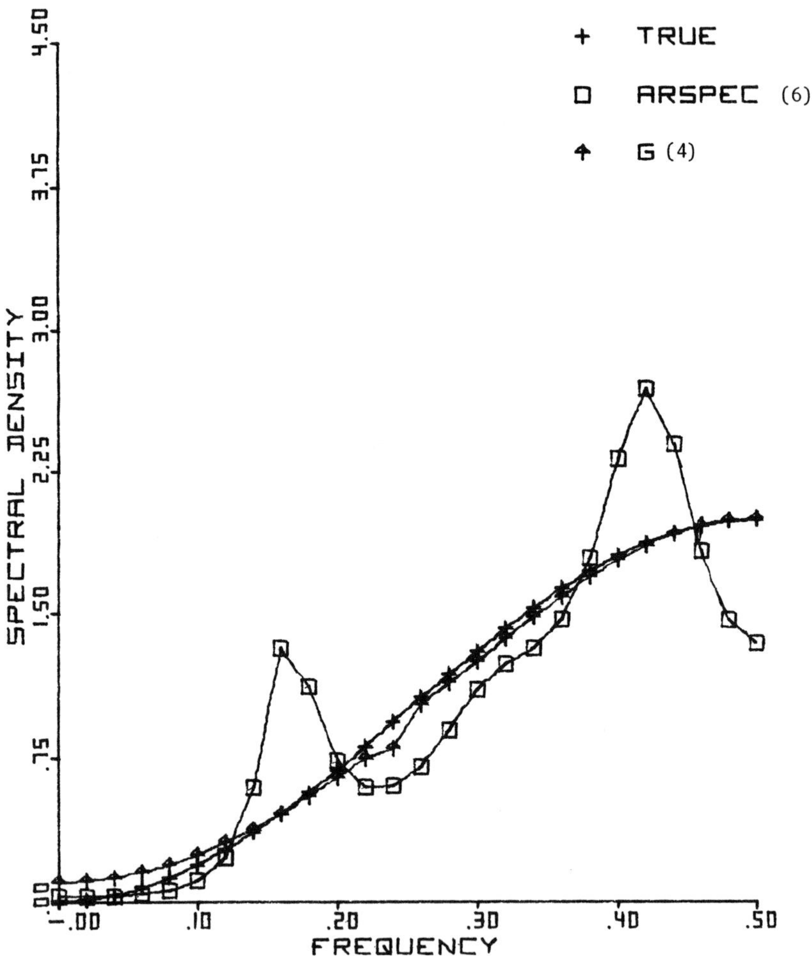

Fig. 8. Spectral estimates based on the sample acf of the MA(1) process $X_t = Z_t - .9Z_{t-1}$.

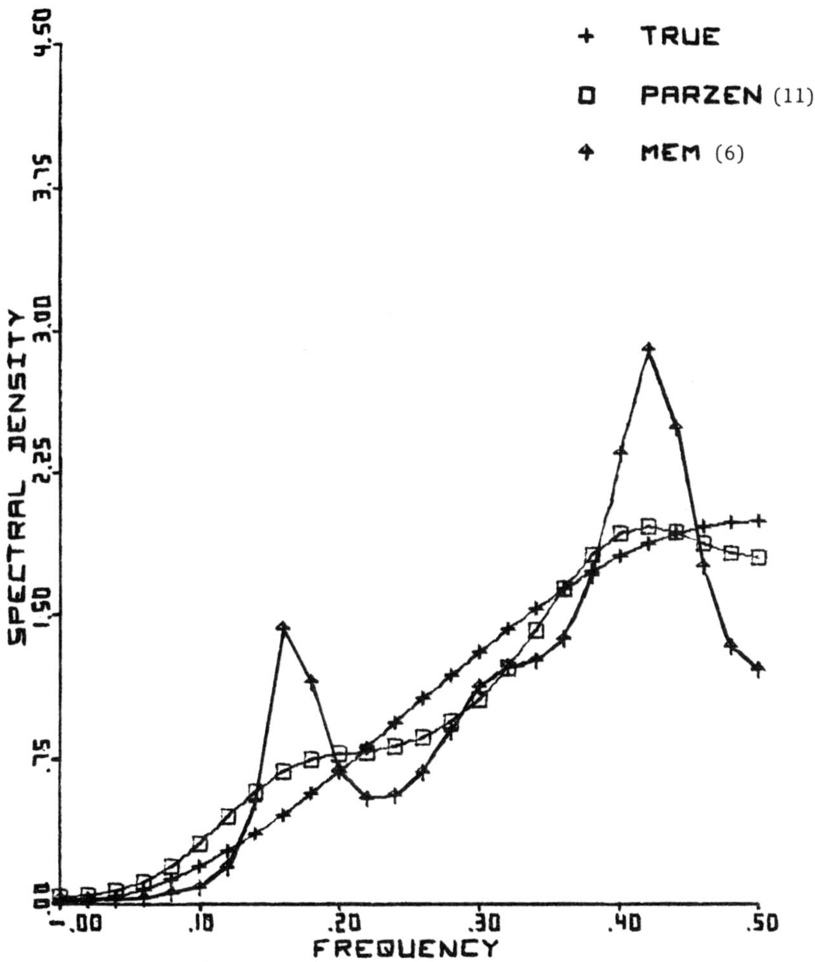

Fig. 9. Spectral estimates based on the sample acf of the MA(1) process $X_t = Z_t - .9Z_{t-1}$.

TABLE 11

Summary Statistics for Spectral Estimates of the Discrete ARMA(1,1) Process $X_t = .25X_{t-1} + Z_t - .95Z_{t-1}$.

Statistics Based on 30 Realizations of 200 Observations Each.

Frequency (ω)	S(ω)	Estimator	Sample Mean	Sample Variance	Sample MSE
0.00	.003	BARTLETT (23)	.062	.00006	.00356
		TUKEY(20)	.019	.00007	.00034
		PARZEN(28)	.021	.00018	.00052
		ARSPEC(4-17)	.034	.00059	.00159
		MEM(6-20)	.022	.00025	.00062
		G(2-14)	.045	.00382	.00558
0.22	1.048	BARTLETT	1.058	.0809	.0810
		TUKEY	1.054	.0806	.0806
		PARZEN	1.056	.0834	.0835
		ARSPEC	1.132	.2446	.2517
		MEM	1.073	.2605	.2611
		G	1.127	.0781	.0842
0.38	1.513	BARTLETT	1.499	.1380	.1382
		TUKEY	1.533	.1421	.1425
		PARZEN	1.529	.1479	.1482
		ARSPEC	1.612	.3398	.3495
		MEM	1.550	.4641	.4654
		G	1.572	.1294	.1330
0.50	1.598	BARTLETT	1.583	.2710	.2712
		TUKEY	1.573	.2721	.2727
		PARZEN	1.583	.2747	.2750
		ARSPEC	1.864	2.5610	2.6315
		MEM	1.782	1.9241	1.9580
		G	1.625	.2718	.2725

about the same results at each frequency. Both ARSPEC and MEM have higher MSE at all frequencies except low frequencies, where there is little power.

The spectral estimates for the randomly selected realization are plotted in Figures 10 and 11. Due to rather poor estimates of the autocorrelation function, the estimators did not estimate the true spectral density function very well. However, the G yielded the smoothest version and the summary statistics of Table 11 indicate this to be true in general. Both ARSPEC and MEM gave many spurious peaks. The minimum FPE criterion selected a 17th-order model for both estimators.

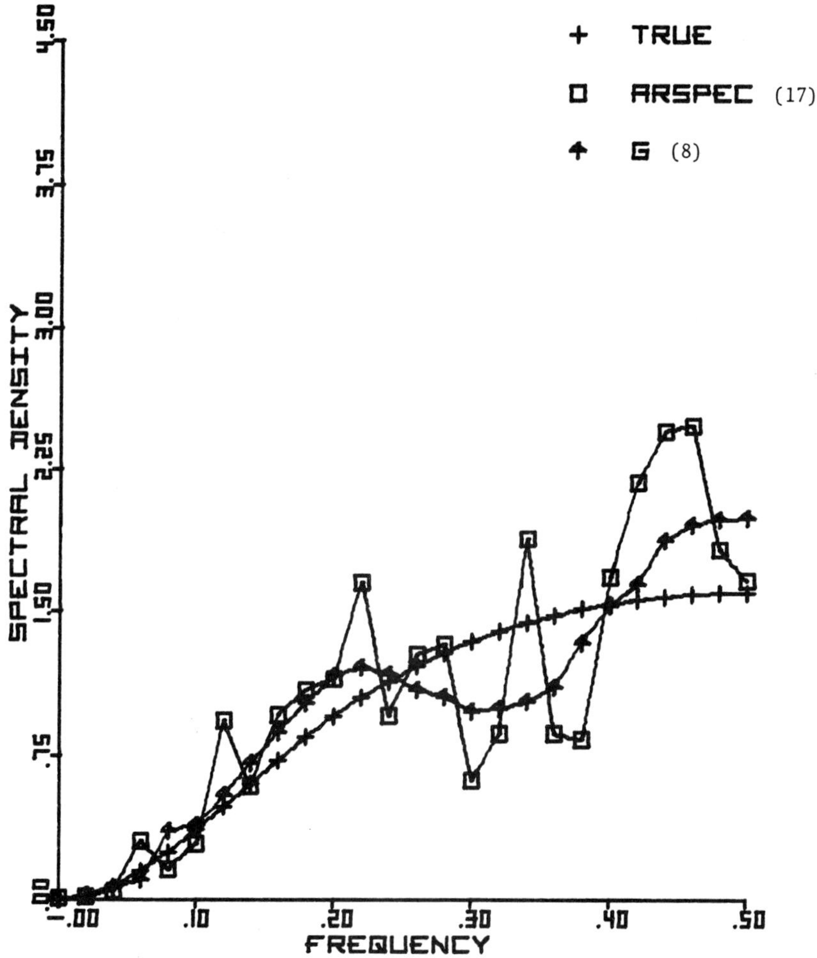

Fig. 10. Spectral estimates based on the sample acf of the discrete ARMA(1,1) process $X_t = .25X_{t-1} + Z_t - .95Z_{t-1}$.

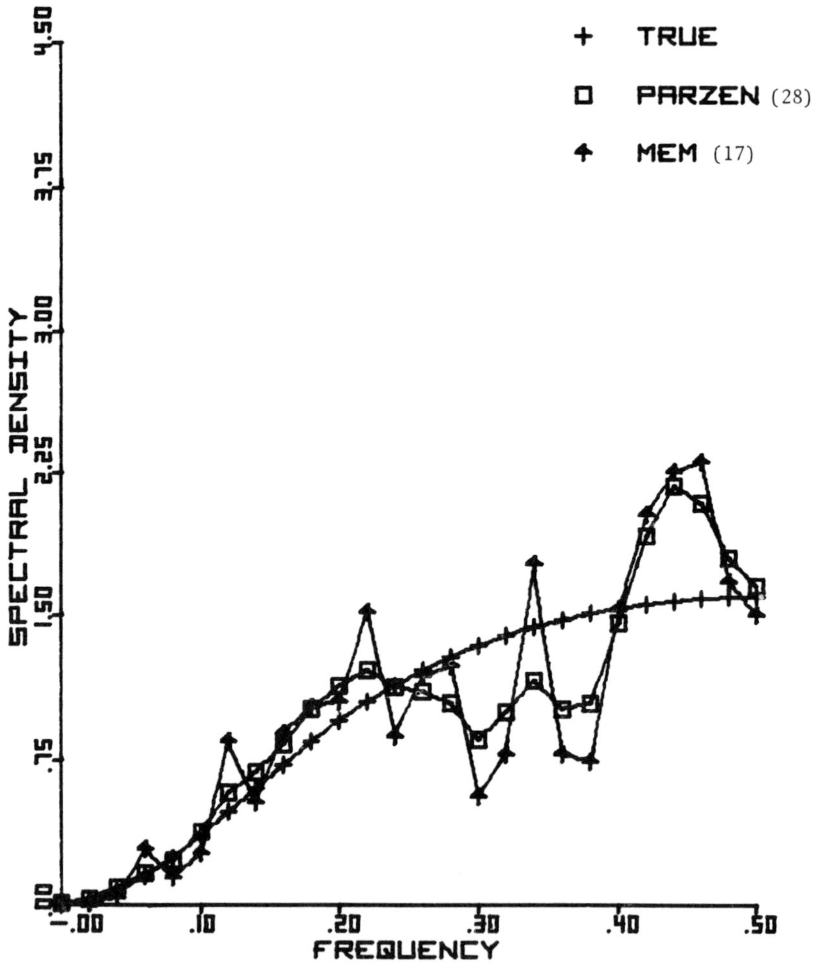

Fig. 11. Spectral estimates based on the sample acf of the discrete ARMA(1,1) process $X_t = .25X_{t-1} + Z_t - .95Z_{t-1}$.

PARZEN was based on 28 lags and the G was based on four to eight lags utilizing the 2nd- to 4th-order transform.

Example 16. An ARMA(1,2) model.

The summary statistics for the spectral estimators obtained from 30 realizations of 300 observations each from the discrete ARMA(1,2) process given by (41) are exhibited in Table 12. Note that the G has the smallest MSE at the

TABLE 12

Summary Statistics for Spectral Estimates of the Discrete ARMA(1,2) Process $X_t = -.5X_{t-1} + Z_t - 1.75Z_{t-1} + .8Z_{t-2}$ *. Statistics Based on 30 Realizations of 300 Observations Each.*

Frequency (ω)	S(ω)	Estimator	Sample Mean	Sample Variance	Sample MSE
0.00	.0001	BARTLETT(23)	.0270	.00001	.00074
		TUKEY(20)	.0022	.00000	.00001
		PARZEN(28)	.0033	.00009	.00010
		ARSPEC(5-17)	.0023	.00000	.00001
		MEM(10-20)	.0005	.00000	.00000
		G(2-14)	.0026	.00001	.00001
0.30	.571	BARTLETT	.624	.0145	.0258
		TUKEY	.583	.0148	.0192
		PARZEN	.586	.0154	.0201
		ARSPEC	.596	.0254	.0317
		MEM	.588	.0389	.0439
		G	.578	.0283	.0320
0.46	3.942	BARTLETT	3.558	.3847	.5232
		TUKEY	3.680	.2744	.3432
		PARZEN	3.675	.3082	.3794
		ARSPEC	3.955	.6671	.6673
		MEM	3.888	1.8032	1.8061
		G	3.720	.1326	.1820
0.50	4.581	BARTLETT	3.885	1.024	1.509
		TUKEY	4.127	.800	1.006
		PARZEN	4.093	.889	1.128
		ARSPEC	4.857	3.955	4.031
		MEM	4.427	6.280	6.304
		G	4.262	.725	.827

peak frequency .5. Again, both ARSPEC and MEM suffer from their underlying assumptions as is reflected in the high variability shown in Table 12. In order to demonstrate the GSPEC procedure at selected frequencies, the s- and r-matrices are presented in Table 13 for frequency $\omega = 0$ for a realization selected at random. One should examine that table at values of $s_n^{(\tau)}$ and $r_n^{(\tau)}$ at and near $\tau = 0$, since it is well known that for large values of τ, $\hat{\rho}_T(\tau)$ is highly variable. From the previous examples, it is clear that inspection of Table 13 would lead one to consider the underlying process as an ARMA(1,1), which is very nearly the case, since the last MA term is not very prominent. In fact, for these data, inspection of the G-matrix showed $G_1^{(1)}$ to yield the best spectral estimates at $\omega = 0$, .5 while $G_2^{(1)}$ was usually best at other frequencies and hence having known the underlying model was actually ARMA(1,2) would have led to no better G estimates.

TABLE 13

The s- and r-Matrices at Frequency 0.00 for a Randomly Selected Realization of the Discrete ARMA(1,2) Process $X_t = -.5X_{t-1} + Z_t - 1.75Z_{t-1} + .8Z_{t-2}$

	s-matrix				r-matrix			
m	$s_0^{(m)}$	$s_1^{(m)}$	$s_2^{(m)}$	$s_3^{(m)}$	$r_1^{(m)}$	$r_2^{(m)}$	$r_3^{(m)}$	$r_4^{(m)}$
-3	1.0000	-3.5490	8.2537	-9.0508	- .3478	-.1793	-.1487	-.0715
-2	1.0000	-2.8311	4.6805	-4.3888	.8865	.3435	.1388	.0294
-1	1.0000	-2.2321	2.9556	-2.4742	-1.6233	-.3767	-.0673	-.0153
0	1.0000	-1.8116	2.1192	.8845	2.0000	.2379	.0113	-.0956
1	1.0000	-1.5461	1.8482	-4.1882	-1.6233	-.0882	.0167	.0054
2	1.0000	-1.3923	2.9154	-1.0932	.8865	.0289	-.0073	-.0069
3	1.0000	-1.2767	3.4879	-2.6045	- .3478	-.0370	-.0050	.0067
4	1.0000	- .7850	3.3513	13.0593	.0962	.1276	-.0011	
5	1.0000	-5.6267	3.2791		.0207	.0516	-.0056	
6	1.0000	-2.5934	4.6127		- .0957	-.0136		
7	1.0000	-2.3614			.1526	.0130		
8	1.0000	-2.3125			- .2077			
9	1.0000				.2521			
10	1.0000							

For this realization, the G-spectral estimates are given in Table 14 for $\omega \in [0, .5]$ in increments of .02 and plotted in Figure 12.

TABLE 14

G-spectral Estimates for the Randomly Selected Realization of the Discrete ARMA(1,2) Process

Frequency	G-spectral estimate	Frequency	G-spectral estimate
0.00	.00354	0.26	.25296
0.02	.00315	0.28	.39250
0.04	.00120	0.30	.58700
0.06	.00174	0.32	.84410
0.08	.00408	0.34	1.07938
0.10	.00490	0.36	1.53896
0.12	.00507	0.38	1.95676
0.14	.00741	0.40	2.40427
0.16	.01438	0.42	2.86167
0.18	.02850	0.44	3.29165
0.20	.05325	0.46	3.64170
0.22	.09374	0.48	4.44973
0.24	.15714	0.50	4.57638

The other spectral estimators for this realization are plotted in Figures 12 and 13. Note that PARZEN, ARSPEC, and MEM put the peak at frequency 0.46 while the G indicated the correct peak frequency, 0.50. The minimum FPE criterion selected a 10th-order model for ARSPEC and a 12th-order model for MEM. PARZEN was determined using 28 lags of the acf.

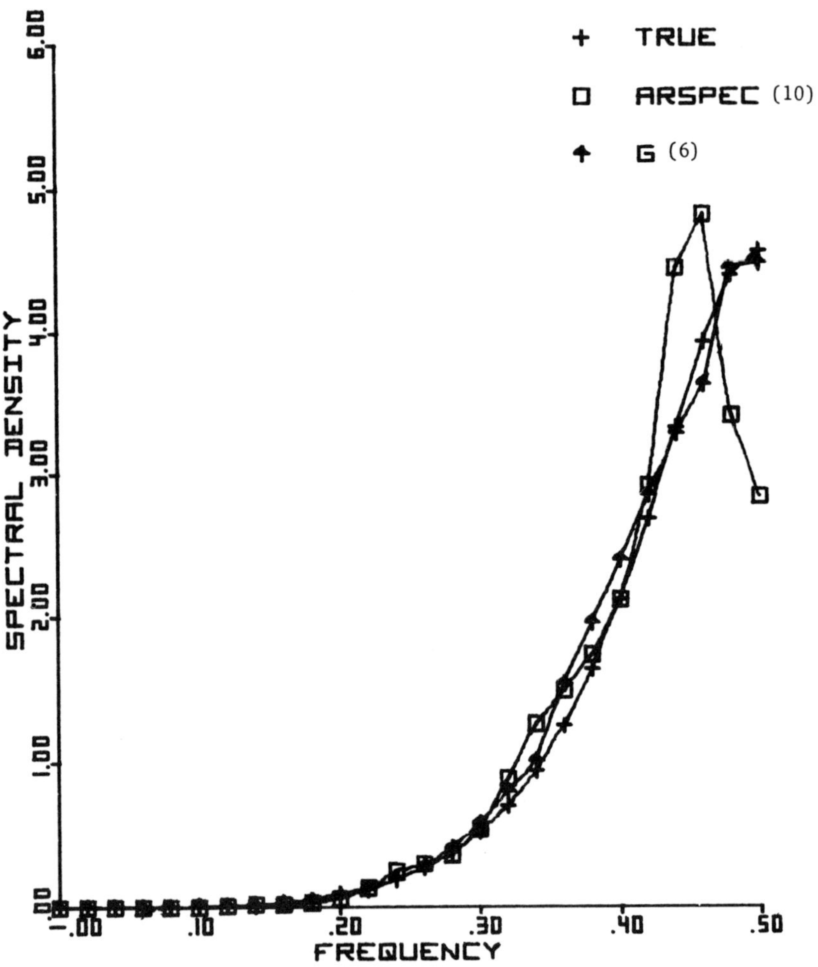

Fig. 12. Spectral estimates based on the sample acf of the discrete ARMA(1,2) process $X_t = -.5X_{t-1} + Z_t - 1.75Z_{t-1} + .8Z_{t-2}$.

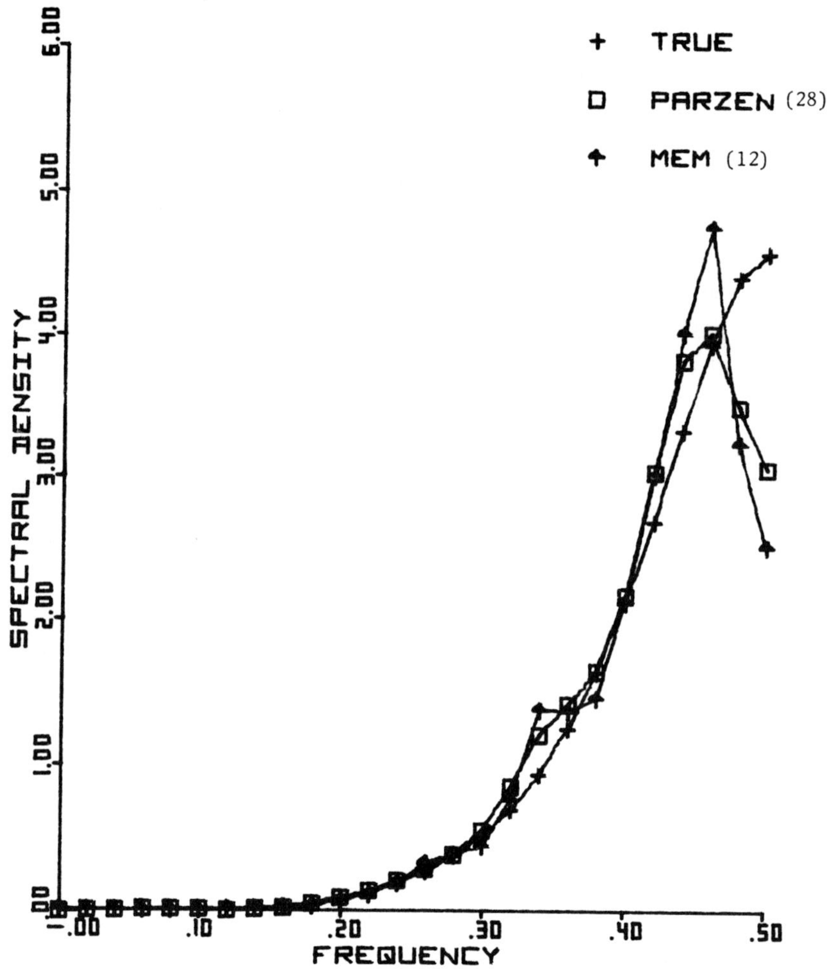

Fig. 13. Spectral estimates based on the sample acf of the discrete ARMA(1,2) process $X_t = -.5X_{t-1} + Z_t - 1.75Z_{t-1} + .8Z_{t-2}$.

Example 17. An invertible ARMA(3,2) model.

The summary statistics for each of the spectral estimators for the discrete ARMA(3,2) process given by (42) are given in Table 15. These results were determined from 30 realizations of 400 observations each from this process. Note that the window estimators, TUKEY and PARZEN, have nearly the same MSE as G at the lower frequencies. Both ARSPEC and MEM were highly variable at and near the peak frequency. The G was based on a 3rd- to 6th-order transform using from 4 to 14 lags of the acf. ARSPEC was calculated from a 7th- to an 18th-order autoregressive model, while MEM was determined from an 8th- to a 20th-order model. BARTLETT, TUKEY, and PARZEN were based on 38, 33, and 46 lags of the acf, respectively.

The spectral estimates are plotted for a randomly selected realization from this process in Figures 14 and 15. Note that the G, ARSPEC, and MEM indicate the peak at the correct frequency, 0.18, while PARZEN indicates the peak at frequency 0.16. Also, PARZEN, ARSPEC, and MEM indicate a smaller peak, that should not be there, near frequency .36. Note that the G also has a slight peak here, but not nearly as pronounced. For this realization, the G was based on a 2nd- to a 4th-order transform utilizing from four to eight lags of the acf. The minimum FPE criterion selected a 9th-order model for ARSPEC and a 10th-order model for MEM. PARZEN was determined using 46 lags of the acf.

Example 18. A "signal plus noise" model.

In order to demonstrate that the G-spectral estimator need not be restricted to ARMA processes, a "signal plus noise" process has also been considered. This process is given by

$$X(t) = \sin(2\pi\omega_0 t + \phi) + Z(t)$$

where ϕ is uniformly distributed over the interval $(-\pi, \pi)$ and $Z(t)$ is white noise.

TABLE 15

Summary Statistics for Spectral Estimates of the Discrete ARMA(3,2) Process $X_t = 1.5X_{t-1} - 1.21X_{t-2} + .455X_{t-3} + Z_t - .175Z_{t-1} + .8Z_{t-2}$. Statistics Based on 30 Realizations of 400 Observations Each.

Frequency (ω)	S(ω)	Estimator	Sample Mean	Sample Variance	Sample MSE
0.00	.017	BARTLETT(38)	.065	.00010	.00238
		TUKEY(33)	.023	.00010	.00014
		PARZEN(46)	.018	.00028	.00028
		ARSPEC(7-18)	.023	.00012	.00015
		MEM(8-20)	.014	.00003	.00004
		G(4-14)	.018	.00018	.00019
0.16	4.063	BARTLETT	3.438	.2819	.6718
		TUKEY	3.615	.2305	.4314
		PARZEN	3.598	.2463	.4619
		ARSPEC	3.907	1.1609	1.1851
		MEM	3.771	1.1351	1.2200
		G	4.022	.1908	.1924
0.18	4.480	BARTLETT	4.079	.6587	.8192
		TUKEY	4.226	.5708	.6353
		PARZEN	4.224	.6054	.6710
		ARSPEC	5.142	3.3303	3.7695
		MEM	4.624	2.4798	2.5005
		G	4.815	.6139	.7261
0.26	1.042	BARTLETT	1.109	.0579	.0624
		TUKEY	1.071	.0611	.0619
		PARZEN	1.078	.0611	.0624
		ARSPEC	1.092	.0630	.0655
		MEM	1.033	.0745	.0746
		G	1.123	.0639	.0705
0.50	.319	BARTLETT	.366	.0150	.0172
		TUKEY	.349	.0145	.0154
		PARZEN	.347	.0151	.0158
		ARSPEC	.376	.0199	.0231
		MEM	.361	.0235	.0252
		G	.336	.0081	.0083

It has mean zero and autocovariance function as follows:

$$\gamma(\tau) = \frac{1}{2}\cos(2\pi\omega_0\tau) + \sigma_z^2\delta(\tau), \quad -\infty < \tau < \infty,$$

where $\sigma_z^2\delta(\tau)$ is the variance of $Z(t)$ and $\delta(\tau)$ is, as mentioned earlier, the Dirac delta function. The spectral density function for this process is given by

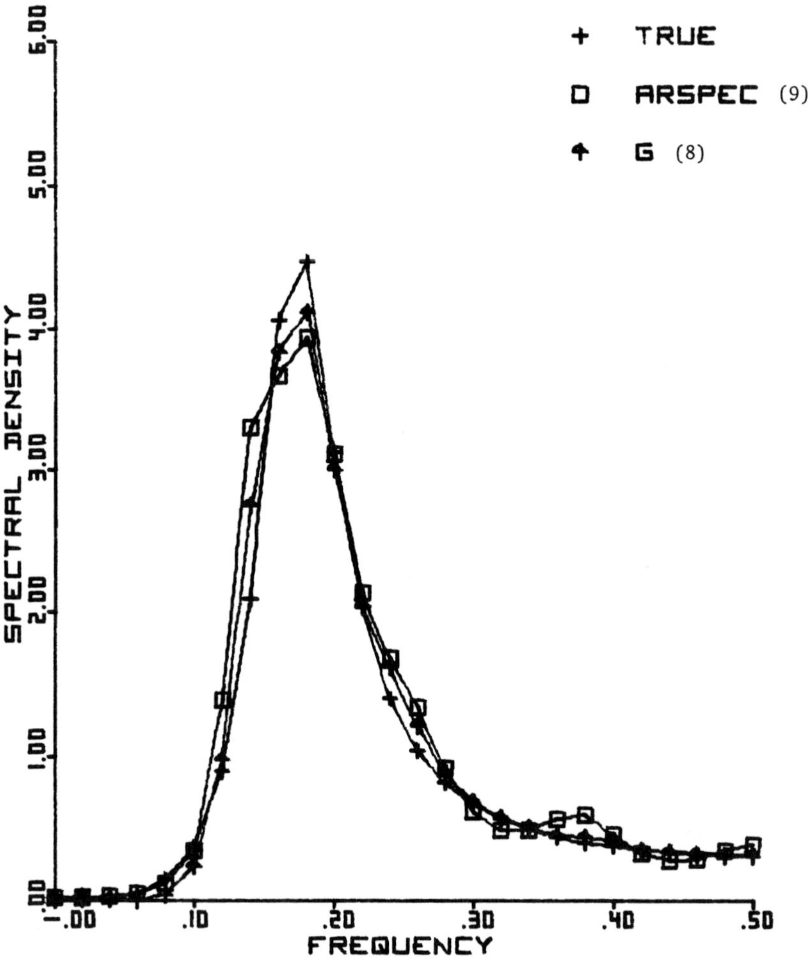

Fig. 14. Spectral estimates based on the sample acf of the discrete ARMA(3,2) process $X_t = 1.5X_{t-1} - 1.21X_{t-2} + .455X_{t-3} + Z_t - 1.75Z_{t-1} + .8Z_{t-2}$.

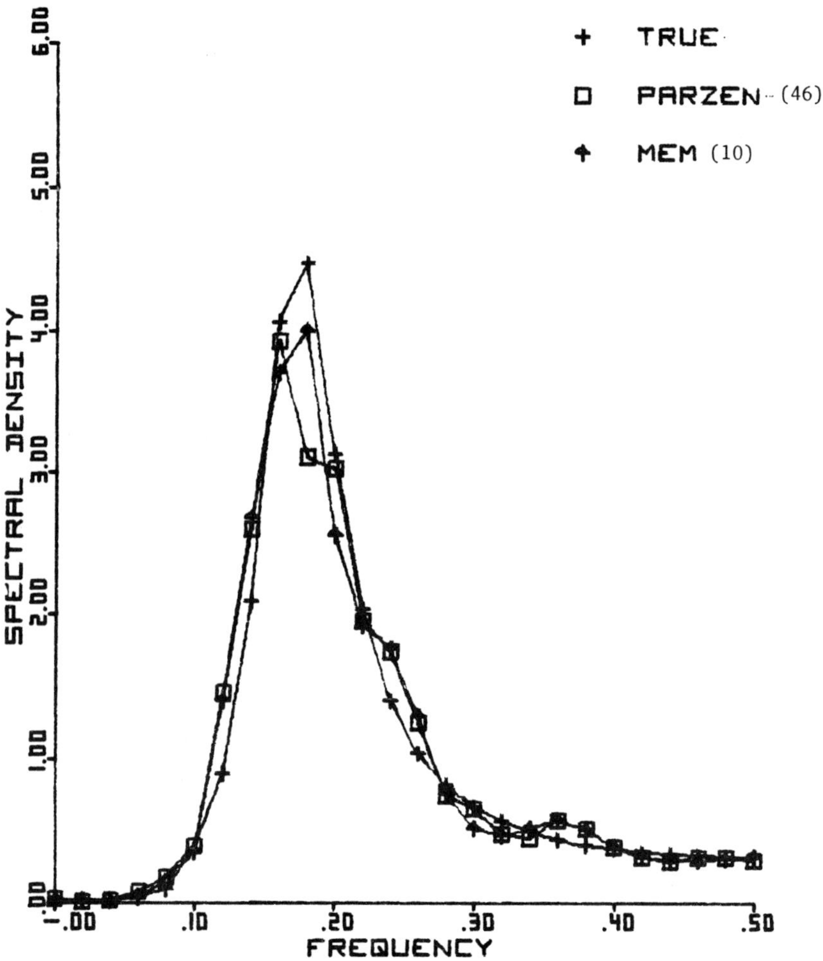

Fig. 15. Spectral estimates based on the sample acf of the discrete ARMA(3,2) process $X_t = 1.5X_{t-1} - 1.21X_{t-2} + .455X_{t-3} + Z_t - 1.75Z_{t-1} + .8Z_{t-2}$.

$$S(\omega) = \frac{\frac{1}{4}[\delta(\omega - \omega_0) + \delta(\omega + \omega_0)] + \sigma_z^2}{\gamma(0)}$$

$$= \frac{\delta(\omega - \omega_0) + \delta(\omega + \omega_0)}{2(2\sigma_z^2 + 1)} + \frac{2\sigma_z^2}{2\sigma_z^2 + 1}$$

for $-\infty < \omega < \infty$. Hence, the spectral density is constant over all frequencies with delta functions at frequencies ω_0 and $-\omega_0$

One realization of length 20 has been generated from the process

$$X(t) = \sin(2\pi t) + Z(t)$$

at sampling interval $\Delta t = .05$ and superimposed with 10% white noise, i.e., $\sigma_z^2 = (.1/\sqrt{2})^2 = .005$. Hence, the realization is of length one period. A spectral analysis should indicate the peak at frequency 1.0. The spectral estimates for this realization are plotted in Figures 16 and 17 for $\omega \in [0, 2.0]$ with an increment of 0.10. The G was determined by a 2nd-, 3rd-, and 4th-order transform using from four to eight lags of the acf. PARZEN was calculated using 15 lags of the acf. The minimum FPE criterion selected a second-order autoregressive model for ARSPEC and a ninth-order model for MEM. Note that the G, ARSPEC, and MEM indicated the correct peak frequency, 1.0, while PARZEN indicated no peak. This is due to the additive effects of the window at frequencies 1.0 and -1.0.

<u>Example 19</u>. A real data example.

The data for this example are the "uncontrolled" temperature readings (every minute for 226 minutes) of a chemical process given in [13] and referred to as "Series C". Figure 18 is a graph representing the data.

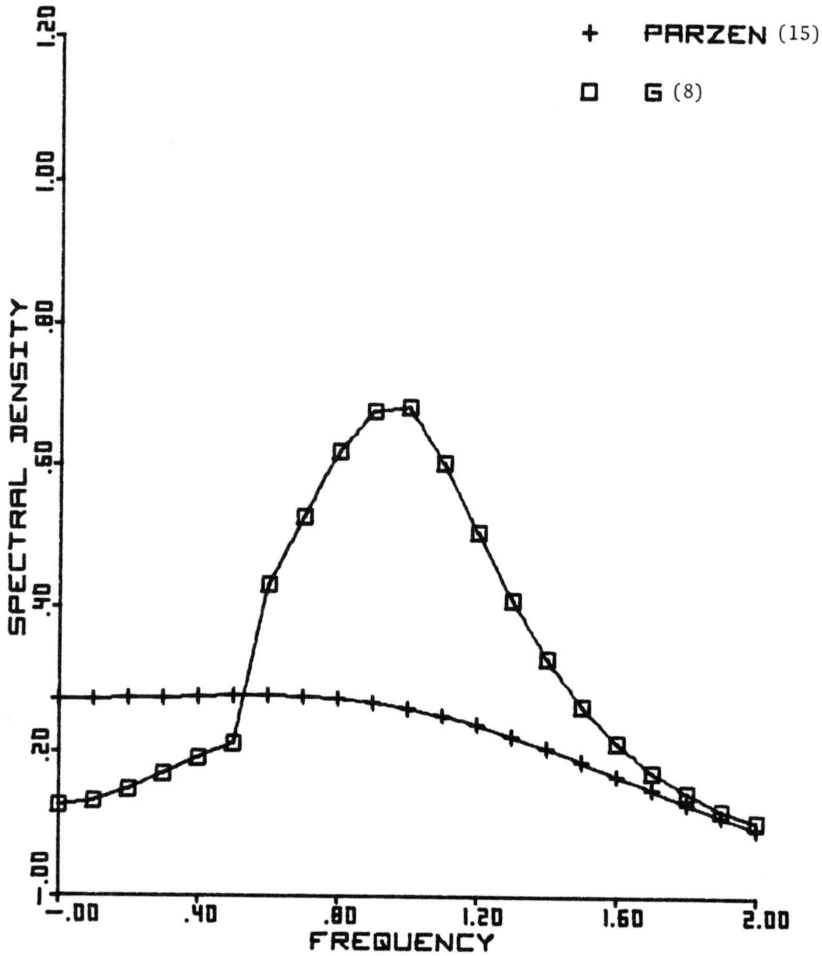

Fig. 16. Spectral estimates based on the sample acf of the "signal plus noise" process $X_t = sin(2\pi t) + Z(t)$.

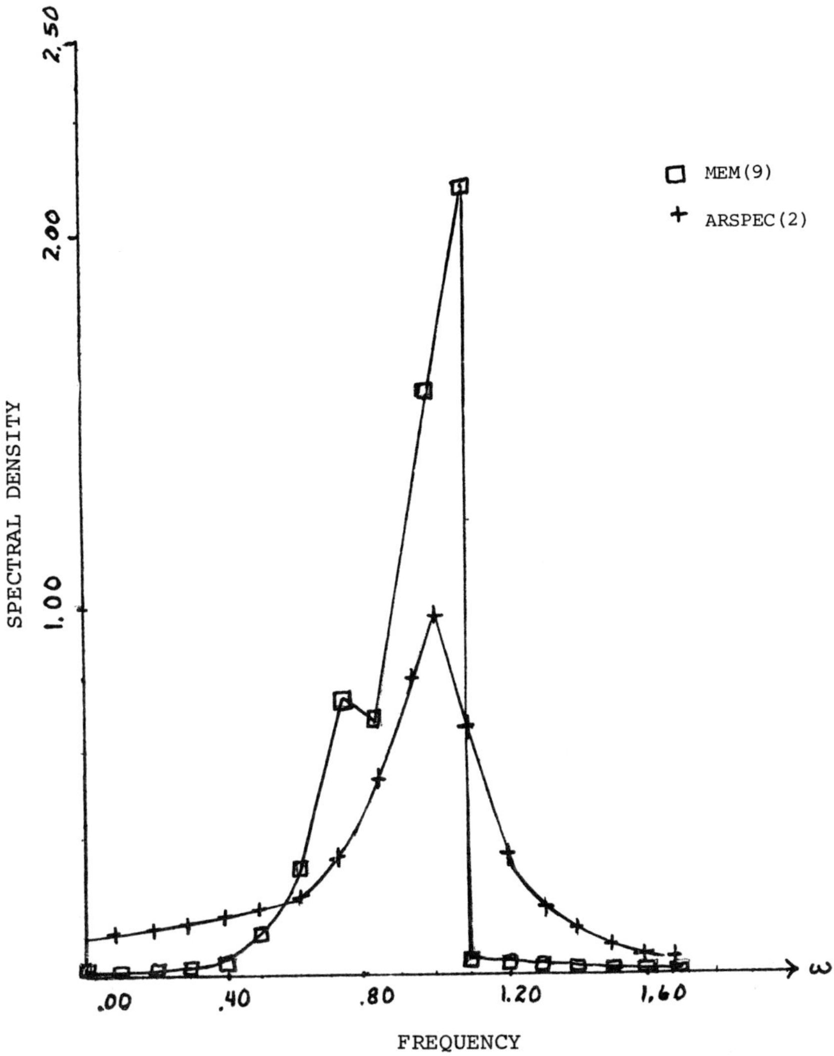

Fig. 17. Spectral estimates based on the sample acf of the "signal plus noise" process $X_t = \sin(2\pi t) + Z(t)$.

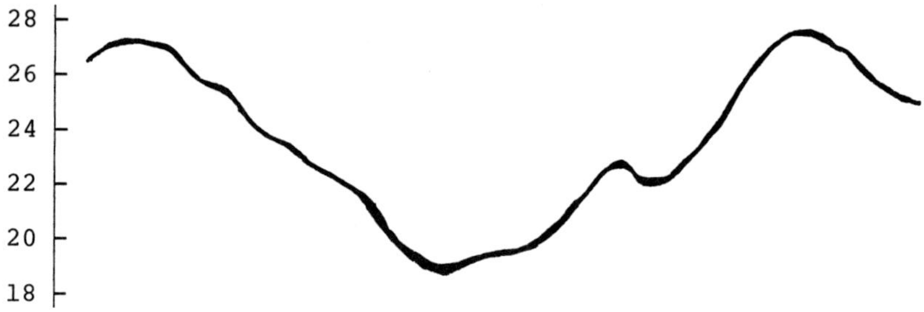

Fig. 18. Series C.

The process has most of its power at zero as is clearly seen by Figure 19, which compares the autoregressive spectral estimator with the G-spectral estimator. As can be seen from that figure, the autoregressive spectral estimator puts considerably more power at zero. As we have seen from our previous examples, it is common for the autoregressive estimator to overestimate at the peak. Moreover, this set of real data looks essentially like our simulated data and, hence, in this case, the G-spectral estimator is probably more correct.

Tables 16, 17, 18, and 19 give the r-, s-, and G-matrices at 0 and .5. From Tables 16 and 17, the process appears to be an ARMA(2,1). The series in fact has been identified by Box and Jenkins as an ARMA(2,0). Either model suffices for our problem. The stabling off procedure utilized in the program given in the Appendix used $G_2^{(1)}$ to $G_6^{(1)}$ as estimates with $G_{(4)}^{(1)}$ being most common. At $\omega = 0, .5$, the estimate selected by the program was $G_2^{(1)}$. This is the same as one would have selected by inspecting the r- and s-matrices. It should be remarked that at $\omega = 0$, inspection of Table 16 suggests that $p = 2$, $q = 1$. Moreover, it should be noted that since all of the elements in Column 2 of the s-matrix are near zero, the process is nearly non-stationary. This is

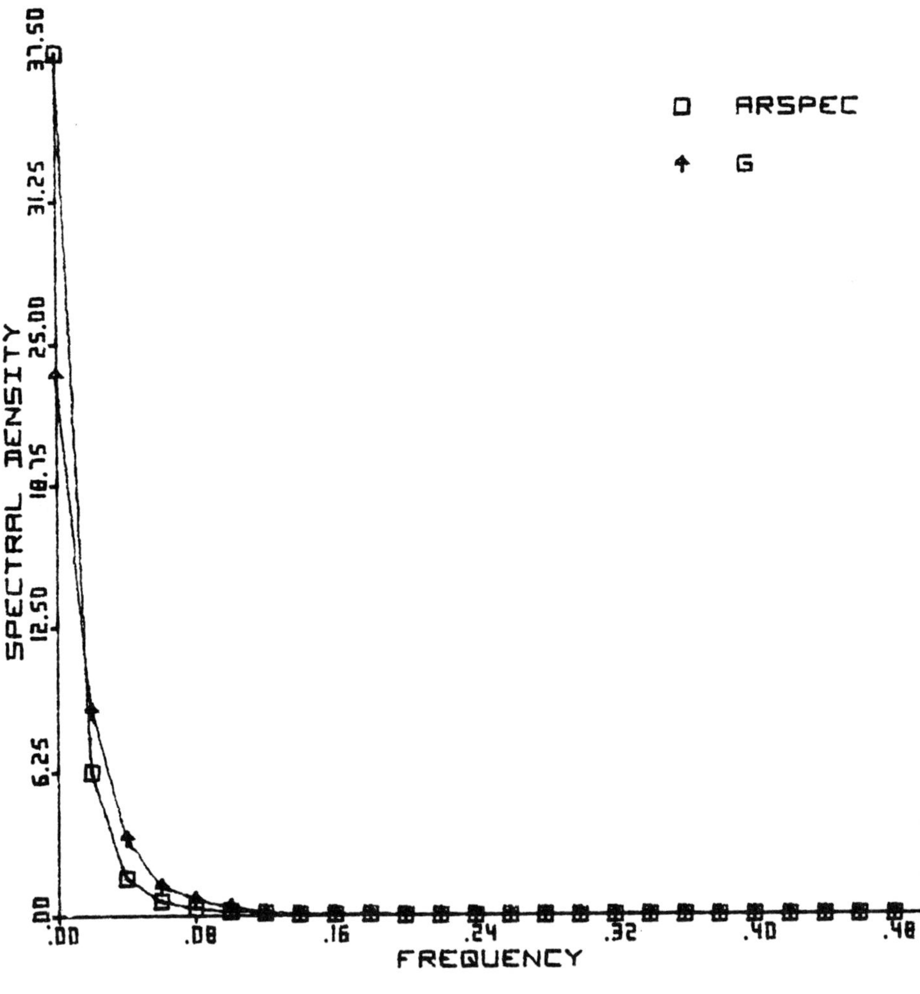

Fig. 19. Spectral estimates of "Series C".

TABLE 16

r-Matrix (ω = 0) "Series C"

m	$r_1^{(m)}$	$r_2^{(m)}$	$r_3^{(m)}$	$r_4^{(m)}$	$r_5^{(m)}$	$r_6^{(m)}$	$r_7^{(m)}$	$r_8^{(m)}$
-15	.50507	- .09980	.30381	- .13331	.47427	.19238	- 1.26744	- .21676
-14	.61087	- .09644	.09756	.20253	.62402	.77013	- .24349	- .01085
-13	.71793	- .10450	.46301	- .98364	1.32912	- .17551	- .08774	1.63206
-12	.82685	- .10516	- .26964	- .53049	- .24649	.12500	.12379	3.66317
-11	.93644	- .10260	.27698	- .12852	- .07505	.22086	- 2.58896	- 2.94141
-10	1.04662	- .09256	.01906	- .17799	- .09949	.00422	- 2.55052	2.26228
- 9	1.15769	- .11218	.14185	.20427	- .02243	3.12729	- 2.10838	-37.52870
- 8	1.27072	- .13384	4.03638	- .00992	- .01810	.64235	- 2.09669	- 3.00756
- 7	1.38384	- .13213	.09598	.02791	-1.72452	2.49871	- .73116	- 1.98716
- 6	1.49511	- .18687	- .07675	- .03085	-3.51733	6.51053	2.01184	.07455
- 5	1.60470	- .22949	- .01088	.93294	-3.13506	-2.10767	- .02468	.09091
- 4	1.70863	- .31017	.02372	2.93463	2.23642	.09583	- .09268	1.31131
- 3	1.80443	- .44647	-3.18850	-2.46647	- .12875	- .19510	- .09598	.05196
- 2	1.88820	- .68976	2.92587	.23005	- .04969	- .01777	- .08921	.31942
- 1	1.95514	-3.95514	- .45940	.00823	- .07953	.06674	- .09702	- .15976
0	2.00000	1.02927	.01990	.03651	- .02994	- .02017	- .09367	.01651
1	1.95514	.55879	- .00844	.08096	- .06635	.09686	.17187	.10679
2	1.88820	.35467	- .04939	- .01621	- .08702	.26078	- .15002	
3	1.80443	.24898	.14166	.09570	- .05323	.07300	- .22676	
4	1.70863	.19706	- .14067	- .09165	- .10422	.25816		
5	1.60470	.13416	-1.98677	.17948	- .17553	.08897		
6	1.49511	.13606	.01861	.15856	- .70370			
7	1.38384	.11210	- .15286	.10547	- .37759			
8	1.27072	.09095	- .07577	- .40553				
9	1.15769	.10178	- .13438	.53883				
10	1.04662	.10460	1.33264					
11	.93644	.10386	- .14450					
12	.82685	.09480						
13	.71793	.09863						
14	.61087							
15	.50507							

s-Matrix (ω = 0) "Series C"

m	$s_0^{(m)}$	$s_1^{(m)}$	$s_2^{(m)}$	$s_3^{(m)}$	$s_4^{(m)}$	$s_5^{(m)}$	$s_6^{(m)}$	$s_7^{(m)}$
-15	1.00000	.20948	- .00589	- .00860	- .00794	- .00838	- .03293	- .01718
-14	1.00000	.17526	.01267	.00315	- .02654	- .01096	.02126	- .03247
-13	1.00000	.15172	.00084	.00453	- .00970	- .01732	.05071	- .03649
-12	1.00000	.13254	- .00286	.02106	.01461	- .02965	.01513	- .51754
-11	1.00000	.11766	- .01039	- .01928	.04263	.01974	.02362	.21474
-10	1.00000	.10612	.02070	.11073	.06060	- .02408	-14.46230	.46239
- 9	1.00000	.09764	.01719	- .02822	.03109	- .01955	- 2.66732	- .03375
- 8	1.00000	.08902	- .00103	- .02965	.10137	3.35682	6.08753	- .63789
- 7	1.00000	.08041	.03037	- .02658	.03561	2.10646	.97944	- 3.26178
- 6	1.00000	.07330	.01477	- .01692	2.02623	.61003	.86944	- .04002
- 5	1.00000	.06477	.01971	- .06486	-5.61296	- .65681	.03953	.08085
- 4	1.00000	.05607	.02040	-2.61608	.38335	- .03781	.02935	.00155
- 3	1.00000	.04643	.01932	- .20829	.03575	- .00967	.04329	- .01956
- 2	1.00000	.03545	.10862	- .03270	.01574	- .04763	.27714	- .00817
- 1	1.00000	.02294	.02827	- .01633	- .07929	- .05827	- .09329	.01872
0	1.00000	- .02243	.01565	- .02308	.09345	.07164	- .54196	.05066

(continued on following page)

TABLE 16

(continued)

s-Matrix ($\omega = 0$) *"Series C"*

m	$s_0^{(m)}$	$s_1^{(m)}$	$s_2^{(m)}$	$s_3^{(m)}$	$s_4^{(m)}$	$s_5^{(m)}$	$s_6^{(m)}$	$s_7^{(m)}$
1	1.00000	-.03424	.01621	.07675	.05889	.09342	-.01787	.05553
2	1.00000	-.04437	.01582	-.04906	.29990	-.01056	-.02965	.01600
3	1.00000	-.05309	.01268	-.04345	.02719	.04117	.03128	
4	1.00000	-.06083	.02180	-.01389	.04298	.01233	.00380	
5	1.00000	-.06830	-.00106	-.01453	.01802	-.00581		
6	1.00000	-.07442	.01440	-.15464	-.00193	-.00507		
7	1.00000	-.08174	.01679	.00576	.01094			
8	1.00000	-.08895	-.01143	-.00226	.02113			
9	1.00000	-.09594	-.00292	-.00907				
10	1.00000	-.10527	.00083	-.01274				
11	1.00000	-.11703	.01150					
12	1.00000	-.13173	-.00603					
13	1.00000	-.14912						
14	1.00000	-.17320						
15	1.00000							
16	1.00000							

TABLE 17

G-Matrix ($\omega = 0$) *"Series C"*

m	$G_0^{(m)}$	$G_1^{(m)}$	$G_2^{(m)}$	$G_3^{(m)}$	$G_4^{(m)}$	$G_5^{(m)}$	$G_6^{(m)}$	$G_7^{(m)}$	$G_8^{(m)}$
0	0.00000	88.16937	22.40053	23.26272	22.87205	23.28994	23.25272	25.10164	25.04729
1	1.97757	58.10624	23.62840	23.73838	22.36371	23.07399	28.49429	25.39891	
3	5.74556	38.83150	19.20115	22.46156	18.94194	20.23474			
4	7.50209	34.73825	25.69913	15.57119	17.70354				
5	9.15875	31.85285	158.61624	21.87912					
6	10.70865	30.05025	20.59812						
7	12.14813	28.38568							
8	13.47540								

compatible with the findings of Box and Jenkins in [13]. Now, at $\omega = .5$ (the model of course has not changed, but the order used in GSPEC may change with ω), the situation is not as clear since the columns after the first in the r-matrix are near zero and the second row of the s-matrix is near constant. However, the entire story is easily seen in the s-matrix. That is, it has been shown by McIntire and Gray in some current research work at Southern Methodist University that an ARMA process is a stationary ARMA(p,q) if and only if p is the smallest integer, such that $r_{p+1}^{(m)} \neq 0$ for $m = q - p, -q - p$, and $r_{p+1}^{(m)} \equiv 0$ for $m \not\in [q - p, -q - p]$. Moreover, they also have shown the process to be a stationary ARMA(p,q) if and only if $s_p^{(m)} \neq$ constant for $m = q - p$, $q - p - 1 \ldots , - q - p + 1$, $s_p^{(m)} = c_1$, for $m \geq q - p + 1$ and $s_p^{(m)} = c_] \neq c_1$, for $m < - q - p + 1$. In addition, they have shown that an ARMA(2,q) process is non-stationary if $s_1^{(m)}$ at $\omega = .5$ is identically 2. These results and others, as mentioned before, concerning the s- and r-matrices will appear in a later paper. For our purposes, they are not essential but at times helpful. Table 18 vividly demonstrates the correctness of those results. In fact, the closeness of the results to the theoretical results stated above is remarkable, since the table is of course a table of estimated $r_n^{(m)}$'s and $s_n^{(m)}$'s.

Notice that in the r-matrix, $r_3^{(-1)}$, $r_3^{(-2)}$, and $r_3^{(-3)}$ are several orders of magnitude larger than the absolute value of the other $r_3^{(m)}$. The proximity to zero of all of these numbers is caused by the near non-stationarity of the process. In the s-matrix, it is clear that $s_2^{(-1)} \neq s_2^{(-2)}$; whereas, nearby the $s_n^{(m)}$'s are approximately constant. The data are, therefore, well modeled by an ARMA(2,1).

The general problem of ARMA(p,q) model fitting using the above techniques is currently being considered by D. McIntire, G. Kelley, and H. Gray, and should appear in the near future. This example certainly suggests the approach should be fruitful.

TABLE 18

r-Matrix (ω = .5) "Series C"

m	$r_1^{(m)}$	$r_2^{(m)}$	$r_3^{(m)}$	$r_4^{(m)}$	$r_5^{(m)}$	$r_6^{(m)}$	$r_7^{(m)}$	$r_8^{(m)}$
-15	- .50507	- .00946	.00045	.00079	.00061	- .00095	- .00517	.00034
-14	.61087	.00777	.00029	- .00012	.00065	- .01083	- .00043	- .00015
-13	- .71793	- .00737	- .00010	- .00037	.00068	.00026	- .00056	- .02003
-12	.82685	.00654	.00019	- .00102	- .00057	.00059	- .00528	.13297
-11	- .93644	- .00570	.00075	.00063	.00492	.00062	.02288	- .00650
-10	1.04662	.00466	.00009	.00062	- .00104	.00004	.02158	- .01542
- 9	-1.15769	- .00522	- .00056	.00165	.00032	- .01940	.02918	.03959
- 8	1.27072	.00570	- .00105	- .00015	.00072	.02644	.03149	- .02656
- 7	-1.38384	- .00511	- .00061	- .00030	.01728	- .02551	.00754	- .03558
- 6	1.49511	.00661	- .00026	- .00010	.01152	- .01584	.03514	.00133
- 5	-1.60470	- .00720	.00005	.01082	.01378	- .03682	.00043	.00080
- 4	1.70863	.00846	.00011	- .01352	.03481	.00167	- .00096	- .00038
- 3	-1.80443	- .01013	.01384	- .03840	.00200	.00047	- .00144	- .00035
- 2	1.88820	.01201	.03318	.00358	- .00022	.00018	- .00123	.00071
- 1	-1.95514	- .04486	.00521	- .00004	- .00026	.00275	- .00082	- .00580
0	2.00000	.01167	.00009	.00042	- .00030	.00013	- .00083	- .00020
1	-1.95514	- .00973	.00004	.00026	- .00265	.00082	.00733	.00148
2	1.88820	.00805	- .00022	.00017	- .00125	.00073	.00096	
3	-1.80443	- .00679	- .00048	.00143	.00055	.00034	.00040	
4	1.70863	.00618	- .00090	- .00074	- .00684	- .00038		
5	-1.60470	- .00474	- .00052	- .00062	.00040	- .00125		
6	1.49511	.00526	.00007	- .00078	.00036			
7	-1.38384	- .00478	.00071	.00020	.00039			
8	1.27072	.00423	.00021	- .00015				
9	-1.15769	- .00513	- .00010	- .00032				
10	1.04662	.00581	.00028					
11	- .93644	- .00646	.00044					
12	.82685	.00668						
13	- .71793	- .00795						
14	.61087							
15	- .50507							

s-Matrix (ω = .5) "Series C"

m	$s_0^{(m)}$	$s_1^{(m)}$	$s_2^{(m)}$	$s_3^{(m)}$	$s_4^{(m)}$	$s_5^{(m)}$	$s_6^{(m)}$	$s_7^{(m)}$
-15	1.00000	-2.20948	3.96160	- 1.45673	6.15644	- 1.69061	8.07456	- 10.92709
-14	1.00000	-2.17526	4.19207	- 5.34549	- 25.37871	.77980	11.92878	- 2.34793
-13	1.00000	-2.15172	4.02425	- 11.93929	18.98242	- 11.64810	- 7.91385	- 2.97376
-12	1.00000	-2.13254	3.96463	10.94603	6.34841	- 6.28964	- .35499	- 14.25727
-11	1.00000	-2.11766	3.82915	- 3.92129	.64969	- 7.04232	2.67187	- 97.10245
-10	1.00000	-2.10612	4.44628	- 31.84150	5.81791	- 2.83606	1708.96556	- 67.83837
- 9	1.00000	-2.09764	4.37071	3.49396	2.16327	- 3.15138	-192.71132	- 31.99257
- 8	1.00000	-2.08902	3.94326	- 1.98500	- 2.54289	81.54503	-405.38725	- 72.22381
- 7	1.00000	-2.08041	4.75556	- 2.46482	3.55445	206.33582	94.96884	182.19500
- 6	1.00000	-2.07330	4.31440	- 5.28836	-618.79679	-250.66350	49.77149	- 2.23527
- 5	1.00000	-2.06477	4.47188	5.59048	**********	37.59936	2.26303	- 9.15457
- 4	1.00000	-2.05607	4.49725	567.95996	24.62592	- 2.16461	2.84589	- 1.46459
- 3	1.00000	-2.04643	4.44947	13.38024	2.29684	- 3.97262	- 2.88299	- 2.87034
- 2	1.00000	-2.03545	9.57672	- 2.10089	3.58234	4.66564	20.02284	- 3.69368
- 1	1.00000	-2.02294	2.49221	- 3.54481	24.21412	1.41559	11.02422	- .51582
0	1.00000	-1.97757	3.60458	- 1.08930	9.32651	- 11.55025	- 61.31363	- 4.12882

(continued on following page)

TABLE 18

(continued)

s-Matrix (ω = .5) "Series C"

m	$s_0^{(m)}$	$s_1^{(m)}$	$s_2^{(m)}$	$s_3^{(m)}$	$s_4^{(m)}$	$s_5^{(m)}$	$s_6^{(m)}$	$s_7^{(m)}$
1	1.00000	-1.96576	3.57258	-23.99536	-1.47730	-11.00389	.41915	-4.01470
2	1.00000	-1.95563	3.58958	4.54958	20.87030	-3.76784	4.62169	-10.17638
3	1.00000	-1.94691	3.70466	2.90924	2.61044	-8.73441	17.54655	
4	1.00000	-1.93917	3.41399	-1.71978	.65510	-8.29485	-.93288	
5	1.00000	-1.93170	4.06055	-4.17814	7.83219	-.41286		
6	1.00000	-1.92558	3.66072	31.45143	3.77402	1.02297		
7	1.00000	-1.91826	3.60463	-2.99440	10.46306			
8	1.00000	-1.91105	4.21072	-5.95026	-16.37932			
9	1.00000	-1.90406	4.04190	-15.38435				
10	1.00000	-1.89473	3.97439	2.15756				
11	1.00000	-1.88297	3.80277					
12	1.00000	-1.86827	4.05133					
13	1.00000	-1.85088						
14	1.00000	-1.82680						
15	1.00000							
16	1.00000							

TABLE 19

G-Matrix (ω = .5) "Series C"

m	$G_0^{(m)}$	$G_1^{(m)}$	$G_2^{(m)}$	$G_3^{(m)}$	$G_4^{(m)}$	$G_5^{(m)}$	$G_6^{(m)}$	$G_7^{(m)}$	$G_8^{(m)}$
0	0.00000	.01134	.00810	.00815	.00810	.00808	.00808	.00788	.00788
1	.02243	.00540	.00813	.00813	.00830	.00806	.00610	.00793	
2	-.01104	.01038	.00814	.00819	.00818	.00785	.00769		
3	.03085	.00625	.00808	.00825	.00770	.00776			
4	-.01705	.00975	.00794	.00741	.00854				
5	.03492	.00655	.00772	.00759					
6	-.01988	.00901	.00757						
7	.03575	.00627							
8	-.02080								

IX. Concluding Remarks.

In this paper, we have introduced a new spectral estimator which appears to be of some value in the problem of spectral estimation. Although we have tied this estimator closely to the ARMA processes, it should be understood that that model is not an underlying assumption. Indeed, the underlying assumption is basically that the spectrum of an associated continuous process satisfies the very general requirement that it is continued fraction integrable. See [61]. Thus, the method should be fairly robust. However, if it is reasonable to consider the underlying model as ARMA(p,q), then the results of this paper actually also suggest another possible alternative approach to the one presented here. That is, in this paper, we have also suggested an approach to estimating p and q in an underlying ARMA(p,q) model. Given p and q, one could then calculate the spectral density by estimating the unknown coefficients in the mixed model by known methods. The result would be an "ARMA spectral estimator" (such an estimator, using a computer algorithm, has been suggested by Akaike). Currently, the authors are investigating this approach to spectral estimation as well. In fact, the problem of estimating p and q utilizing the r- and s-matrices has essentially been solved and incorporated in current computer program routines at Southern Methodist University. The implications of these results in model fitting as well as their application to spectral estimation in the manner just described will be discussed in a later paper. However, it is expected that as a spectral estimator, the "ARMA spectral estimator" will give better definition near the peak power frequencies for data which are well fit by an ARMA model, but will in general be less robust to deviations from this model than the G estimator. Thus, for example, if $\rho(\tau) = e^{-\alpha\tau^2}$, the G-spectral estimator should still yield a good approximation of the spectrum, whereas the "ARMA spectral estimator" might not. For data which are nearly error free, this latter property should be clear. That is, for this type of data, the G-estimator can simply be considered as a technique for calculating Fourier transforms of an arbitrary function in such a way as to reduce the leakage error. For this purpose, of course, the other methods are not applicable. However, if the data points are taken as simply the values of the function to be transformed, i.e., the data are input into the program in place of values of $\rho(\tau)$, the G estimator will yield an approximation of the Fourier transform which normally will have smaller leakage effects than say the discrete Fourier transform. In essence in this application, the G-transform is a G_n-transform of a discrete Fourier transform, where n is taken as large as the data allow, or until stability is reached.

Appendix 1

Proof of Theorem 2.

Since $e_m[A_k;\ell] = A_{k-1} + e_m[B_k;\ell]$, where $B_{k+j\ell} = A_{k+j\ell} - A_{k-1}$, it follows that we can take $h=0$ w.l.o.g. Then since $a_i \in L(n,\Delta)$, there exists (assuming no repeated roots in the characteristic equation) a set of constants $b_1, b_2 \ldots b_n$ such that

$$a_i = \sum_{m=1}^{n} C_m \pi_m^{\ i} \quad , \quad i = 0,1,2,\ldots ,$$

where the π_m are roots of the characteristic equation

$$r^n + b_1 r^{n-1} + \ldots + b_n = 0.$$

Consequently, the elements of the sequence $\{a_{i+j\ell}\}$, $i = 0,1,2,\ldots$ can be written

$$a_{i+j\ell} = \sum_{m=1}^{n} C_m \pi_m^{\ i} (\pi_m^{\ \ell})^j \ .$$

Then, for fixed i and ℓ $a_{i+j\ell}$, as a function of j, is an element of $L(n,\Delta)$ with its characteristic equation having the roots $\{\pi_m^{\ \ell}\}$. Now let $A_\infty = L$. Then

$$L - e_n[A_k;\ \ell] = \frac{H_{n+1}(L - A_{k+j\ell};\ a_{k+(i+j)\ell})}{H_{n+1}(1;\ a_{k+(i+j)\ell})} .$$

But

$$L - A_{k+j\ell} = \sum_{i=k+j\ell+1}^{\infty} a_i$$

$$= \sum_{i=k+j\ell+1}^{\infty} (c_1\pi_1{}^i + c_2\pi_2{}^i + \ldots + c_n\pi_n{}^i) .$$

But since $L < \infty$, $|\pi_i| < 1$ for $i = 1, \ldots n$ and, therefore,

$$L - A_{k+j\ell} = \frac{c_1\pi_1{}^{k+1}}{1 - \pi_1}(\pi_1{}^\ell)^j + \ldots + \frac{c_n\pi_n{}^{k+1}}{1 - \pi_n}(\pi_n{}^\ell)^j .$$

Consequently, $L - A_{k+j\ell}$ and $a_{i+j\ell}$ satisfy the same difference equation. Thus (see [24])

$$H_{n+1}(L - A_{k+j\ell};\ a_{k+(i+j)\ell}) \equiv 0$$

and

$$H_{n+1}(1;\ a_{k+(i+j)\ell}) \neq 0 .$$

Since

$$H_{n+1}(L - A_{k+j\ell};\ a_{k+(i+j)\ell}) = L\ H_{n+1}(1;\ a_{k+(i+j)\ell}) -$$

$$H_{n+1}(A_{k+j\ell};\ a_{k+(i+j)\ell}) ,$$

the result follows if $m = n$. If $m > n$, then clearly

$$e_m[A_k;\ \ell] \equiv 0/0$$

and the drop back rule again gives

$$e_m[A_k; \ell] \equiv L.$$

<u>Proof of Theorem 8</u>.

Proof. Suppose $f_m \in L(n,\Delta)$ for $m \geq m_0 + n$. Then

$$f_m - a_1 f_{m-1} - \dots - a_n f_{m-n} = 0 \tag{1}$$

for some set of a_i and by elementary row operations

$$H_{n+1}^{(m)}[1; f_{i+j}] = (-1)^n (1 - \sum_{i=1}^{n} a_i) H_n^{(m)}[f_j] .$$

But since n is the smallest integer for which f satisfies an equation such as (1), $H_n^{(m)}[f_j] \neq 0$. Thus

$$s_n^{(m)} = (-1)^n (1 - \sum_{i=1}^{n} a_i) , \tag{2}$$

which is independent of m. Now suppose for $m \geq m_0 + n$

$$H_{n+1}^{(m)}[1; f_{i+j}] = C H_n^{(m)}[f_j] . \tag{3}$$

But

$$H_n^{(m)}[f_j] = (-1)^n H_{n+1}^{(m)}[\beta_j; f_{i+j}] ,$$

where $\beta_{n+1} = 1$ and $\beta_1 = \beta_2 = \dots = \beta_n = 0$.

Therefore by (3) if $m \geq m_0 + n$

$$H_{n+1}^{(m)}[1; f_{i+j}] - (-1)^n C H_{n+1}^{(m)}[\beta_j; f_{i+j}] = 0$$

or taking $A = 1 - (-1)^n C$, we have

$$\begin{vmatrix} 1 & 1 & \cdots & 1 & A \\ f_m & f_{m+1} & \cdots & f_{m+n-1} & f_{m+n} \\ \vdots & \vdots & & & \\ f_{m+n-1} & f_{m+n} & \cdots & f_{m+2n-2} & f_{m+2n-1} \end{vmatrix} = 0 . \tag{4}$$

From (4) it follows that for some set of $c_i(m)$ that

$$\begin{bmatrix} A \\ f_{m+n} \\ f_{m+n+1} \\ \vdots \\ f_{m+2n-1} \end{bmatrix} = c_1(m) \begin{bmatrix} 1 \\ f_m \\ f_{m+1} \\ \vdots \\ f_{m+n-1} \end{bmatrix} + \cdots + c_n(m) \begin{bmatrix} 1 \\ f_{m+n-1} \\ f_{m+n} \\ \vdots \\ f_{m+2n-2} \end{bmatrix} \tag{5}$$

But by differencing the first row in (5), and applying the operator defined by $Bf_t = f_{t-1}$ to the j+l st row and subtracting from the jth row, we obtain

$$\begin{bmatrix} 1 & 1 & \cdots & 1 \\ f_m & f_{m+1} & \cdots & f_{m+n-1} \\ \vdots & & & \\ f_{m+n-2} & f_{m+n-1} & \cdots & f_{m+2n-3} \end{bmatrix} \begin{bmatrix} \Delta c_1(m) \\ \Delta c_2(m) \\ \vdots \\ \Delta c_n(m) \end{bmatrix} = 0 . \tag{6}$$

Since $s_n^{(m)}$ is defined and not zero (31) and (6) imply $\Delta c(m) = 0$ if $m \geq m_0 + n$, i.e., the $c_i(m)$ are constant. Thus, $f \in L(n,\Delta)$, where n is the smallest such integer since $s_n^{(m)}$ is defined, i.e., $H_n^{(m)}[f_j] \neq 0$.

<u>Theorem 12.</u>

Suppose X_t is a discrete MA(q) process with autocorrelation function ρ_k and $\theta_i \neq 0$, $1 = 1,2,\cdots q$.

(i) The for $N_o \geq q$ and $n = 0$,

$$e_n\left[1 + 2 \sum_{k=1}^{N_o} \rho_k \cos(2\pi\omega k);\ h\right] = 1 + 2 \sum_{k=1}^{\infty} \rho_k \cos(2\pi\omega k) \qquad (7)$$

where h is a positive integer, and ω is such that $-1/2 \leq \omega \leq 1/2$.

(ii) If $1 \leq N_o < q$ and m and h are positive integers such that $N_o + mh = q$, then (7) holds for $n \geq (q - N_o)(1/h) + 1$ and all $\omega \in [-1/2, 1/2]$ such that ω is not an odd integer multiple of $1/4(N_o + mh) = 1/4\,q$.

(iii) If $0 < q - mh < N_o < q - (m - 1)h$, then (7) holds for $n \geq (q - N_o)\ 1/h$ and all $\omega \in [-1/2, 1/2]$ such that ω is not an odd integer multiple of $1/4[N_o + (m - 1)\,h]$.

Proof.

(i) Since X_t is a discrete MA(q) process, its autocorrelation function satisfies (since $\theta_i \neq 0$)

$$\rho_0 = 1$$

$$\rho_k \neq 0, \quad 0 < k \leq q$$

$$\rho_k = 0, \quad k \geq q + 1 .$$

Therefore, its spectral density function is given by

$$S(\omega) = 1 + 2 \sum_{k=1}^{\infty} \rho_k \cos(2\pi\omega k)$$

$$= 1 + 2 \sum_{k=1}^{q} \rho_k \cos(2\pi\omega k) .$$

By the definition of the e_0-transformation, it follows that

$$e_0\left[1 + 2 \sum_{k=1}^{N_o} \rho_k \cos(2\pi\omega k); h\right] \equiv S(\omega)$$

for $N_o \geq q$ and any positive integer h. Hence, (i) holds.

Now, suppose $1 \leq N_o < q$. Then for h a positive integer, there exists an integer $m \geq 1$ such that $N_o + (m - 1)h < q$ and $N_o + mh \geq q$. Let

$$S_{N_o} = 1 + 2 \sum_{k=1}^{N_o} \rho_k \cos(2\pi\omega k)$$

and

$$s_k = 2\rho_k \cos(2\pi\omega k)$$

for a fixed $\omega \in [-1/2, 1/2]$.

(ii) Suppose $N_o + mh = q$ for some $h \geq 1$. Also, suppose that ω is not an integer multiple of $1/4(N_o + mh)$. Then, clearly,

$$S_{N_o+(m-1)h} \neq S_{N_o+mh} = S_{N_o+(m+1)h} = \cdots = S(\omega)$$

and

$$s_k = 0 \quad \text{for } k > N_o + mh = q.$$

Consider $e_{m+1}[S_{N_o}; h]$. It is given by

$$e_{m+1}[S_{N_o};\ h]$$

$$= \frac{\begin{vmatrix} S_{N_o} & S_{N_o+h} & \cdots & S_{N_o+(m-1)h} & S_{N_o+mh} & S_{N_o+(m+1)h} \\ s_{N_o} & s_{N_o+h} & \cdots & s_{N_o+(m-1)h} & s_{N_o+mh} & s_{N_o+(m+1)h} \\ \vdots & \vdots & & & & \vdots \\ s_{N_o+mh} & s_{N_o+(m+1)h} & \cdots & & & s_{N_o+2(m+1)h-h} \end{vmatrix}}{\begin{vmatrix} 1 & 1 & \cdots & 1 & 1 & 1 \\ s_{N_o} & s_{N_o+h} & \cdots & s_{N_o+(m-1)h} & s_{N_o+mh} & s_{N_o+(m+1)h} \\ \vdots & \vdots & & & & \vdots \\ s_{N_o+mh} & s_{N_o+(m+1)h} & \cdots & & & s_{N_o+2(m+1)h-h} \end{vmatrix}}$$

$$= \frac{\begin{vmatrix} S_{N_o} & S_{N_o+h} & \cdots & S_{q-1} & S_q & S_q \\ s_{N_o} & s_{N_o+h} & \cdots & s_{q-1} & s_q & 0 \\ & \vdots & & & & \\ s_{N_o+h} & & \cdots & s_q & 0 & 0 \\ \vdots & s_q & & & \vdots & \vdots \\ s_q & 0 & \cdots & & 0 & 0 \end{vmatrix}}{\begin{vmatrix} 1 & 1 & \cdots & 1 & 1 & 1 \\ s_{N_o} & s_{N_o+h} & \cdots & s_{q-1} & s_q & 0 \\ & \vdots & & & & \\ s_{N_o+h} & & & s_q & 0 & 0 \\ \vdots & s_q & & & \vdots & \vdots \\ s_q & 0 & \cdots & & 0 & 0 \end{vmatrix}}$$

$$= \frac{S_q(-1)^m s_q^{m+1}}{(-1)^m s_q^{m+1}}$$

$$= S_q = S(\omega)$$

for $s_q \neq 0$, i.e., for $2\rho_q \cos(2\pi\omega q) \neq 0$. But $2\rho_q \cos(2\pi\omega q) = 0$ only if $\cos(2\pi\omega q) = 0$, that is, if ω is an odd integer multiple of $1/4q = 1/4(N_o + mh)$. Also, it is clear that $e_n[S_{N_o}; h] \equiv 0/0$ for $n > m + 1$ so that by the drop back rule

$$e_n[S_{N_o}; h] = e_{m+1}[S_{N_o}; h]$$

for $n > m + 1$. Hence, if $N_o + mh = q$ and $h \geq 1$, then

$$e_n[S_{N_o}; h] \equiv S(\omega)$$

for $n \geq (q - N_o)1/h + 1$ and all $\omega \in [-1/2; 1/2]$ such that ω is not an integer multiple of $1/4q$. Therefore, (ii) holds.

(iii) Suppose $N_o + mh > q$, $N_o + (m - 1)h < q$, and $h > 1$. Then $S_{N_o+mh-h} \neq S_{N_o+mh} = S_q$ and $s_{N_o+mh} = 0$ so that $e_m[S_{N_o}; h]$ is given by

$$\frac{\begin{vmatrix} s_{N_o} & s_{N_o+h} & \cdots & s_{N_o+(m-1)h} & s_{N_o+mh} \\ s_{N_o} & s_{N_o+h} & \cdots & s_{N_o+(m-1)h} & 0 \\ \vdots & \vdots & & \vdots & \vdots \\ s_{N_o+(m-1)h} & 0 & \cdots & 0 & 0 \end{vmatrix}}{\begin{vmatrix} 1 & 1 & \cdots & 1 & 1 \\ s_{N_o} & s_{N_o+h} & \cdots & s_{N_o+(m-1)h} & 0 \\ \vdots & \vdots & & \vdots & \vdots \\ s_{N_o+(m-1)h} & 0 & \cdots & 0 & 0 \end{vmatrix}}$$

which is equal to

$$s_{N_o+mh} \frac{(-1)^{m-1}[s_{N_o+(m-1)h}]^m}{(-1)^{m-1}[s_{N_o+(m-1)h}]^m} \quad .$$

Hence,

$$e_m[s_{N_o}; h] \equiv s_{N_o+mh} = S(\omega)$$

for $N_o+mh > q$ and $s_{N_o+(m-1)h} \neq 0$. But $s_{N_o+(m-1)h} \neq 0$ since ω is not an odd integer multiple of $1/4[N_o + (m - 1)h]$. Clearly,

$$e_n[S_{N_o}; h] \equiv 0/0$$

for $n > m$ and the drop back rule applies. Therefore, if $h > 1$ and $N_o+mh > q$, then

$$e_n[S_{N_o}; h] = S(\omega)$$

for $n \geq (q - N_o)l/h$ and all $\omega \in [-1/], 1/2]$ and the proof is complete. The assumption $\theta_i \neq 0$ was for convenience in this proof and not necessary for its validity if h is properly chosen.

Appendix 2

The following subroutine is a reasonably self-contained program for calculating G-spectral estimates. It is intended as an aid to understanding of G-spectral estimation and no attempt has been made at computational efficiency.

The program was written by Dr. A. G. Houston. Questions concerning it should be directed to Dr. Houston at the following address:

Dr. A. G. Houston
Mail Code TF2
NASA Johnson Space Center
Houston, Texas 77058.

The user is cautioned that this program contains an automatic check for stability in order to select a specific G-spectral estimate. The measure used here for stability is rather ad hoc and a visual inspection of the G-matrix is recommended. Large values of α in the program are suggestive of the need for closer inspection.

```
      SUBRØUTINE GSPEC (SACF,WØRK,NLAG,KG,K,H,W,IRS,IG)
CCCCCCCCCCCCCCCCCCCCCCCCCCCCCCCCCCCCCCCCCCCCCCCCCCCCCCCCCCCCCCCCCCCCCCC
C     SACF - INPUT VECTØR ØF LENGTH NLAG CØNTAINING ACF (INCLUDING
C            ZERØ LAG)
C     WØRK - ØUTPUT VECTØR ØF LENGTH NLAG
C            WØRK(I) = 2.0 * SACF(I) * CØS(2-PI*W*(I-1)*H)
C     NLAG - NUMBER ØF LAGS INCLUDING THE ZERØ LAG
C     KG   - INPUT INTEGER - HIGHEST ØRDER TRANSFØRM DESIRED
C            (KG LESS THAN ØR EQUAL TØ 29 AND LESS THAN NLAG)
C     K    - INPUT INTEGER - INCREMENT ØF TRANSFØRM (USUALLY 1)
C     H    - INPUT REAL - INCREMENT ØF ACF (USUALLY 1.) -- CAN
C            ALSØ BE USED AS LENGTH ØF SAMPLING INTERVAL.
C     W    - INPUT REAL - FREQUENCY AT WHICH SPECTRAL ESTIMATE IS
C            DESIRED
C            -1/(2*H) .LE. W .LE. 1/(2*H)
C            ØUTPUT  G-SPECTRAL ESTIMATE AT FREQUENCY
C     IRS  - INPUT INTEGER - IRS .EQ. 0 - PRINT ØUT S-MATRIX AND
C            R-MATRIX AT FREQUENCY W.  IRS .NE. 0 - PRINTING
C            SUPPRESSED
C     IG   - INPUT INTEGER - IG .EQ. 0 - PRINT ØUT G-MATRIX AT
C            FREQUENCY W.  IG .NE. 0 - PRINTING SUPPRESSED
C
C     ØUTPUT:  G-SPECTRAL ESTIMATE AT FREQUENCY W IS G(N,M), WHERE
C              N = ØRDER ØF TRANSFØRM SELECTED BY STABLING ØFF
C                  PRØCEDURE
C              M = RØW PØSITIØN IN CØLUMN N ØF THE G-MATRIX
C              NUMBER ØF LAGS ØF ACF USED FØR ESTIMATE IS 2*N + M - 1
C
CCCCCCCCCCCCCCCCCCCCCCCCCCCCCCCCCCCCCCCCCCCCCCCCCCCCCCCCCCCCCCCCCCCCCCC
      DIMENSIØN WØRK(1), S(29,116), R(29,116), G(30,30), SACF(1)
      DATA S,R,G/3364*1.E-5,3364*2.E-5,900*3.E-5/
      TWØPI=6.283185307
C
      GØ TØ 15
   10 W=W-.001/H
   15 CØNTINUE
      F=W*TWØPI
      WØRK(1)=2.0*SACF(1)
      DØ 20 I=2,NLAG
      FI=F*H*FLØAT(I-1)
   20 WØRK(I)=2.0*SACF(I)*CØS(FI)
C
      KG1=KG+1
      N=2*KG
      N1=N+1
      NM1=N-1
      NP=2*N-1
```

```
      NPP1=NP+1
C
C     CHECK FOR LIMITING FREQUENCY.
C
   30 NMAX=(2*KG1*K+1)/2
      DO 60 I=1,NMAX
      DO 50 J=1,KG1
      WCHK=(2.0*FLOAT(I)-1.0)/(4.0*H*FLOAT(J)*FLOAT(K))
      IF(ABS(W-WCHK)/WCHK.GT.1.E-6) GO TO 50
   40 WRITE (6,490) W,K
      GO TO 10
   50 CONTINUE
   60 CONTINUE
C
C     INITIALIZE FIRST COLUMN OF S-MATRIX.
C
      DO 65 I=1,NPP1
      S(1,I)=1.0
   65 CONTINUE
C
C     INITIALIZE FIRST COLUMN OF R-MATRIX.
C
      J=(N-1)*K
      DO 70 I=1,NM1
      R(1,I)=WORK(J+1)
      NPMIP1=NP-I+1
      R(1,NPMIP1)=WORK(J+1)
      J=J-K
   70 CONTINUE
      R(1,N)=WORK(1)
C
C     CALCULATE REMAINING KG-1 COLUMNS OF R- AND S-MATRICES.
C
      MP=NP-1
      DO 120 I=2,KG
      DO 90 J=1,MP
      IF (R(I-1,J).EQ.0.)  GO TO 80
      RATIO=DBLE(R(I-1,J+1))/DBLE(R(I-1,J))-1.0D0
      IF(ABS(RATIO).GT.1.E-6)  GO TO 75
      S(I,J)=0.
      GO TO 90
   75 S(I,J)=S(I-1,J+1)*RATIO
      IF(S(I-1,J+1).EQ.0.) S(I,J)=0.
      GO TO 90
   80 S(I,J)=-S(I-1,J+1)
   90 CONTINUE
      MPP=MP-1
```

```
      DO 110 J=1,MPP
      IF (S(I,J).EQ.0.)  GO TO 100
      RATIO=DBLE(S(I,J+1))/DBLE(S(I,J))-1.DO
      IF(ABS(RATIO).GT.1.E-6)  GO TO 95
      R(I,J)=0.
      GO TO 110
   95 R(I,J)=R(I-1,J+1)*RATIO
      IF(R(I-1,J+1).EQ.0.) R(I,J)=0.
      GO TO 110
  100 R(I,J)=-R(I-1,J+1)
  110 CONTINUE
      MP=MP-2
  120 CONTINUE
C
C     INITIALIZE FIRST COLUMN OF G-MATRIX.
C
      LL=1
      DO 160 J=1,KG1
      IF (LL.GT.1) GO TO 130
      R2=0.0
      GO TO 150
  130 R2=(WORK(1)+WORK(LL))/2.0
      L1=LL-1
      IF (L1.LT.2) GO TO 150
      DO 140 I=2,L1
      R2=R2+WORK(I)
  140 CONTINUE
  150 CONTINUE
      G(1,J)=R2*H
      LL=LL+K
  160 CONTINUE
C
C     CALCULATE REMAINING KG-1 COLUMNS OF G-MATRIX.
C
      NM=KG
      DO 220 I=2,KG1
      DO 210 J=1,NM
      IM1=I-1
      JPNM1=J+N-1
      JP1=J+1
      JPN=J+N
      UN=R(IM1,JPNM1)*(DBLE(G(IM1,JP1))-DBLE(G(IM1,J)))
      UD=R(IM1,JPNM1)-R(IM1,JPN)
      IF(ABS(UD).GT.1.E-6*AMAX1(ABS(R(IM1,JPNM1)),ABS(R(IM1,JPN))))
     1GO TO 200
      IF(R(IM1,JPNM1).EQ.0.0) GO TO 190
  170 WRITE (6,500) IM1,JPNM1,IM1,JPN,R(IM1,JPN)
```

```
      K=K+1
      IF(2*KG*K.LE.NLAG) GO TO 30
  180 WRITE (6,510)
  190 G(I,J)=G(I-1,J)
      GO TO 210
  200 G(I,J)=G(I-1,J)+UN/UD
  210 CONTINUE
      NM=NM-1
  220 CONTINUE
      IF(IRS .NE. 0) GO TO 279
C
C    PRINT R-MATRIX
C
      WRITE(6,520)
      KGM = MINO(KG,15)
      WRITE(6,530) (I,I=1,KGM)
      WRITE(6,540)
      KP = K*(1-2*KG)
      DO 230 J = 1,N1
      WRITE(6,550) KP,(R(I,J),I=1,KGM)
      KP = KP + K
  230 CONTINUE
      N2=2*KG+2
      N4 = KG
      DO 240 J = N2,NP,2
      N4=N4-1
      N4M=MINO(N4,15)
      WRITE(6,550) KP,(R(I,J),I=1,N4M)
      KP = KP + K
      WRITE(6,550) KP,(R(I,J+1),I=1,N4M)
      KP = KP + K
  240 CONTINUE
C
      IF (KG .LE. 15) GO TO 242
      WRITE(6,521)
      WRITE(6,530) (I,I=16,KG)
      WRITE(6,540)
      KP = K*(1-2*KG)
      DO 231 J = 1,N1
      WRITE(6,550) KP,(R(I,J),I=16,KG)
      KP = KP + K
  231 CONTINUE
      N2 = 2*KG+2
      N4 = KG
      DO 241 J = N2,NP,2
      N4 = N4-1
      IF(N4 .LT. 16) GO TO 242
```

```
      WRITE(6,550) KP,(R(I,J),I=16,N4)
      KP = KP + K
      WRITE(6,550) KP,(R(I,J+1),I=16,N4)
  241 KP = KP + K
  242 CONTINUE
C
C
C     PRINT S-MATRIX
C
      WRITE(6,560)
      IF(KG .GT. 1) GO TO 250
      WRITE(6,570)
      GO TO 255
  250 KGM1=MINO(KG-1,14)
      WRITE(6,570) (I,I=1,KGM1)
  255 WRITE(6,540)
      KGM = MINO(KG,15)
      KP = K-2*KG*K
      DO 260 J = 1,N2
      WRITE(6,550) KP,(S(I,J),I=1,KGM)
      KP = KP + K
  260 CONTINUE
      N4 = KG
      N5 = N2 + 1
      N6 = NP + 1
      DO 270 J = N5,N6,2
      N4=N4-1
      N4M = MINO(N4,15)
      WRITE(6,550) KP,(S(I,J),I=1,N4M)
      KP = KP + K
      WRITE(6,550) KP,(S(I,J+1),I=1,N4M)
  270 KP = KP + K
C
      IF(KG .LE. 15) GO TO 279
      WRITE(6,561)
      KGM1 = KG-1
      WRITE(6,571),(I,I=15,KGM1)
      WRITE(6,540)
      KP=K-(2*KG)*K
      DO 261 J = 1,N2
      WRITE(6,550) KP,(S(I,J),I=16,KG)
      KP=KP+K
  261 CONTINUE
      N4 = KG
      N5 = N2 + 1
      N6=NP+1
      DO 271 J = N5,N6,2
```

```
      N4=N4-1
      IF(N4 .LT. 16) GO TO 279
      WRITE(6,550) KP,(S(I,J),I=16,N4)
      KP = KP + K
      WRITE(6,550) KP,(S(I,J+1),I=16,N4)
  271 KP = KP + K
  279 CONTINUE
C
C     PRINT G-MATRIX
C
      IF(IG .NE. 0) GO TO 340
      KGM=MINO(KG-1,14)
      IF (IRS.EQ.0) GO TO 300
      WRITE(6,590)
      GO TO 310
  300 WRITE(6,600)
  310 IF(KG1 .EQ. 1) GO TO 320
      WRITE(6,570) (I,I=1,KGM)
      GO TO 325
  320 WRITE(6,570)
  325 CONTINUE
      WRITE(6,540)
      KP = 0
      DO 330 J = 1,KG1
      JJ = MINO(KG1-J+1,15)
      WRITE(6,550) KP,(G(I,J),I=1,JJ)
      KP = KP + K
  330 CONTINUE
C
      IF(KG1 .LE. 15) GO TO 339
      WRITE(6,601)
      WRITE(6,571) (I,I=15,KG)
      WRITE(6,540)
      KP = 0
      DO 331 J = 1,KG1
      JJ = KG1 - J + 1
      IF(JJ .LT. 16) GO TO 339
      WRITE(6,550) KP,(G(I,J),I=16,JJ)
      KP = KP + K
  331 CONTINUE
  339 CONTINUE
C
C     DETERMINATION OF THE G-SPECTRAL ESTIMATE AT FREQUENCY W FROM
C     THE APPROXIMATE CONVERGENCE OF THE SEQUENCE OF G-MATRIX
C     ENTRIES.
C
CCCCCCCCCCCCCCCCCCCCCCCCCCCCCCCCCCCCCCCCCCCCCCCCCCCCCCCCCCCCCCCCCCCCCCCC
```

```
C      AUTOMATIC ORDER SELECTION:  ORDER SELECTED WHEN THE DIFFERENCE
C      BETWEEN SUCCESSIVE ORDER TRANSFORMS .LE. ALPHA * (SMALLER OF THE
C      TWO)
C      ALPHA INITIALLY SET TO 0.0 AND INCREMENTED BY .005 TILL ORDER IS
C      SELECTED OR ALPHA .EQ. 0.35
CCCCCCCCCCCCCCCCCCCCCCCCCCCCCCCCCCCCCCCCCCCCCCCCCCCCCCCCCCCCCCCCCCCCCCCC
  340 NM=KG
      ALPHA=0.0
      J=2
  350 I=3
  360 IF (G(I,J).LT.0.0) GO TO 400
  370 IF (G(I,J).GE.G(I-1,J)) GO TO 390
  380 IF (G(I,J).LT.G(I-1,J)-ALPHA*G(I,J)) GO TO 410
      GO TO 440
  390 IF (G(I,J).GT.G(I-1,J)+ALPHA*G(I-1,J)) GO TO 410
      GO TO 440
  400 I=I+2
      GO TO 420
  410 I=I+1
  420 IF (I.LE.NM) GO TO 360
  430 ALPHA=ALPHA+.005
      IF (ALPHA.LE.0.35) GO TO 350
      GO TO 450
  440 GOUT=G(I-1,J)
      IOUT=I-2
      JOUT=J-1
      WRITE (6,610) ALPHA
      GO TO 470
  450 J=J+1
      ALPHA=0.00
      IF (J.GE.KG) GO TO 460
      NM=NM-1
      GO TO 350
  460 WRITE (6,620) W
      GO TO 480
  470 WRITE (6,630) W,IOUT,JOUT,GOUT
      W=GOUT
  480 CONTINUE
      RETURN
C
  490 FORMAT  (///,45X,25H*************************,//,19H  ///   FREQUEN
     1CY  ,F10.6,34H IS A LIMITING FREQUENCY WITH K = ,I3,48H -- IT HAS
     2BEEN REPLACED BY A NEARBY FREQUENCY.  ///,/)
  500 FORMAT (1HO,3H R(,I2,1H,,I2,6H) = R(,I2,1H,,I2,4H) = ,F10.6,7H NE
     10 .,//,50H K IS NOW INCREMENTED BY 1 IF NLAG IS LARGE ENOUGH,/)
  510 FORMAT (1HO,45H NLAG IS NOT LARGE ENOUGH TO INCREMENT K BY 1,//,63
     1H NLAG SHOULD BE INCREASED, K SHOULD BE INCREMENTED BY 1 AND THE,/
     2/,62H G-SPECTRAL ESTIMATOR SHOULD BE RECALCULATED AT THIS FREQUENC
     3Y,/)
  520 FORMAT(///,45X,25(1H*),//53X,8HR-MATRIX,/)
  530 FORMAT(4H0  N,I7,14I8)
  540 FORMAT(3H  M,/)
  521 FORMAT(//,57X,18HR-MATRIX CONTINUED)
  550 FORMAT(1X,I3,1X,15F8.3)
  560 FORMAT(//,53X,8HS-MATRIX,/)
  570 FORMAT(11H0  N      0,14I8)
  561 FORMAT(//,57X,18HS-MATRIX CONTINUED)
  571 FORMAT(4H0  N,I7,14I8)
  590 FORMAT(///,45X,25(1H*),//53X,8HG-MATRIX,/)
  600 FORMAT(//,53X,8HG-MATRIX,/)
  601 FORMAT(//,57X,18HG-MATRIX CONTINUED)
  610 FORMAT (1H0,17H   ///   ALPHA = ,F5.3,6H   ///)
  620 FORMAT (25(1H*),/,81H   ///BEWARE: STABLING OFF NOT REACHED. CHECK
     1 R-MATRIX AND S-MATRIX AT FREQUENCY ,F8.5,36H ,OR TRY HIGHER ORDER
     2 TRANSFORMS ///)
  630 FORMAT (///,30X,34H G-SPECTRAL ESTIMATE AT FREQUENCY ,F8.5,5H : G(
     1,I2,1H,,I2,4H) = ,F12.6,/)
      END
```

Bibliography

[1] Akaike, H. (1969a). "Fitting Autoregressive Models for Prediction." *Ann. Inst. Statist. Math. 21*, 243-247.

[2] Akaike, H. (1969b). "Power Spectrum Estimation through Autoregressive Model Fitting." *Ann. Inst. Statist. Math. 21*, 407-419.

[3] Akaike, H. (1970a). "Statistical Predictor Identification." *Ann. Inst. Statist. Math. 22*, 203-217.

[4] Akaike, H. (1970b). "A Fundamental Relation between Predictor Identification and Power Spectrum Estimation." *Ann. Inst. Statist. Math. 22*, 219-223.

[5] Akaike, H. (1974). "A New Look at the Statistical Model Identification." *IEEE Trans. Auto. Control AC-19*, (6), 716-723.

[6] Bartlett, M. S. (1946). "On the Theoretical Specification and Sampling Properties of Autocorrelated Time-Series." *J. R. Statist. Soc., Ser. B 8*, (1), 27-41.

[7] Bartlett, M. S. (1948). "Smoothing Periodograms from Time Series with Continuous Spectra." *Nature 161*, 686-687.

[8] Bartlett, M. S. (1950). "Periodogram Analysis and Continuous Spectra." *Biometrika 37*, 1-16.

[9] Bartlett, M. S. (1955). *An Introduction to Stochastic Processes with Special Reference to Methods and Applications.* Cambridge University Press, Cambridge.

[10] Bartlett, M. S., and Medhi, J. (1955). "On the Efficiency of Procedures for Smoothing Periodograms from Time Series with Continuous Spectra." *Biometrika 42*, 143-150.

[11] Berk, K. N. (1974). "Consistent Autoregressive Spectral Estimates." *Ann. Statist. 2*, (3), 489-502.

[12] Blackman, R., and Tukey, J. (1958). *The Measurement of Power Spectra* (from the Point of View of Communications Engineering). Dover, New York.

[13] Box, G. E. P., and Jenkins, G. M. (1970). *Time Series Analysis: Forecasting and Control.* Holden-Day, San Francisco.

[14] Brigham, E. O. (1974). *The Fast Fourier Transform.* Prentice-Hall, Inc., Englewood Cliffs, N. J.

[15] Burg, J. P. (1967). "Maximum Entropy Spectral Analysis." Presented to the 37th Meeting, Soc. Explor. Geophys., Oklahoma City, Oklahoma, Oct. 31, 1967.

[16] Burg, J. P. (1975). *Maximum Entropy Spectral Estimation.* Doctoral Thesis, Stanford University, Stanford, California.

[17] Chen, W. Y., and Stegen, G. R. (1974). "Experiments with Maximum Entropy Power Spectra of Sinusoids." *J. Geophys. Res. 79*, (20), 3019-3022.

[18] Cogburn, R., and Davis, H. T. (1974). "Periodic Splines and Spectra Estimation." *Ann. Statist. 2*, (6), 1108-1126.

[19] Cooley, J. W., and Tukey, J. W. (1965). "An Algorithm for the Machine Calculation of Complex Fourier Series." *Mathematics of Computation 19*, (90), 297-301.

[20] Fryer, G. J., Odegard, M. E., and Sutton, G. H. (1975). "Deconvolution and Spectral Estimation using Final Prediction Error." *Geophysics 40*, (3), 411-425.

[21] Gray, H. L. (1975). "On a Unification of Bias Reduction and Numerical Approximation." Unpublished manuscript.

[22] Gray, H. L., and Atchison, T. A. (1967). "Nonlinear Transformations related to the Evaluation of Improper Integrals, I." *SIAM J. Numer. Anal. 5*, (2), 363-371.

[23] Gray, H. L., and Atchison, T. A. (1968). "The Generalized G-Transform." *Math. Comp. 22*, (103), 595-606.

[24] Gray, H. L., Atchison, T. A., and McWilliams, G. V. (1971). "Higher Order G-Transformations." *SIAM J. Numer. Anal. 8*, (2), 365-381.

[25] Gray, H. L., and Clark, W. D. (1969). "On a Class of Nonlinear Transformations and Their Applications to the Evaluation of Infinite Series." *J. Res. Nat. Bur. Standards, Sec. B 73B*, (3), 251-274.

[26] Gray, H. L., and Foster, M. R. (1975). "Autoregressive, Maximum Entropy, and G-Spectral Estimation." *The Search for Oil: Some Statistical Methods and Techniques* (Ed., D. B. Owen). Marcel Dekker, Inc., New York, 169-189.

[27] Grenander, U. (1951). "On Empirical Spectral Analysis of Stochastic Processes." *Ark. Mat. 1*, 503-531.

[28] Grenander, U., and Rosenblatt, M. (1953). "Statistical Spectral Analysis Arising from Stationary Stochastic Processes." *Ann. Math. Statist. 24*, 537-558.

[29] Grenander, U., and Rosenblatt, M. (1957). *Statistical Analysis of Stationary Time Series.* John Wiley, New York.

[30] *IEEE Transactions on Audio and Electroacoustics AU-15*, (2), June, 1967.

[31] Jenkins, G. M. (1961). "General Considerations in the Analysis of Spectra." *Technometrics 3*, (2), 133-166.

[32] Jenkins, G. M. (1965). "A Survey of Spectral Analysis." *Appl. Statist. 14*, (1), 2-32.

[33] Jenkins, G. M., and Watts, D. G. (1968). *Spectral Analysis and its Applications.* Holden-Day, San Francisco.

[34] Jones, R. H. (1974). "Identification and Autoregressive Spectrum Estimation. *IEEE Trans. Auto Control AC-19*, (6), 894-897.

[35] Koopmans, L. H. (1974). *The Spectral Analysis of Time Series.* Academic Press, New York.

[36] Kromer, R. E. (1970). *Asymptotic Properties of the Autoregressive Spectral Estimator.* Ph.D. Dissertation, Stanford University, Stanford, California.

[37] Lacoss, R. T. (1971). "Data Adaptive Spectral Analysis Methods." *Geophysics 36*, (4), 661-675.

[38] Lubkin, S. (1952). "A Method of Summing Infinite Series." *J. Res. Nat. Bur. Standards, Sect. B 48*, 228-254.

[39] McWilliams, G. V. (1969). *The e_m and Higher Order G-Transformations.* Doctoral Thesis, Texas Tech University, Lubbock, Texas.

[40] Morgan, F. W. (1975). *The G-Spectral Estimator.* Doctoral Thesis, Southern Methodist University, Dallas, Texas.

[41] Papoulis, A. (1965). *Probability, Random Variables, and Stochastic Processes.* McGraw-Hill, New York.

[42] Parzen, E. (1957). "On Consistent Estimates of the Spectrum of a Stationary Time Series." *Ann. Math. Statist. 28*, 329-348.

[43] Parzen, E. (1961). "Mathematical Considerations in the Estimation of Spectra." *Technometrics 3*, (2), 167-190.

[44] Parzen, E. (1964). "An Approach to Empirical Time Series Analysis." *J. Res. Nat. Bur. Standards, Sect. D 68D*, 937-951.

[45] Parzen, E. (1967). *Stochastic Processes.* Holden-Day, San Francisco.

[46] Parzen, E. (1969). "Multiple Time Series Modeling." *Multivariate Analysis II* (Ed., P. R. Krishnaiah). Academic Press, New York, 389-409.

[47] Parzen, E. (1972). "Some Recent Advances in Time Series Analysis." *Statistical Models and Turbulence* (Ed., M. Rosenblatt). Springer-Verlag, New York.

[48] Parzen, E. (1974). "Some Recent Advances in Time Series Modeling." *IEEE Trans. Auto. Control AC-19*, (6), 723-730.

[49] Phadke, M. S., and Wu, S. M. (1974). "Modeling of Continuous Stochastic Processes from Discrete Observations with Applications to Sunspots Data." *J. Amer. Statist. Assoc. 69*, (346), 325-329.

[50] Pye, W. C. (1972). *Pseudoinverses and Higher Order G-Transformations.* Doctoral Thesis, Texas Tech University, Lubbock, Texas.

[51] Pye, W. C., and Atchison, T. A. (1973). "An Algorithm for the Computation of the Higher Order G-Transformation." *SIAM J. Numer. Anal. 10*, (1), 1-7.

[52] Schuster, A. (1898). "Investiation of Hidden Periodicities." *Terrestrial Magnetism 3*, 13-41.

[53] Shanks, D. (1955). "Nonlinear Transformation of Divergent and Slowly Convergent Sequences." *J. Math. and Phys. 34*, 1-42.

[54] Steiglitz, K. (1964). "Power Spectrum Identification for Adaptive Systems. *IEEE Trans. Appl. Industry 83*, 195-197.

[55] Tukey, J. W. (1949). "The Sampling Theory of Power Spectrum Estimates." *Symposium on Applications of Autocorrelation Analysis to Physical Problems.* Woods Hole, June 13-14.

[56] Tukey, J. W. (1961). "Discussion, Emphasizing the Connection between Analysis of Variance and Spectrum Analysis." *Technometrics 3*, (2), 191-219.

[57] Tukey, J. W. (1967). "An Introduction to the Calculations of Numerical Spectral Analysis." *Spectral Analysis of Time Series* (Ed., B. Harris). John Wiley and Sons, New York.

[58] Ulrych, T. J. (1972). "Maximum Entropy Power Spectrum of Truncated Sinusoids." *J. Geophys. Res.* *77*, (8), 1396-1400.

[59] Ulrych, T. J., and Bishop, T. N. (1975). "Maximum Entropy Spectral Analysis and Autoregressive Decomposition." *Reviews of Geophysics and Space Physics* *13*, 183-200.

[60] Wynn, P. (1956). "On a Device for Computing the $e_m(s_n)$ Transformation." *Math. Tables and Aids to Comp.* *X*, 91-96.

[61] Wynn, P. (1968). "Upon the Definition of an Integral as a Limit of a Continued Fraction." *Arch. Rational Mech. Anal.* *28*, 83-147.

[62] Yule, G. U. (1927). "On a Method of Investigating Periodicities in Disturbed Series, with Special Reference to Wolfer's Sunspot Numbers." *Phil. Trans. Royal Soc.* *226A*, 267-298.

MULTIVARIATE AUTOREGRESSION ESTIMATION USING RESIDUALS

Richard H. Jones
University of Colorado Medical Center
Denver, Colorado

Abstract.

Autoregression estimation techniques based on the Yule-Walker equations are briefly reviewed and compared to Burg's maximum entropy algorithm. The problems encountered in generalizing Burg's algorithm to multivariate time series are discussed, and generalizations presented. The selection of the order of the autoregression also is considered.

1. Introduction.

The classical method of fitting autoregressions to stationary time series is to first estimate the covariance sequence from the data and to fit the autoregressions using the Yule-Walker equations. Burg (1967) introduced another algorithm referred to as maximum entropy estimation which is asymptotically equivalent to the Yule-Walker estimates for long data sequences, but has quite different small sample properties. A comparison of the two algorithms is given in Ulrych and Bishop (1975).

This paper reviews the univariate algorithms, the multivariate generalization of the Yule-Walker estimates, and the problems associated with generalizing Burg's algorithm. Generalizations are discussed with a FORTRAN subroutine attached as an Appendix for a procedure suggested by Morf, Viera, and Kailath (1976) and independently by Robinson (1976). This procedure does not directly generalize Burg's algorithm, but estimates the normalized reflection coefficient matrix as a correlation matrix between forward and backward residuals and produces a stable estimate. In the univariate

ISBN 0-12-257250-5

case, the procedure simply estimates the partial autocorrelation coefficient as the usual correlation estimate between forward and backward residuals. The paper concludes with a discussion of the autoregression order selection problem.

2. Univariate Time Series.

Let

$$x_1 , x_2 , \dots x_n ,$$

be equally spaced observations of a univariate weakly stationary time series which has zero mean or has the sample mean or trends subtracted. The usual positive definite estimate of the covariance sequence for lag j is

$$\hat{R}_j = \frac{1}{n} \sum_{t=1}^{n-j} x_{t+j} x_t . \tag{2.1}$$

If the time series is an autoregression of order p,

$$x_t = \sum_{k=1}^{p} a_k x_{t-k} + \varepsilon_t , \tag{2.2}$$

where ε_t are uncorrelated random variables with zero mean and variance σ^2, the Yule-Walker equations are

$$\sum_{k=1}^{p} a_k R_{j-k} = R_j \qquad j > 0 . \tag{2.3}$$

The autoregression coefficients, a_k, are estimated from the first p equations ($j = 1,p$) by replacing the covariances by the estimate (2.1) and solving the system of p equations with p unknowns. The one-step prediction error variance is estimated by substitution into the Yule-Walker equation for $j = 0$,

$$\sigma^2 = R_o - \sum_{k=1}^{p} a_k R_k . \tag{2.4}$$

The usual procedure is to fit autoregressions of increasing orders in a stepwise fashion using the Levinson (1947) - Durbin (1960) recursion. Denoting the estimate of the kth coefficient when fitting an autoregression of order p as $a_k^{(p)}$, and the estimate of σ^2 as S_p, the recursion begins with

$$S_o = R_o \tag{2.5}$$

$$a_1(1) = R_1/R_o$$

$$S_1 = [\ 1 - (a_1^{(1)})^2\ ]\ S_o\ .$$

The general step is

$$a_p^{(p)} = [R_p - \sum_{k=1}^{p-1} a_k^{(p-1)} R_{p-k}]/S_{p-1}\ . \tag{2.6}$$

This is referred to as the partial autocorrelation coefficient or reflection coefficient of order p.

$$S_p = [\ 1 - (a_p^{(p)})^2\ ]\ S_{p-1}\ . \tag{2.7}$$

The other coefficients are then updated,

$$a_k^{(p)} = a_k^{(p-1)} - a_p^{(p)} a_{p-k}^{(p-1)} \quad k=1,\ p-1, \tag{2.8}$$

completing the recursion.

It can be shown that the Yule-Walker estimate is equivalent to annexing p additional zeros at the beginning of the time series and obtaining least squares estimates based on predictions from t = 1 to n. For small n, this tends to bias the estimated autocorrelation coefficients toward zero. Burg's (1967) algorithm avoids this source of bias by working entirely within the data span. This algorithm works with the data rather than with an estimated covariance function. An estimated covariance function can be produced as a byproduct of the algorithm. Define residuals from a one-step prediction based on an autoregression of order p,

$$e_t^{(p)} = x_t - \sum_{k=1}^{p} a_k^{(p)} x_{t-k} \ , \ t=p+1, n \ . \qquad (2.9)$$

Assume that coefficients for order p-1 have been estimated, and substitute Equation (2.8) into Equation (2.9),

$$e_t^{(p)} = x_t - \sum_{k=1}^{p-1} \left(a_k^{(p-1)} - a_p^{(p)} a_{p-k}^{(p-1)} \right) x_{t-k} - a_p^{(p)} x_{t-p} \ . \qquad (2.10)$$

This equation now contains only one unknown parameter, $a_p^{(p)}$, and can be simplified by noting that embedded in the equation are residuals from a backward prediction based on the autoregression of order p-1. Let these backward residuals be

$$\eta_t^{(p)} = x_t - \sum_{k=1}^{p} a_k^{(p)} x_{t+k} \ , \quad t=1, n-p \ . \qquad (2.11)$$

Equation (2.10) can then be simplified,

$$e_t^{(p)} = e_t^{(p-1)} - a_p^{(p)} \eta_{t-p}^{(p-1)} \ , \ t=p+1, n \ . \qquad (2.12)$$

and the recursion for the backward residuals is

$$\eta_t^{(p)} = \eta_t^{(p-1)} - a_p^{(p)} e_{t+p}^{(p-1)} \ , \quad t=1, n-p \ . \qquad (2.13)$$

$a_p^{(p)}$ is then estimated from

$$\sum_{t=p+1}^{n} \left[e_t^{(p)} \right]^2 + \sum_{t=1}^{n-p} \left[\eta_t^{(p)} \right]^2 = \min \ , \qquad (2.14)$$

giving

$$a_p^{(p)} = 2\sum_{t=p+1}^{n} e_t^{(p-1)} \eta_{t-p}^{(p-1)} \Big/ \left\{ \sum_{t=1}^{n-p} \left[\eta_t^{(p-1)} \right]^2 + \sum_{t=p+1}^{n} \left[e_t^{(p-1)} \right]^2 \right\}. \tag{2.15}$$

The recursion is then completed using Equations (2.7) and (2.8), and started with

$$S_o = R_o \, . \tag{2.16}$$

The equations in this section are given by Burg (1975) with proofs of many of the properties of the estimates. A main property is that the roots of the equation

$$\sum_{k=1}^{p} a_k^{(p)} z^{-k} = 1 \tag{2.17}$$

cannot be outside of the unit circle, so that the spectrum estimate

$$s(f) = hS_p / \left| 1 - \sum_{k=1}^{p} a_k^{(p)} e^{2\pi ikhf} \right|^2 , \tag{2.18}$$

where f is frequency in cycles per unit time and h is the time interval between observations, cannot be negative. For a finite sample size, the denominator of this expression can be closer to zero than when estimates are based on the Yule-Walker equations and this causes the apparent increase in resolution when there are strong peaks in the spectrum.

3. Multivariate Time Series.

Whittle (1963) and Robinson (1963) generalized the Levinson-Durbin recursion to the multivariate case by fitting both forward and backward autoregressions in a stepwise fashion (see also Wiggins and Robinson, 1965). Burg's algorithm does not generalize directly since the forward and backward autoregression matrices are not the same in the multivariate case, and the forward and backward one-step prediction error covariance matrices are different, although they have the same determinant (Akaike, 1973).

Let

$$X_1, X_2, \ldots X_n$$

denote n equally spaced observations of a multivariate weakly stationary time series of dimension d. Each X_t is a d × 1 column vector. Assume that the process has zero mean or that the sample mean has been subtracted from each observation. The d × d covariance sequence matrices are estimated, for lag j, as

$$\hat{R}_j = \frac{1}{n} \sum_{t=1}^{n-j} X_{t+j} X_t' \quad , \tag{3.1}$$

where the prime denotes the transposed vector. The forward predicting autoregression of order p is

$$\hat{X}_t^{(f)} = \sum_{k=1}^{p} A_k^{(p)} X_{t-k} \quad , \tag{3.2}$$

and the backward predicting autoregression of order p is

$$\hat{X}_t^{(b)} = \sum_{k=1}^{p} B_k^{(p)} X_{t+k} \quad , \tag{3.3}$$

where $A_k^{(p)}$ and $B_k^{(p)}$ are d × d matrices. The stepwise procedure starts by letting

$$S_o^{(f)} = S_o^{(b)} = R_o \quad . \tag{3.4}$$

These are the forward and backward one-step prediction error covariance matrices for an autoregression of order zero, i.e., the mean (zero) is the prediction. The first step is

$$\begin{aligned} A_1^{(1)} &= R_1 \left[S_o^{(b)} \right]^{-1} \\ B_1^{(1)} &= R_{-1} \left[S_o^{(f)} \right]^{-1} \end{aligned} \quad . \tag{3.5}$$

Since

$$R_{-1} = R_1' \quad ,$$

for computation, the backward matrix would be calculated

$$B_1^{(1)} = R_1' \left[S_o^{(f)} \right]^{-1} \quad . \tag{3.6}$$

Assuming that the recursion has been completed for order p-1, the general step calculates

$$\begin{aligned} A_p^{(p)} &= (R_p - \sum_{k=1}^{p-1} A_k^{(p-1)} R_{p-k}) \left[S_{p-1}^{(b)} \right]^{-1} \\ B_p^{(p)} &= (R_p' - \sum_{k=1}^{p-1} B_k^{(p-1)} R'_{p-k}) \left[S_{p-1}^{(f)} \right]^{-1} \quad , \end{aligned} \tag{3.7}$$

and the other matrices are updated

$$\begin{aligned} A_k^{(p)} &= A_k^{(p-1)} - A_p^{(p)} B_{p-k}^{(p-1)} \\ B_k^{(p)} &= B_k^{(p-1)} - B_p^{(p)} A_{p-k}^{(p-1)} \quad . \end{aligned} \tag{3.8}$$

The one-step forward and backward prediction error covariance matrices are

$$\begin{aligned} S_p^{(f)} &= (I - A_p^{(p)} B_p^{(p)}) S_{p-1}^{(f)} \\ S_p^{(b)} &= (I - B_p^{(p)} A_p^{(p)}) S_{p-1}^{(b)} \quad . \end{aligned} \tag{3.9}$$

Equations (3.9) are a simplification of the equations given by Whittle which express the change in the prediction error covariance matrices in terms of the partial autocorrelation or reflection matrices (see, for example, Burg, 1975). This procedure has been programmed by Jones (1964), Robinson (1967) and others.

When attempting to generalize Burg's algorithm, it is noted that the recursion on the forward and backward prediction residuals does generalize. Letting

$$e_t^{(p)} = X_t - \sum_{k=1}^{p} A_k^{(p)} X_{t-k}, \quad t=p+1, n$$
$$\eta_t^{(p)} = X_t - \sum_{k=1}^{p} B_k^{(p)} X_{t+k}, \quad t=1, n-p \ , \tag{3.10}$$

and substituting from Equation (3.8), gives

$$e_t^{(p)} = e_t^{(p-1)} - A_p^{(p)} \eta_{t-p}^{(p-1)} \ , \quad t=p+1, n$$
$$\eta_t^{(p)} = \eta_t^{(p-1)} - B_p^{(p)} e_{t+p}^{(p-1)} \ , \quad t=1, n-p \ . \tag{3.11}$$

Since the forward and backward autoregression matrices are different, if $A_p^{(p)}$ and $B_p^{(p)}$ are estimated separately from the two equations, the least squares estimates would be

$$A_p^{(p)} = U\, V^{-1}$$
$$B_p^{(p)} = U' W^{-1} \tag{3.12}$$

where

$$U = \sum_{t=1}^{n-p} e_{t+p}^{(p-1)} (\eta_t^{(p-1)})' \tag{3.13}$$

is the sum of cross products of forward and backward residuals at lag p, and

$$V = \sum_{t=1}^{n-p} \eta_t^{(p-1)} (\eta_t^{(p-1)})'$$

and (3.14)

$$W = \sum_{t=1}^{n-p} e_{t+p}^{(p-1)} (e_{t+p}^{(p-1)})'$$

estimate

$$(n-p)\ S_{p-1}^{(b)}$$

and

$$(n-p)\ S_{p-1}^{(f)} \quad .$$

A difficulty with these separate estimates is that the spectra calculated from the forward and backward estimates are different. Although the forward and backward autoregressive matrices and the prediction error covariance matrices are different, the multivariate spectra should be identical when calculated from the forward and backward fits by

$$S(f) = h[A(f)]^{-1}\ S_p^{(f)} [A^*(f)]^{-1}$$

or by

$$S(f) = h[B(f)]^{-1}\ S_p^{(b)} [B^*(f)]^{-1}$$

where

$$A(f) = I - \sum_{k=1}^{p} A_k^{(p)}\ e^{2\pi ikhf} \quad ,$$

$$B(f) = I - \sum_{k=1}^{p} B_k^{(p)}\ e^{-2\pi ikhf} \quad ,$$

and $*$ denotes complex conjugate transpose.

4. Multivariate Generalizations.

A possible generalization, independently derived by Jones (1976a), Nuttall (1976a), and Lacoss (1976), involves specifying the partial autocorrelation matrices to be of the form

$$A_p^{(p)} = \Gamma \left[S_{p-1}^{(b)} \right]^{-1}$$
$$B_p^{(p)} = \Gamma' \left[S_{p-1}^{(f)} \right]^{-1} \tag{4.1}$$

where $S_{p-1}^{(f)}$ and $S_{p-1}^{(b)}$ are known from the previous step of the recursion, and estimating the $d \times d$ matrix Γ by minimizing the sum of the traces of the forward and backward sum of squares and cross products matrices of one-step prediction residuals. Now

$$e_t^{(p)} = e_t^{(p-1)} - \Gamma \left[S_b^{(p-1)} \right]^{-1} \eta_{t-p}^{(p-1)}$$
$$\eta_t^{(p)} = \eta_t^{(p-1)} - \Gamma' \left[S_f^{(p-1)} \right]^{-1} e_{t+p}^{(p-1)} . \tag{4.2}$$

Using the notation of Equations (3.13) and (3.14), it is necessary to solve the following system of equations for Γ,

$$\Gamma A + B\Gamma = C \tag{4.3}$$

where all entries are $d \times d$ matrices and A, B, and C are known constant matrices,

$$A = \left[S_b^{(p-1)}\right]^{-1} V \left[S_b^{(p-1)}\right]^{-1}$$

$$B = \left[S_f^{(p-1)}\right]^{-1} W \left[S_f^{(p-1)}\right]^{-1} \qquad (4.4)$$

$$C = U \left[S_b^{(p-1)}\right]^{-1} + \left[S_f^{(p-1)}\right]^{-1} U .$$

Equation (4.3) can be solved by unraveling the $d \times d$ matrix Γ into a $d^2 \times 1$ vector and solving a system of $d^2 \times d^2$ equations. There is a large body of literature on the solution of Equation (4.3). See, for example, Jameson (1968) and Hartwig (1972).

In the univariate case,

$$S_b^{(p-1)} = S_f^{(p-1)} = S^{(p-1)}$$

and (4.3) becomes

$$\Gamma V + \Gamma W = 2\ US^{(p-1)}$$

$$\Gamma = 2\ US^{(p-1)}/(V + W) \qquad (4.5)$$

$$A_p^{(p)} = 2\ U/(V + W)$$

giving Burg's estimate.

The multivariate spectral density estimate is

$$S(f) = h \left[A(f)\right]^{-1} S_f^{(p)} \left[A^*(f)\right]^{-1} \qquad (4.6)$$

where

$$A(f) = I - \sum_{k=1}^{p} A_k^{(p)} e^{2\pi ikhf} \qquad (4.7)$$

and * denotes complex conjugate transpose.

It has been pointed out to the author by Nuttall (1976b) that there are examples where the above estimation scheme does not produce a stationary solution. Meanwhile, another procedure suggested by Morf, Viera, and Kailath (1976) and independently by Robinson (1976), which was proved to produce stationary estimates, was programmed. The FORTRAN IV program is attached as an appendix.

Using the notation for residual sum of squares and cross products from Equations (3.13) and (3.14), a step in the recursion begins by factoring W and V into the product of lower triangular matrices times their transposes using the Cholesky decomposition,

$$W = L_W L_W'$$
$$V = L_V L_V' \quad . \tag{4.8}$$

The normalized reflection coefficient matrix, which is a matrix of correlations is calculated as

$$\rho = L_W^{-1} \, U(L_V')^{-1} \tag{4.9}$$

where the matrix products are calculated using forward elimination. This procedure is stable since the singular values of ρ, which are the canonical correlations between the forward and backward residuals (Akaike, 1976) are less than one. The forward and backward prediction error covariance matrices from the previous step also are factored,

$$S_f^{(p-1)} = L_f L_f'$$
$$S_b^{(p-1)} = L_b L_b' \quad , \tag{4.10}$$

and the partial autocorrelation matrices calculated as

$$A_p^{(p)} = L_f \rho L_b^{-1}$$
$$B_p^{(p)} = L_b \rho' L_f^{-1} \tag{4.11}$$

where postmultiplication by the inverse of a lower triangular matrix is carried out using backward elimination. Equation (3.9) for updating the forward and backward prediction error covariance matrices can now be written

$$S_p^{(f)} = S_{p-1}^{(f)} - (L_f\rho)(L_f\rho)'$$
$$S_p^{(b)} = S_{p-1}^{(b)} - (L_b\rho')(L_b\rho')' \quad . \tag{4.12}$$

where the matrix products in parentheses are available from the previous step.

In the univariate case, this procedure reduces to estimating the partial autocorrelation coefficient as the usual correlation estimate between the forward and backward residuals,

$$a_p^{(p)} = U/(WV)^{1/2} \quad , \tag{4.13}$$

where Burg's estimate is

$$a_p^{(p)} = U\Big/\left(\frac{W + V}{2}\right) \quad . \tag{4.14}$$

Treitel (1976) has been using estimate (4.13) and pointed out to the author an earlier reference (Itakura and Saito, 1971). It is interesting to note that (4.13) is the maximum likelihood estimate of a correlation coefficient for Gaussian variables with zero mean and unknown variances, and (4.14) is the maximum likelihood estimate if the unknown variances are known to be the same as is the case for the forward and backward one-step prediction error variance for a univariate time series. However, in the time series case, these are not the exact maximum likelihood estimates.

In the univariate situation, there are many possible estimates (Makhoul, 1976) and the multivariate case is worse. The above discussion suggests another possible estimate. If the forward residuals, $\varepsilon_t^{(p-1)}$, are premultiplied by L_f^{-1} and the backward residuals, $\eta_t^{(p-1)}$, are premultiplied by L_b^{-1}, the modified residuals should then have the identity matrix as the covariance matrices. Since this will not be true exactly because of estimation error, assume that the covariance matrices of the forward and backward modified residuals are the same and use the usual pooled estimate. Let this pooled estimate be

$$M = 1/2\ [L_f^{-1} W(L_f')^{-1} + L_b^{-1} V(L_b')^{-1}] \ . \tag{4.15}$$

Factoring this matrix,

$$M = L_m L_m' \ . \tag{4.16}$$

The estimate of the normalized reflection coefficient matrix is

$$\rho = L_m^{-1} L_f^{-1} U(L_b')^{-1} (L_m')^{-1} \ . \tag{4.17}$$

The estimation procedure is completed using Equations (4.11) and (4.12).

5. Order Selection.

The problem of selecting the order of the autoregression to be fit to data depends on the purpose of the data analysis. In the univariate case, Burg's algorithm was developed to give high resolution spectrum estimates. Fitting autoregressions of order n-1 to time series of length n gives estimates of maximum resolution and maximum variability, and the term maximum entropy was introduced. This led some researchers to use Burg's algorithm to estimate the spectrum of data which was not highly structured and conclude that spectral peaks due only to sampling variability were real. Therefore, in many application, it is necessary to have a method for selecting the order of the autoregression. In the multivariate case, this becomes more critical, since to fit order n-1 involves estimating more parameters for predicting each channel of data than there are data points in the channel.

In a recent simulation study (Jones, 1976b), it became obvious that certain biases are present in the various algorithms when the order of the autoregression is not small compared to the length of the time series. These simulations involved fitting autoregressions to white noise and calculating for each order the mean square error for one-step predictions. Burg's algorithm was fairly unbiased until the order of the autoregression reached n/2 then became biased toward zero. On the other hand, the Yule-Walker algorithm started unbiasedly for small orders, then showed an increasing bias as the order increased. When the fitted autoregressions were used to predict independent data, Burg's algorithm showed a drastic increase in prediction error variance for orders greater than n/2 indicating that if the purpose of fitting autoregressions is prediction, orders greater than n/2 would rarely be appropriate.

At the present time, my preference for order selection is to use Akaike's (1973, 1974) Information Criterion (AIC) with the maximum order searched being

$$p_{max} \leq n/2d \qquad (5.1)$$

so that the number of coefficients estimated to predict each channel of data is not greater than half the number of observations. The order is then selected for which

$$AIC_p = n \log |S_p| + 2pd^2 = \min . \qquad (5.2)$$

6. Conclusion.

This paper has discussed multivariate generalizations of

algorithms for fitting autoregressions working with residuals within the data span. It must be noted that a time series of dimensions d with n time points observed contains a total of nd observations. Fitting an autoregression of order p involves estimating pd^2 coefficients. This conflict is removed by limiting the maximum order searched to n/2d and using an order selecting criterion. Given these considerations, in the multivariate case, the algorithm does not seem to have any advantages over the Yule-Walker algorithm, and the name maximum entropy does not seem appropriate. The advantages and disadvantages of the two algorithms may not be clear until more experience is gained.

References.

Akaike, H. (1973). Information theory and an extension of the maximum likelihood principle. *Second International Symposium on Information Theory* (B. N. Petrov and F. Csaki, Eds.). Budapest: Akademia Kaido, 267-281.

Akaike, H. (1973). Block Toeplitz matrix inversion. *SIAM J. Appl. Math. 24*, 234-241.

Akaike, H. (1974). A new look at the statistical model identification. *IEEE Trans. Automat. Contr. AC-19*, 716-723.

Akaike, H. (1976). Personal communication.

Burg, J. P. (1967). Maximum entropy spectral analysis. Paper presented at the 37th Annual International Meeting. Soc. of Explor. Geophys., Oklahoma City, October 31, 1967.

Burg, J. P. (1975). Maximum entropy spectral analysis. Ph.D. dissertation, Dept. of Geophysics, Stanford University.

Durbin, J. (1960). The fitting of time series models. *Rev. Int. Inst. Statist. 28*, 233-289.

Hartwig, R. E. (1972). Resultants and the solutions of AX-XB = -C. *SIAM J. Appl. Math. 23*, 104-117.

Itakura, F., and S. Saito (1971). Digital filtering techniques for speech analysis and synthesis. *Seventh International Congress on Acoustics*, Budapest, 261-264.

Jameson, A. (1968). Solution of the equation AX + XB = C by inversion of an M × M or N × N matrix. *SIAM J. Appl. Math. 16*, 1020-1023.

Jones, R. H. (1964). Prediction of multivariate time series. *J. Appl. Meteorol.* *3*, 255-289.

Jones, R. H. (1976a). Multivariate maximum entropy spectral analysis. Applied Time Series Analysis Symposium, Tulsa, Oklahoma, May 14-15, 1976.

Jones, R. H. (1976b). Autoregression order selection. *Geophysics* *41*, August, 1976.

Lacoss, R. (1976) Personal communication.

Levinson, H. (1947). The Wiener RMS (root mean square) error criterion in filter design and prediction. *J. Math. Phys.* *25*, 261-278.

Makhoul, J. (1976). New lattice methods for linear prediction. Presented at the 1976 International Conference on Acoustics, Speech, and Signal Processing, Philadelphia, April 12-14, 1976.

Morf, M., A. Viera, and T. Kailath (1976). The multichannel maximum entropy method. Unpublished manuscript. ILS - Stanford University.

Nuttall, A. H. (1976a). FORTRAN program for multivariate linear predictive spectral analysis, employing forward and backward averaging. NUSC Technical Document 5419, Naval Underwater Systems Center, New London, Connecticut.

Nuttall, A. H. (1976b). Personal communication.

Robinson, E. A. (1963). Mathematical development of discrete filters for the detection of nuclear explosions. *J. Geophys. Res.* *68*, 5559-5567.

Robinson, E. A. (1967). *Multichannel Time Series Analysis with Digital Computer Programs*. Holden-Day, San Francisco.

Robinson, E. A. (1976). Personal communication.

Treitel, S. (1976). Personal communication.

Ulrych, T. J., and T. N. Bishop (1975). Maximum entropy spectral analysis and autoregressive decomposition. *Rev. Geophysics and Space Physics* *13*, 183-200.

Whittle, P. (1963). On the fitting of multivariate autoregressions, and the approximate canonical factorization of a spectral density matrix. *Biometrika* *50*, 129-134.

Wiggins, R. A., and E. A. Robinson (1965). Recursive solution to the multichannel filtering problem. *J. Geophys. Res.* *70*, 1885-1891.

Appendix.

```
      SUBROUTINE MRAUTO(N,NFD,ND,M,X,IA,V,A,B,PF,PB,Y,U,W,C,D)
      DIMENSION X(NFD,ND),Y(NFD,ND),V(ND,ND),A(ND,ND,M)
      DIMENSION B(ND,ND,M),U(ND,ND),W(ND,ND),PF(ND,ND,1)
C     DIMENSION PB(ND,ND,1),C(ND,ND),D(ND,ND)
C   MULTIVARIATE STEPWISE AUTOREGRESSION USING RESIDUALS
C   WITHIN THE DATA SPAN AND ORDER SELECTION USING AKAIKE'S
C   INFORMATION CRITERION
C   PROGRAMMED JUNE, 1976,  BY RICHARD H JONES
C                              DEPARTMENT OF BIOMETRICS
C                              BOX B-119
C                              UNIV. OF COLORADO MEDICAL CENTER
C                              DENVER, COLORADO  80220
C   INPUT TO SUBROUTINE
C      N   SAMPLE SIZE
C      NFD   FIRST DIMENSION OF X AND Y ARRAYS IN CALLING
C            PROGRAM
C      ND   DIMENSION OF MULTIVARIATE TIME SERIES
C      M   MAXIMUM ORDER AUTOREGRESSION SEARCHED FOR BEST FIT
C      X   ARRAY OF TIME SERIES DATA
C   OUTPUT FROM SUBROUTINE
C      IA   ORDER SELECTED BY AKAIKE'S INFORMATION CRITERION
C      V   ESTIMATED ONE STEP PREDICTION COVARIANCE MATRIX
C      A   ESTIMATED AUTOREGRESSION MATRICES
C      B   ESTIMATED BACKWARD AUTOREGRESSION MATRICES
C      PF  ESTIMATED ONE STEP PREDICTION COVARIANCE MATRIX
C          FOR EACH ORDER FIT. LAST DIMENSION OF PF MUST BE
C          M+1 IN CALLING PROGRAM SINCE ORDER ZERO IS STORED
C          IN FIRST POSITION
C      PB  SAME FOR BACKWARD PREDICTION
C      U   ESTIMATED ONE STEP BACKWARD PREDICTION COVARIANCE
C   MATRIX WORKSPACE
C      Y   ARRAY WITH SAME DIMENSION AS X USED FOR RESIDUALS
C          FROM BACKWARD PREDICITIONS. RESIDUALS FROM FOREWARD
C          PREDICTIONS ARE STORED IN X. ORIGINAL DATA ARE
C          DESTROYED.
C      W,C,D
C
      IF(M.GT.N/(2*ND))M=N/(2*ND)
      FN=N
      DO 11 I=1,N
      DO 11 J=1,ND
   11 Y(I,J)=X(I,J)
```

```
C     CALCULATE SUM OF SQUARES AND CROSS PRODUCT MATRIX
      DO 14 I=1,ND
      DO 14 J=I,ND
      W(I,J)=0.0
      DO 13 K=1,N
   13 W(I,J)=W(I,J)+X(K,I)*X(K,J)
      PF(I,J,1)=W(I,J)/FN
      W(J,I)=W(I,J)
      PF(J,I,1)=PF(I,J,1)
      PB(I,J,1)=PF(I,J,1)
   14 PB(J,I,1)=PB(I,J,1)
      CALL DET(PF,U,ND,AICM)
      AICM=FN*ALOG(AICM)
      IA=0
      PRINT 1
    1 FORMAT(6H0ORDER,7X,3HAIC)
      PRINT 2,IA,AICM
    2 FORMAT(1X,I4,F15.2)
C
C     START STEPWISE PROCEDURE
      DO 26 L=1,M
      NP=N-L
      NP1=NP+1
      L1=L+1
C     CALCULATE LAG PRODUCT SUMS
      DO 15 I=1,ND
      DO 15 J=1,ND
      U(I,J)=0.0
      DO 15 K=1,NP
   15 U(I,J)=U(I,J)+X(K+L,I)*Y(K,J)
      DO 17 I=1,ND
      DO 17 J=I,ND
      C(I,J)=0.0
      D(I,J)=0.0
      DO 19 K=1,NP
      C(I,J)=C(I,J)+X(K+L,I)*X(K+L,J)
   19 D(I,J)=D(I,J)+Y(K,I)*Y(K,J)
      C(J,I)=C(I,J)
   17 D(J,I)=D(I,J)
C     FACTOR MATRICES AND CALCULATE NORMALIZED REFLECTION
C     COEFFICIENTS IN U
      CALL CHOLES(C,ND,ND)
      CALL CHOLES(D,ND,ND)
      CALL FORWARD(C,ND,ND,U,ND)
      CALL POST(D,ND,ND,U,ND,ND)
```

```
C     CALCULATE FORWARD AND BACKWARD PARTIAL AUTOREGRESSION
C     MATRICES IN A AND B
      DO 18 I=1,ND
      DO 18 J=1,ND
      W(I,J)=PF(I,J,L)
   18 V(I,J)=PB(I,J,L)
      CALL CHOLES(W,ND,ND)
      CALL CHOLES(V,ND,ND)
      DO 16 I=1,ND
      DO 16 J=1,ND
      C(I,J)=0.0
      D(I,J)=0.0
      DO 23 K=1,I
      C(I,J)=C(I,J)+W(I,K)*U(K,J)
   23 D(I,J)=D(I,J)+V(I,K)*U(J,K)
      A(I,J,L)=C(I,J)
   16 B(I,J,L)=D(I,J)
      CALL BACK(V,ND,ND,A(1,1,L),ND,ND)
      CALL BACK(W,ND,ND,B(1,1,L),ND,ND)
C     CALCULATE FORWARD AND BACKWARD PREDICTION COVARIANCE
C     MATRICES IN PF AND PB
      DO 41 I=1,ND
      DO 41 J=1,ND
      PF(I,J,L1)=PF(I,J,L)
      PB(I,J,L1)=PB(I,J,L)
      DO 41 K=1,ND
      PF(I,J,L1)=PF(I,J,L1)-C(I,K)*C(J,K)
   41 PB(I,J,L1)=PB(I,J,L1)-D(I,K)*D(J,K)
C     CALCULATE RESIDUALS
      DO 24 K=L1,N
      KK=K-L
      DO 20 I=1,ND
      U(I,1)=0.0
      DO 20 J=1,ND
   20 U(I,1)=U(I,1)+A(I,J,L)*Y(KK,J)
      DO 21 I=1,ND
      DO 21 J=1,ND
   21 Y(KK,I)=Y(KK,I)-B(I,J,L)*X(K,J)
      DO 22 I=1,ND
   22 X(K,I)=X(K,I)-U(I,1)
   24 CONTINUE
C     CALCULATE AIC
      CALL DET(PF(1,1,L1),U,ND,AIC)
      IF(AIC.LE.0.0)GO TO 34
      AIC=FN*ALOG(AIC)+ND*L*(1+2*ND)
```

```
      PRINT 2,L,AIC
      IF(AIC.GE.AICM)GO TO 26
      AICM=AIC
      IA=L
   26 CONTINUE
C
C    CALCULATE ONE STEP PREDICTION COVARIANCE MATRIX
   34 DO 27 I=1,ND
      DO 27 J=1,ND
   27 V(I,J)=PF(I,J,IA+1)
C    GENERATE PREDICTION MATRICES FOR ORDER SELECTED
      PRINT 3,IA
    3 FORMAT(6H0ORDER,I3,9H SELECTED)
      IF(IA.LE.1)GO TO 33
      DO 31 I=2,IA
      JJ=I/2
      DO 31 J=1,JJ
      KK=I-J
      DO 28 K=1,ND
      DO 28 L=1,ND
      U(K,L)=B(K,L,J)
      W(K,L)=A(K,L,J)
      DO 28 LL=1,ND
      U(K,L)=U(K,L)-B(K,LL,I)*A(LL,L,KK)
   28 W(K,L)=W(K,L)-A(K,LL,I)*B(LL,L,KK)
      IF(KK.LE.J)GO TO 30
      DO 29 K=1,ND
      DO 29 L=1,ND
      DO 29 LL=1,ND
      B(K,L,KK)=B(K,L,KK)-B(K,LL,I)*A(LL,L,J)
   29 A(K,L,KK)=A(K,L,KK)-A(K,LL,I)*B(LL,L,J)
   30 DO 31 K=1,ND
      DO 31 L=1,ND
      B(K,L,J)=U(K,L)
   31 A(K,L,J)=W(K,L)
C    CALCULATE BACKWARD PREDICTION COVARIANCE MATRIX
   33 DO 35 I=1,ND
      DO 35 J=1,ND
   35 U(I,J)=PB(I,J,IA+1)
      RETURN
      END
```

```
      SUBROUTINE CHOLES(R,N,ND)
      DIMENSION R(ND,1)
C   IN PLACE CHOLESKY FACTORIZATION OF A POSITIVE DEFINITE
C   MATRIX R INTO THE PRODUCT OF A LOWER TRIANGULAR MATRIX
C   AND ITS TRANSPOSE. ONLY THE LOWER TRIANGULAR PART OF R
C   IS USED AND REPLACED BY THE FACTOR.
C   INPUT TO SUBROUTINE
C      R  N BY N POSITIVE DEFINITE MATRIX TO BE FACTORED
C      ND  FIRST DIMENSION OF R IN CALLING PROGRAM
      IF(R(1,1).LE.0.0)GO TO 14
      R(1,1)=SQRT(R(1,1))
      IF(N.LE.1)RETURN
      DO 13 K=2,N
      K1=K-1
      DO 12 I=1,K1
      IF(I.EQ.1)GO TO 11
      I1=I-1
      DO 10 J=1,I1
   10 R(K,I)=R(K,I)-R(K,J)*R(I,J)
   11 R(K,I)=R(K,I)/R(I,I)
   12 R(K,K)=R(K,K)-R(K,I)**2
      IF(R(K,K).LE.0.0)GO TO 14
   13 R(K,K)=SQRT(R(K,K))
      RETURN
   14 PRINT 1
    1 FORMAT(39H0MATRIX NOT POSITIVE DEFINITE IN CHOLES)
      STOP
      END
C
C
C
      SUBROUTINE FORWARD(R,N,ND,S,M)
      DIMENSION R(ND,1),S(ND,1)
C   PREMULTIPLIES S BY R INVERSE USING FORWARD ELIMINATION,
C   WHERE R IS LOWER TRIANGULAR
C   INPUT TO SUBROUTINE
C      R   LOWER TRIANGULAR N BY N MATRIX
C      ND  FIRST DIMENSION OF R IN CALLING PROGRAM
C      S   N BY M RIGHT HAND SIDE MATRIX WHICH IS REPLACED
C          BY R INVERSE TIMES S
      DO 12 J=1,M
      S(1,J)=S(1,J)/R(1,1)
      IF(N.LE.1)GO TO 12
      DO 11 I=2,N
      I1=I-1
```

```
      DØ 10 K=1,I1
   10 S(I,J)=S(I,J)-R(I,K)*S(K,J)
   11 S(I,J)=S(I,J)/R(I,I)
   12 CØNTINUE
      RETURN
      END
C
C
C
      SUBRØUTINE PØST(R,N,ND,S,M,MD)
      DIMENSIØN R(ND,1),S(MD,1)
C    PØST MULTIPLIES S BY R INVERSE TRANSPØSED BY FØRWARD
C    ELIMINATIØN WHERE R IS LØWER TRIANGULAR
C    INPUT TØ SUBRØUTINE
C       R    LØWER TRIANGULAR N BY N MATRIX
C       ND   FIRST DIMENSIØN ØF R IN CALLING PRØGRAM
C       S    M BY N RIGHT HAND SIDE MATRIX WHICH IS REPLACED
C            BY S TIMES R INVERSE TRANSPØSED
C       MD   FIRST DIMENSIØN ØF S IN CALLING PRØGRAM
      DØ 12 J=1,M
      S(J,1)=S(J,1)/R(1,1)
      IF(N.LE.1)GØ TØ 12
      DØ 11 I=2,N
      I1=I-1
      DØ 10 K=1,I1
   10 S(J,I)=S(J,I)-R(I,K)*S(J,K)
   11 S(J,I)=S(J,I)/R(I,I)
   12 CØNTINUE
      RETURN
      END
C
C
C
C
      SUBRØUTINE BACK(R,N,ND,S,M,MD)
      DIMENSIØN R(ND,1),S(MD,1)
C    PØST MULTIPLIES S BY R INVERSE BY BACKWARD ELIMINATIØN,
C    WHERE R IS LØWER TRIANGULAR
C    INPUT TØ SUBRØUTINE
C       R    LØWER TRIANGULAR N BY N MATRIX
C       ND   FIRST DIMENSIØN ØF R IN CALLING PRØGRAM
C       S    M BY N RIGHT HAND SIDE MATRIX WHICH IS REPLACED
C            BY S TIMES R INVERSE
C       MD   FIRST DIMENSIØN ØF S IN CALLING PRØGRAM
      DØ 12 I=1,M
```

```
      S(I,N)=S(I,N)/R(N,N)
      IF(N.LE.1)GO TO 12
      DO 11 J=2,N
      NN=N+1-J
      J1=NN+1
      DO 10 K=J1,N
   10 S(I,NN)=S(I,NN)-S(I,K)*R(K,NN)
   11 S(I,NN)=S(I,NN)/R(NN,NN)
   12 CONTINUE
      RETURN
      END
C
C
C
      SUBROUTINE DET(B,A,N,D)
      DIMENSION A(N,N),B(N,N)
C
C    EVALUATES DETERMINANT OF N BY N MATRIX B
C    FIRST DIMENSION OF B IN CALLING PROGRAM MUST BE N
C    A IS N BY N MATRIX USED FOR WORK SPACE
C
      IF(N.GT.1)GO TO 5
      D=B(1,1)
      RETURN
    5 DO 10 I=1,N
      DO 10 J=1,N
   10 A(I,J)=B(I,J)
      DO 12 L=2,N
      K=L-1
      DO 7 I=L,N
    7 A(I,K)=A(I,K)/A(K,K)
      J2=K-1
      IF(J2.EQ.0)GO TO 11
      DO 9 J=L,N
      DO 9 I=1,J2
    9 A(K,J)=A(K,J)-A(K,I)*A(I,J)
   11 DO 12 I=L,N
      DO 12 J=1,K
   12 A(I,L)=A(I,L)-A(I,J)*A(J,L)
      D=1.
      DO 13 I=1,N
   13 D=D*A(I,I)
      RETURN
      END
```

TWO-DIMENSIONAL RECURSIVE FILTERING IN THEORY AND PRACTICE

James H. Justice
The University of Tulsa

I. Introduction.

The application of digital data processing techniques to seismic data had its inception with the Geophysical Analysis Group at MIT in the 1950s. Prior to that time, little in the way of sophisticated computational analysis could be applied to large quantities of seismic data, since all interpretive work was carried out by hand using photographically recorded paper records.

The advent of the digital computer changed our ways of working with seismic data. We began to record data in the field on magnetic tape which could be digitized and processed by computer. Much of the pioneering work in developing new processing techniques was done by Sven Treitel and Enders Robinson, whose statistical analyses greatly advanced our abilities to extract information from the data obtained by the seismic method.

In particular, the concept of filtering assumed a position of paramount importance in this work. It was inevitable that eventually we would take advantage of both the temporal and spatial nature of seismic data in designing filters to process this data. Multichannel Wiener filters underwent rapid development [54], and in 1969 the first paper [56] appeared on two-dimensional recursive filtering, an extension to two dimensions of the recursive filter introduced by Shanks in 1967 [57].

The extension to two dimensions of recursive filter design techniques is not automatic, and various approaches to the problem have been considered.

ISBN 0-12-257250-5

It is the purpose of this paper to survey many aspects of two-dimensional filtering and to attempt to shed some light on these inquiries via the author's research in this area.

A. Data Acquisition in Seismic Exploration.

Let us review briefly the origin of seismic data with which we shall be primarily concerned. Figure 1 illustrates the principal features of the seismic method of geophysical exploration. A source of high energy (dynamite, vibrator, air gun, sparker, etc.) sends a signal of some kind into the subsurface (on or offshore). The signal encounters variations in velocity due to changes in lithology, density, etc., as it travels into the subsurface. These changes in acoustic impedance cause a portion of the input signal to be reflected back to the surface where it may be recorded through suitable detectors. In the early years of the seismic method, this recording was done on photosensitive paper. Those days are gone, and today we record the signal on magnetic tape suitable for computer processing.

One of the interesting problems brought about by the computer revolution was the problem of finding a way to transfer perhaps 40 years of data accumulated on paper tape to magnetic tape. The problem was never solved to my knowledge due to the complexities of many overlapping traces (see Figure 2).

One further point is to be made before proceeding. In Figure 1, I have tried to indicate the two lines of thought currently in vogue for working with the evolution of the seismic signal in time. The one line of thought is ray tracing--a derivative of geometric optics in which we try to follow a point on the signal wavefront along a "ray". The other approach is through analytic or approximate solution of the wave equation. Currently, each approach has its advantages. Ray tracing offers us the ability to look at large models inexpensively, but does not give us anything we did not purposely build in. Solutions of the wave equation on the other hand give us everything we would ever want to see, but only at great expense in both computer running time

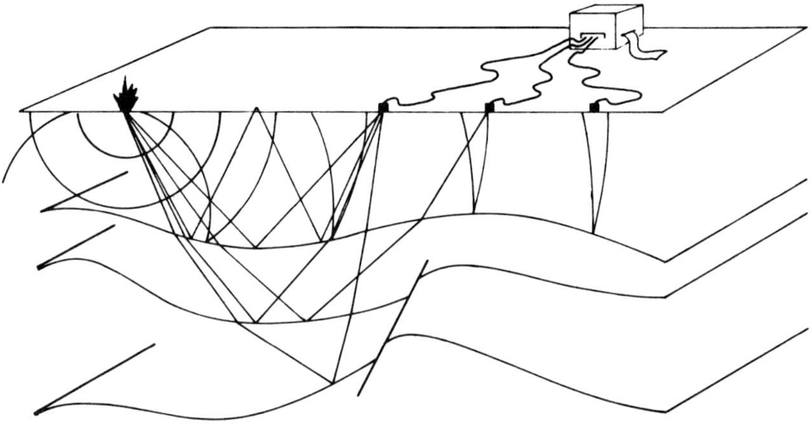

Fig. 1. Seismic Data Acquisition.

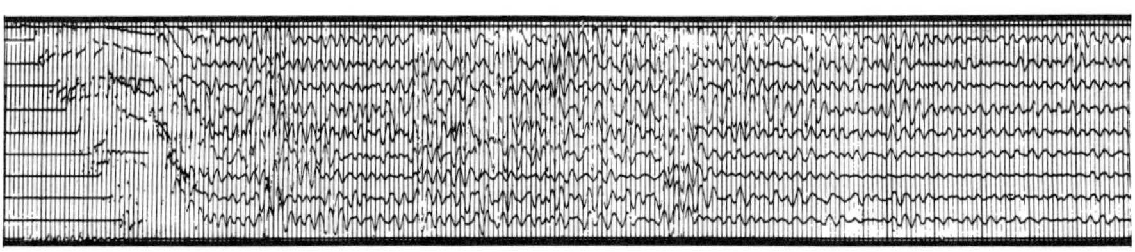

Fig. 2. An example of an old seismic recording on paper.

and storage. Great advances are being made through the latter approach; however, significantly enhancing our ability to interpret the data which we receive.

B. Short Course in Transforms and Sampling.

We shall not get far in our discussion of filtering techniques if we do not know a little about transforms and sampling. I call this my short course and it contains the essence of what one needs to know.

Let f(t) be a function of time which we shall interpret as a signal of some sort.

We define the Fourier Transform of f by the equation

$$F(\omega) = \int_{-\infty}^{\infty} f(t) \exp(2\pi i\omega t)\, dt . \tag{1}$$

If, now, we were to sample the signal f(t) at points in time spaced ΔT seconds apart, we would obtain the sampled version of f given by

$$f_s(t) = \sum_k f_k\, \delta(t - k\Delta T) , \tag{2}$$

where

$$f_k = f(k\Delta T) .$$

Taking the transform of (2) using Equation (1) we obtain

$$F_s(\omega) = \sum_k f_k \int_{-\infty}^{\infty} \delta(t - k\Delta T) \exp(2\pi i\omega t)\, dt$$

$$= \sum_k f_k \exp(2\pi ik\omega\Delta T) . \tag{3}$$

If we now define a complex variable, $z = \exp(i\theta)$ where $\theta = 2\pi\omega\Delta T$, then Equation (3) becomes

$$F_s(\omega) = \sum_k f_k z^k \quad , \tag{4}$$

and this representation is what engineers call the Z-transform of the sampled signal. We should point out that electrical engineers commonly define (4) with z^{-k} in place of z^k, but we shall not use that notation.

We can now observe that as long as f is a real signal, there is no new information in Equation (4) outside the range

$$0 \leq \theta < \pi \quad . \tag{5}$$

That is, we have no information concerning frequencies which lie outside the interval

$$0 \leq \omega < \frac{1}{2\Delta T} \quad , \tag{6}$$

where

$$\omega = \frac{\theta}{2\pi\Delta T} \quad .$$

The number $1/2\Delta T$ is called the Nyquist frequency and our observation is in effect, the Shannon Sampling Theorem. Stated another way, it says that in order to reconstruct accurately a frequency of Ω Hz from our data, we must have sampled that data at a rate no less than 2Ω samples/sec.

The advantage of the Z-transform is that a product of Z-transforms is the Z-transform of the convolution of the respective sequences, and it is easy to work with since it resembles a power series or even a polynomial (for finite sequences) though in fact, it is just another notation for the Fourier transform as long as Z remains on the unit circle ($|Z| = 1$).

C. Filtering and Convolution.

We are now ready to discuss filtering of sampled data. Given an input set of sampled data which we shall think of as a sequence (of sampled values) $\{i_n\}$ and a "filter" sequence, $\{a_n\}$, we can operate on the input with the filter to produce an output sequence, $\{o_n\}$, according to the equation

$$\sum_k a_k \, i_{n-k} = o_n \quad . \tag{7}$$

Equation (7) is called a convolution equation; the output sequence $\{o_n\}$ being the convolution of the input sequence $\{i_n\}$ with the filter sequence $\{a_n\}$. If all the sequences involved are infinite in length, then we generally require them to be absolutely summable so that everything converges and makes sense.

The filter referred to here is called a "convolutional" filter and has a Z-transform given by

$$A(z) = \sum_k a_k \, z^k \quad . \tag{8}$$

(Henceforth, capital letters will stand for transforms where appropriate.)

There is a particular type of filter of great importance known as the recursive filter introduced by Shanks [57].

The recursive filter is characterized by a Z-transform of the form

$$F(z) = \frac{A(z)}{B(z)} \quad . \tag{9}$$

Since a product of Z-transforms is the Z-transform of the convolution of the respective sequences, we may apply the recursive filter to an input sequence and obtain an output sequence according to the equation

$$I(z) \cdot \frac{A(z)}{B(z)} = O(z) \quad . \tag{10}$$

Multiplying by B(z) gives

$$I(z) \cdot A(z) = B(z) \cdot O(z) \quad . \tag{11}$$

If we now normalize B(z) so that

$$B(z) = 1 + \sum_{k=1}^{m} b_k z^k \quad , \tag{12}$$

and let

$$A(z) = \sum_{k=0}^{n} A_k z^k \quad , \tag{13}$$

then (1) is equivalent to

$$o_j = \sum_{k=0}^{n} A_k i_{j-k} - \sum_{\ell=1}^{m} b_\ell o_{j-\ell} \quad . \tag{14}$$

This is the equation for applying the recursive filter. It is called recursive because the current output value is being calculated from previous output values and the input.

The type of filter which we have been discussing here is called a one-dimensional filter since a single index is needed to specify its values (as a sequence).

A two-dimensional filter, then, may be thought of as an array requiring two indices for specifying its values.

D. Filter Design Problems.

There is a variety of philosophies of filtering currently extant and each philosophy of filtering carries with it the statement of a design problem and, hopefully, a solution to that problem.

We shall be concerned with one type of filter design problem, but it is by no means the only one.

I would like to mention just one other which is of great interest in seismic data processing so that we may see recursive filtering in a proper perspective.

The other type of filtering which I would like to mention is called "Wiener" filtering for Norbert Wiener who first made good use of the idea.

In Wiener filtering, we assume that we are given sequences $b \equiv \{b_k\}$ and $c \equiv \{c_k\}$ satisfying

$$||b||^2 \equiv \sum_{k=0}^{\infty} |b_k|^2 < \infty$$

$$||c||^2 \equiv \sum_{k=0}^{\infty} |c_k|^2 < \infty \quad . \tag{15}$$

The design problem is to find a filter sequence $a \equiv \{a_k\}$ satisfying the norm condition (15) and such that the convolution of a with b is the best possible approximation to c in the ℓ_2 norm defined in (15). That is,

$$||b * a - c|| \quad , \tag{16}$$

is a minimum for all $a \varepsilon \ell_2$.

In practice, we may be trying to "shape" the sequence b into the form of the sequence c by using a filter.

The solution to the design problem as stated is given by the "normal" equations

$$\sum \phi_{r-j} a_j = g_r \quad , \tag{17}$$

where ϕ is the autocorrelation of the b sequence, and g is the cross-correlation of the b and c sequences.

Highly optimized procedures such as the Levinson algorithm [54] may be used to solve the normal equations. This type of algorithm can be extended to two dimensions. The author gave a derivation of this type of algorithm using Szegö polynomials [33]. The existence of such an algorithm guarantees that we can design Wiener filters in two dimensions in much the same way that we design them in one dimension. Analyses of this type may be extended to even higher dimensions, and it seems prudent to always choose methods, when possible, which are capable of the greatest possible degree of generalization and extension.

As an example of the use of Wiener filters, let us consider the problem of deconvolution of the seismic wavelet. We may imagine that a signal of some definite form is produced by our energy source and that this "wavelet" travels through the earth undergoing attenuation and reflection, returning to us in modified form. Assuming that the earth behaves in a linear fashion, we may imagine that the input signal is convolved with an "earth filter" (see Figure 3) and the resulting signal is recorded as a seismogram. If we knew the input wavelet shape, we could design a Wiener filter which, when convolved with the wavelet would approximate a δ-distribution. Applying this filter to the seismogram would "deconvolve" the wavelet to the extent that this is possible. The result would be an approximation to the earth filter. We could then hopefully interpret from this the structure of the subsurface in an effort to locate hydrocarbon traps. In practice, of course, we seldom know the wavelet well enough to carry this out.

Related to the problem above is a technique which is called "predictive deconvolution" discussed in the paper [46] by Peacock and Treitel. The appendix to that paper contains an interesting derivation of the fact that if the earth filter is uncorrelated, the seismic wavelet could in fact be estimated. In any event, these techniques have proven to be successful in removing periodic phenomena such as ringing from seismic traces, and Wiener filtering has found great application in seismic data processing.

I should point out that the extraction of the "earth filter" is the ultimate seismic problem which all of our data processing aims at solving. The problem is not an easy one and has certainly stimulated the development of a wide variety of constantly improving techniques in the attempt to obtain a solution.

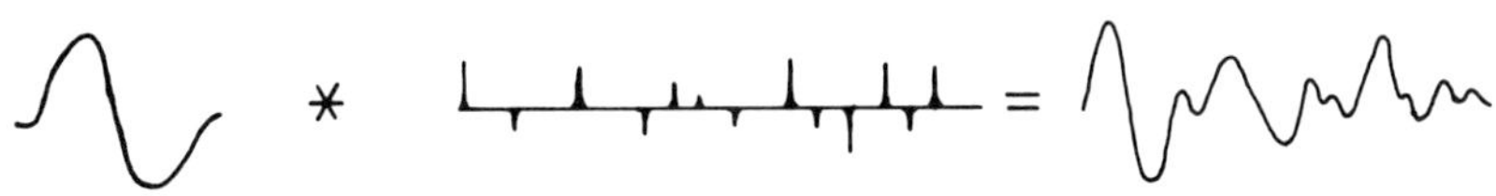

Fig. 3. Construction of an idealized seismogram.

E. Two-Dimensional Transforms and Sampling.

Our one-dimensional sampling and transform theory may be extended in natural ways to two and higher dimensions. For convenience, we shall assume the same sample interval, ΔT, in each dimension and so the Nyquist frequency is simply $1/2\Delta T$ for each dimension. Paralleling our discussion of the one-dimensional case, we may define the Fourier transform of our sampled data $\{a_{nm}\}$ by the equation

$$A\ (f_1, f_2) = \sum \sum a_{nm} \exp\ [2\pi i\ (nf_1 + mf_2)\ \Delta T]\ . \tag{18}$$

If we define

$$\begin{aligned} \theta &= 2\pi f_1 \Delta T \\ \psi &= 2\pi f_2 \Delta T\ , \end{aligned} \tag{19}$$

then (18) becomes

$$A\ (\theta, \psi) = \sum \sum a_{nm} \exp\ (in\theta)\ \exp\ (im\psi)\ , \tag{20}$$

where

$$0 \leq \theta \leq \pi,$$

$$-\pi \leq \psi \leq \pi \text{ (for real coefficients)}$$

corresponds to the frequency range

$$0 \leq f_1 \leq 1/2\Delta T, \; -1/2\Delta T \leq f_2 \leq 1/2\Delta T \; .$$

Defining complex variables Z, w by

$$Z = \exp(i\theta)$$

$$w = \exp(i\psi) \; , \tag{21}$$

we obtain the two-dimensional Z-transform

$$A(z,w) = \sum \sum a_{nm} \, z^n \, w^m \; . \tag{22}$$

The relation between the two-dimensional Z-transform and its generating sequence is the same as in the one-dimensional case. A product of two-dimensional Z-transforms is the Z-transform of the convolution of the corresponding arrays.

F. Two-Dimensional Recursive Filters and Stability.

A two-dimensional recursive filter is a filter whose Z-transform has the form

$$F(z,w) = A(z,w)/B(z,w) \; , \tag{23}$$

where we assume that $B(0,0) = 1$. In this case, the recursion relation for applying the filter becomes

$$O_{m,n} = \sum_{i=0}^{N1} \sum_{j=0}^{N2} a_{ij} \, S_{m-i,n-j} - \sum_{\substack{k=0 \\ k+\ell\neq 0}}^{M1} \sum_{\ell=0}^{M2} b_{k\ell} \, O_{m-k,n-\ell} \; , \tag{24}$$

where $\{O_{mn}\}$ is the output array and $\{S_{m,n}\}$ is the input array.

Filters of recursive type offer both advantages and disadvantages over their convolutional counterparts.

Of primary importance is the fact that such filters can achieve infinite impulse responses due to the denominator term. From the approximation point of view, a quotient of polynomials offers greater ability to fit curves than a single polynomial. Finally, such filters have a "memory" of past events, the current output sample taking into account previous output samples offering a feedback mechanism in filtering. However, recursive filters may be unstable resulting in an unbounded output from a bounded input sequence or array.

As a result, one of the most critical considerations in recursive filter design is filter stability.

In the one-dimensional case, it was known (Robinson) that a recursive filter of the form $F(z) = 1/B(z)$ would be stable provided $B(z) \neq 0$ for $|z| \leq 1$. Shanks conjectured correctly that two-dimensional recursive filters would be stable provided $B(z,w) \neq 0$ for $|z| \leq 1$ and $|w| \leq 1$.

The first proof of this result was given by the author in 1969. It was first published by others to whom it was communicated. The author and Shanks published it in a general form in [35].

Let us state the theorem and sketch its proof. The authors' approach was to find a reasonable setting in which one could obtain a proof of the Shanks conjecture. The proper setting turned out to be the algebra ℓ_1 of absolutely summable sequences (arrays). A sequence $\{b_{nm}\}$ belongs to the algebra ℓ_1 (with convolution as multiplication) if and only if

$$\sum \sum |b_{mn}| < \infty \quad .$$

Once in this setting, a generalized form of a famous Tauberian theorem due to Wiener yields the required result.

The Wiener-Tauberian Theorem: An element b in ℓ_1 is invertible if and only if no continuous homomorphism in the dual of ℓ_1 vanishes at b.

We now need only observe that an arbitrary filter of the form A(z,w)/B(z,w) for $\{a_{nm}\}$ and $\{b_{nm}\}$ in ℓ_1 and $\{b_{nm}\}$ fixed, is always the Z-transform of an ℓ_1 sequence if and only if $\{b_{nm}\}$ is ℓ_1-invertible. (Note that we are actually guaranteeing that our filter output will be absolutely summable rather than just bounded, if the input sequence is also in ℓ_1. However, if we require only that the input sequence be bounded, then this condition still guarantees a bounded output.)

We then establish that the continuous homomorphisms in the dual of ℓ_1 are those functionals of the form

$$\phi_{z_1,z_2}(b) = \sum_{m=0}^{\infty} \sum_{n=0}^{\infty} b_{mn}\, z_1{}^n\, z_2{}^m$$

for

$$|z_1| \leq 1 \quad , \quad |z_2| \leq 1 \quad ,$$

and this yields the required result.

This result is generalized to arrays involving different sets of indices in [35]. (These generalized cases are finding interest today due to recent work in wave equation migration.)

Any good recursive filter design procedure should consider the question of stability of the resulting filter.

Many papers have been written developing tests for stability, but it is the author's opinion that filter stability is most properly considered in the design stage rather than as an afterthought.

G. <u>The Filter Design Problem</u>.

We shall be concerned in this paper with the design of (two-dimensional) recursive filters. As we have seen, these filters have transforms of the form

$$F(z,w) = A(z,w)/B(z,w) \quad , \tag{25}$$

where we assume

$$A(z,w) = \sum_{n=0}^{\infty} \sum_{m=0}^{\infty} a_{nm} z^n w^m$$

$$B(z,w) = 1 + \sum_{\substack{k=0 \\ k+j\neq 0}}^{\infty} \sum_{j=0}^{\infty} b_{kj} z^K w^j . \tag{26}$$

If we use only real coefficients $\{a_{nm}\}$, $\{b_{kj}\}$, then our filter is completely specified if we know its transform for

$$0 \leq \theta < \pi \text{ and } -\pi \leq \psi < \pi$$

where

$$z = e^{i\theta} \text{ and } w = e^{i\psi} .$$

We may imagine this region graphically as the rectangular region of the θ-ψ plane given in Figure 4.

If we wish to use complex coefficients in (26), then we need two more quadrants in Figure 4 to specify the filter transform.

We shall be concerned only with the case of real coefficients in (26). As a result, our filters will exhibit conjugate symmetry of the transform through the origin and the lower two quadrants are not required to specify the filter response. In addition, we shall be concerned only with the specification of the filter amplitude response; i.e., $|F(z,w)|$, the phase being of no interest to us until later when we shall discuss "zero-phase" filters.

Referring now to Figure 5, let us consider the type of data which we are interested in. A simulated seismogram is shown (see also Figure 2), and we see that our two variables of interest are labeled "time" and "space". Each of these variables corresponds to one of the variables Z or W in the transform domain. It follows that the transform variables have the frequency specification "cycles per second" and

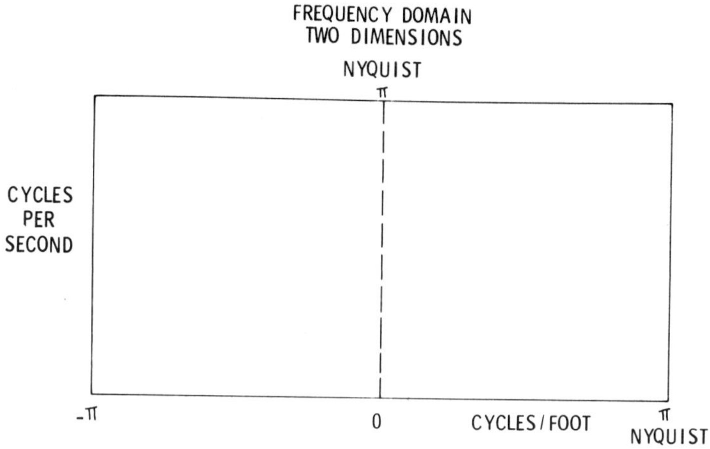

Fig. 4. Frequency domain, two dimensions.

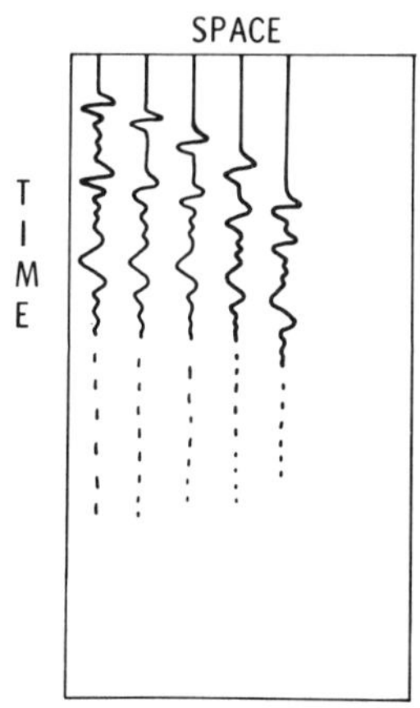

Fig. 5. Space - time domain.

"cycles per foot", respectively. That is, we are dealing with both temporal and spatial frequencies.

Let us now make an observation which we shall make reference to later. Events which lie on straight lines in the time-space domain corespond to events which lie on straight lines in the frequency domain. We notice that the slope of a straight line in the frequency domain has the units feet/second; i.e., velocity. As a result, seismic events which appear to arrive with fixed velocity will be displayed in the frequency domain along a fixed straight line. This observation leads to a technique known as velocity filtering in which we may easily remove seismic events traveling with specified apparent velocities.

Looking at this another way tells us that we can easily remove events which exhibit a linear trend at a specified dip angle. This then, represents one of many possible criteria for filter design in two dimensions.

We may now state the design problem with which we shall be concerned.

Given a desired filter amplitude response $f(\theta,\psi) \geq 0$ for $0 \leq \theta < \pi$ and $-\pi \leq \psi < \pi$, find a recursive filter of predetermined size whose Z-transform $F(z,w)$ approximates $f(\theta,\psi)$ in magnitude.

Many different approaches to this problem could be found depending on how one wished to approximate f with $|F|$.

The technique which we have chosen is the technique of Padé approximation primarily because the technique can be extended to any number of dimensions.

II. Padé Technique in Recursive Filtering.

To begin, let us review the underlying principle for Padé approximation. Let the function f(x) be defined on an interval I. We shall seek polynomials $P_n(x)$, $Q_m(x)$ of degree n,m, respectively, such that

$$|f(x) - P_n(x)/Q_m(x)| \leq M\,|x|^N , \tag{27}$$

where we seek to make N as large as possible. If f has a

Taylor series expansion on I, then we may take $N = n + m + 1$, $Q_m(0) = 1$ and solve for the coefficients of P_n, Q_m in such a way that the equation

$$f(x)\, Q_m(x) - P_n(x) = \sum_{k=n+m+1}^{\infty} C_k\, x^k\,, \tag{28}$$

is satisfied.

It is this concept which we wish to extend to higher dimensions. A moment's reflection, however, may convince us that Tyalor expansions are not necessarily the best approximating mechanisms available.

Because of the polynomial form of the Z-transform, we shall do well to consider some kind of polynomial expansion. In particular, orthogonal polynomials suggest themselves.

Let $\{\phi_k\}$ be a sequence of orthogonal polynomials and suppose that P_n, Q_m, and f are to have the expansions

$$P_n = \sum_1^N p_k\, \phi_k\,,$$

$$Q_m = \sum_1^M q_k\, \phi_k\,, \tag{29}$$

$$f = \sum_1^{\infty} f_k\, \phi_k\,.$$

Then we may generalize (28) and calculate P_n, Q_m to satisfy

$$fQ_m - P_n = \sum_{N+M}^{\infty} C_k\, \phi_k\,. \tag{30}$$

In this case, it is clear that we must satisfy the equation

$$\langle f\, Q_M - P_N,\, \phi_k \rangle = 0,\ 1 \leq k \leq N+M-1\,, \tag{31}$$

and this may be done by solving in order:

$$< f\, Q,\ \phi_k > = 0,\ N + 1 \leq K \leq N+M - 1\ , \tag{32}$$

$$< f\, Q - P,\ \phi_k > = 0,\ 1 \leq K \leq N\ . \tag{33}$$

Recalling that the Tchebycheff polynomials of given degree give almost the best polynomial fit obtainable on the interval [-1,1], we shall use the Tchebycheff polynomials, T_K, for the ϕ_K above.

Since we wish to extend the Padé technique to two or more dimensions, we find that the Tchebycheff polynomials are still a good choice for the extension. This may be exemplified by the following theorem [14].

If f(x,y) is continuous and of bounded variation in the square $[-1 - \delta \leq x,y \leq 1+\delta]$ for some $\delta > 0$ and if one of its partial derivatives is bounded in the square, then f(x,y) has a uniformly convergent double Tchebycheff series expansion on $[-1 \leq x,y \leq 1]$ of the form

$$f(x,y) = \sum_{0}^{\infty} \sum_{0}^{\infty} a_{k\ell}\ T_k(x)\ T_\ell(y)\ .$$

This is the type of expansion which we shall use in two dimensions.

The Padé technique given by Equations (29) through (33) will take the following form in two dimensions:

$$f = \sum_{0}^{\infty} \sum_{0}^{\infty} f_{nm}\ T_n\ T_m\ ,$$

$$P_N = \sum_{0}^{\infty} \sum_{0}^{\infty} p_{nm}\ T_n\ T_m\ ,$$

$$Q_M = \sum_{0}^{\infty} \sum_{0}^{\infty} q_{nm}\ T_n\ T_m\ .$$

We then calculate Q_M from the equation

$$< Q_M f, T_n T_m > = 0 , \tag{35}$$

for

$$N + 1 \leq n \leq Y , \quad 1 \leq m \leq Y ,$$

and

$$N + 1 \leq m \leq Y , \quad 1 \leq n \leq N ,$$

where Y is chosen based on the number of undetermined coefficients to be calculated.

We may now calculate P_N from the equations

$$p_{nm} = < Q_M f, T_n T_m > , \tag{36}$$

for

$$1 \leq n \leq N , \quad 1 \leq m \leq N .$$

Several remarks are in order concerning Equations (34) through (36).

First, we do not have to assume that P_N, Q_M have square coefficient arrays. We have merely done that for convenience in writing. The equations (35) reduce to a fairly simple matrix equation for Q_M because of the nice relationship for a product of Tchebycheff polynomials:

$$T_n T_m = \frac{1}{2} (T_{n+m} + T_{|n-m|}) , \tag{37}$$

which is utilized in calculating fQ_M. The other simplification results from the fact that there is an exact summation formula for the inner products in (35) and (36) so that they are straightforward and easy to calculate. The result is that each element in the matrix for Equation (35) is a sum of nine coefficients of the expansion of f. The details of these calculations will be published in a technical paper elsewhere. Because of the fact that the Tchebycheff polynomials are given by

$$T_n(\theta) = \cos n\theta , \tag{38}$$

on the unit circle, it follows that a filter designed according to the procedure given above would necessarily be symmetric. Since we are also interested in nonsymmetric filters, we must account for sine as well as cosine terms. As a result, although our reasoning remains the same, further modifications are required in detail to calculate nonsymmetric filters. An example of the need for nonsymmetric filters will be given later.

Finally, we point out that the filter which we will ultimately obtain, being an expansion in cosines (and sines) on the unit circle is easily converted to Z-transform format.

A. Stability Procedures.

At this point, we have developed the rudiments of the approximation problem in recursive filter design. Let us recall, now, that stability is just as great a problem and ought to be considered at the design stage.

Let us look at some stability techniques which are available. Again, we point out that the important tests for stability will not be considered here, our interest being directed toward techniques which guarantee stability.

1. Double Least-Squares Inverse.

Let us consider a recursive filter with denominator whose Z-transform is given by B(z,w). According to our stability theorem stated earlier, if B(z,w) has zeroes in the region $|Z| \leq 1$ and $|w| \leq 1$, the filter will be unstable. It was known in one dimension that even if B(z) has zeroes inside or on the unit ball $|z| \leq 1$, the best least-squares approximation of any length to the inverse of B would have no zeroes in this region. As a result, the least-squares approximation of the inverse of the least-squares approximation to the inverse of B should be an approximation to B and would yield stable filters when used as the filter denominator.

Symbolically, we may express this as the following:

Let B be given and let B', B" satisfy

$$B' * B \sim \delta$$

$$B'' * B' \sim \delta \ , \qquad (39)$$

where $\sim$ means best least-squares approximation. The B" is stable and approximates B.

Shanks [58] conjectured that the same result is true in two dimensions. The conjecture remained unresolved for some time as no counterexamples were reported in practice.

However, it has been established that the conjecture is not always true in two dimensions [25]. The construction of a counterexample utilized the same bivariate Szegö polynomials referred to earlier in connection with Wiener filtering (which is clearly involved in the calculation of B' and B" above).

In practice, because truncation of infinite expansions is involved in the calculation of the double least-squares inverse, the amplitude response of B" may not be the same as the response of B.

2. Two-Dimensional Spectral Factorization.

It is well-known for polynomials in one variable that if a given polynomial, p, has no zeroes on the unit circle, then there is a polynomial q which has no zeroes for $|Z| \leq 1$ and satisfies $|q(z)| = |p(z)|$ for $|Z| = 1$. That is, q has the same amplitude spectrum as p, but is stable. The polynomial q can be obtained directly from a factorization of p and there are several ways to do this.

The question is, if we are given a polynomial $p(z,w)$ in two variables, can we obtain a factorization from p such that all the factors are stable and the magnitude of their product agrees with $|p(z,w)|$ (nonzero) for $|z| = |w| = 1$?

Efforts have been made in this direction, and the recent paper by Ekstrom and Woods [20] seems to be the most significant contribution to this problem, indicating that this possibility does exist with stability suitably interpreted for the factors. Their technique does require truncating expansions, however, which may lead to instability, and procedures are given for correcting this condition.

3. Hilbert Transform.

It has long been known that for minimum-phase functions, the phase and log-magnitude of the transform were related. An equation illustrating this relationship is given by

$$\theta(e^{iw}) = -\frac{1}{2\pi}\int_{0}^{2\pi} \operatorname{Log} F(e^{i\Omega}) \cot\left(\frac{w-\Omega}{2}\right) d\Omega \,. \tag{40}$$

It follows that if we have the amplitude response of a filter, we can calculate a phase for the transform of the filter such that the inverse transform is a minimum phase, hence stable filter. This is done in practice using a discrete version of (40).

The advantage of this approach is that we obtain a stable filter without distorting the amplitude response.

Again, the question is, does a two-dimensional version of this technique exist? Read and Treitel [51] investigated the use of a two-dimensional discrete Hilbert Transform for stabilizing filters in two dimensions. The theoretical basis for this work still needs to be developed.

One of the great problems in extending any one-dimensional technique to two dimensions is what Huang has called the fundamental curse - that we cannot factor polynomials in more than one variable. The difficulties introduced by this fact seem almost insurmountable.

4. Spectral Transformation.

Shanks and others have investigated transformations (called Spectral Transformations) which preserve stability of a filter but transform its amplitude spectrum from one form into another. This amounts to a technique for obtaining new stable filters from old ones. The papers of these authors should be consulted for details.

5. Separable Decompositions.

Separable decompositions seem to have found numerous applications in image processing. The idea is that a given doubly-indexed array, G (such as the array of coefficients of a two-dimensional filter), can be decomposed exactly into a sum of outer products of (one-dimensional) sequences in the form

$$G = U\, \Lambda^{1/2}\, V^T \,, \tag{41}$$

or, equivalently

$$G = \Sigma\, \lambda_i^{\ 1/2}\, U_i\, V_i^{\ T} \,, \tag{42}$$

where, as we see, $\Lambda^{1/2}$ is diagonal, and U_i, V_i are ordinary sequences.

Again, Treitel and Shanks [62] introduced separable decompositions to the literature of geophysics. If we attempt to apply this technique to recursive filtering (approximating each term by a separable recursive filter, for example) we run into potential problems in obtaining stability. If each filter in a sum is stabilized (with a possible change in phase spectrum), the amplitude response of the sum may easily be destroyed.

As a result, another type of separation for two-dimensional filters was developed by this author for obtaining stability when it can be used.

The idea is to approximate a given two-dimensional array, which again may be the array of coefficients of a recursive filter denominator, by an outer product of two sequences. The result is that the two-dimensional Z-transform is approximated by a product of two one-dimensional transforms.

These, in turn, can be stabilized by standard procedures. The separation technique is contained in the following theorem [34]:

The orthogonal expansions

$$A(x) = \sum_0^N a_n \phi_n(x) , \quad \text{(with inner product } \langle\cdot,\cdot\rangle_1) \quad \text{for } x \in [a,b]$$

$$B(y) = \sum_0^M b_m \psi_m(y) , \quad \text{(with inner product } \langle\cdot,\cdot\rangle_2) \quad \text{for } y \in [c,d]$$

whose product is the best approximation to the expansion

$$C(x,y) = \sum\sum C_{nm} \phi_n(x) \psi_m(y) , \quad \text{for } (x,y) \in [a,b] \times [c,d]$$

(with inner product defined by

$$\langle\phi_k(x)\ \psi_\ell(y),\ \phi_s(x)\ \psi_t(y)\rangle \equiv \langle\phi_k(x),\ \phi_s(x)\rangle_1 \cdot \langle\phi(y),\ \psi_t(y)\rangle_2)$$

have coefficient vectors given by

$$b = \sum_0^M d_K C_K$$

where $C_K = (C_{k0}, C_{k1}, \ldots, C_{km})$ and $d = \alpha_\nu/(\lambda_\nu)^{1/2}$ where λ_ν is the largest eigenvalue of the matrix:

$$\Phi = (\langle C_K, C_\ell \rangle) ,$$

and α_ν is the normalized eigenvector.

Further, $a_K = \langle b, C_K\rangle$. If $C(x,y)$ is separable, the result is exact.

It is an advantage of the Padé technique that the denominator may be calculated and stabilized before the numerator is calculated. As a result, the numerator assumes a new role of great importance in helping to correct any errors introduced in stabilizing the denominator. This fact lends great power to the Padé technique.

B. Smoothing.

We have now established our basic design criteria and have discussed some of the principles and techniques which can be applied to meet those criteria. If we were to stop at this point, the filters designed by any combination of the preceding techniques would in general fall short of the performance obtainable by exercising a little further thought about the design process in general.

Referring back to our earlier discussion of transforms and sampling (IB), we recall that a Z-transform is nothing more than a Fourier transform or Fourier Series, as you like, when Z is restricted to our design domain, the unit circle. This being true, we must fully expect that a filter designed by almost any process will have a Z-transform which exhibits the well-known properties of Fourier Series such as Gibbs phenomenon.

Fourier Series have been studied for a very long time and several techniques designed to deal with problems exhibited by Fourier Series speak directly to some important aspects of filter design.

There are three basic aspects of filter design which we wish to treat in this section. The first is the "flattening" of the filter response where this response is supposed to be flat. This may involve removal of Gibbs phenomenon or simply smoothing of our response. The second is the elimination of excessively steep roll-off or taper in the response curve. This effect may introduce what is called "ringing" in the filtered output, that is, excessive oscillation of a periodic nature which may damp slowly and obscure detail. The third relates to a question of economics. If we can accelerate in any way the convergence of our designed filters' response to the desired amplitude response curve, we may hope to obtain a filter which exhibits an acceptable response with fewer coefficients. In the authors' opinion, one of the greatest difficulties faced in two-dimensional filtering is the large number of coefficients required to achieve more complicated designs.

There is one technique which addresses itself in some way

to all of the problems mentioned above, and that is the concept of smoothing. The philosophy is simple. If the function used to specify the desired amplitude response (that is, the function which we are concerned with approximating) could be smoothed in some suitable way before the approximating procedure is begun, then we could rely on some basic principles of Fourier Series to operate on our side. Specifically, these principles state that Fourier Series approximations to "smooth" functions do not exhibit Gibbs phenomenon and tend to converge with greater rapidity depending on the degree of "smoothness" of the function being approximated.

Some years ago, the mathematician Lanczos looked into this aspect of Fourier Series and devised a technique which not only deals effectively with Gibbs phenomenon and accelerated convergence, but has the particularly important feature, as far as we are concerned, that it can be easily extended to Fourier Series in any number of dimensions [34]. Let us consider Lanczos' technique. We shall assume that we are given an integrable function f(x) (which may exhibit discontinuities) and a partial sum of its Fourier Series

$$f(x) \sim \sum_{-n}^{n} C_k \exp(ikx) , \tag{43}$$

(which may exhibit Gibbs phenomenon).

One way to obtain a smoothed version of f would be to use an integral operator which would "average" the values of f near each point. Such an operator is given by the following:

$$f^*(x) = \frac{n}{2\pi} \int_{-\pi/n}^{\pi/n} f(x+t)\, dt , \tag{44}$$

where f* is a "smoothed" version of f. It is a fortunate circumstance, now, that the Fourier Series approximation to f* (partial sum) is simply related to the approximation to f given by (43). Specifically, we obtain

$$f^*(x) \sim \sum_{-n}^{n} C_k \, \sigma_k \exp(ikx) \, , \tag{45}$$

where $\{C_k\}$ are as in (43) and the factors σ_k are called Lanczos sigma factors and are given by the following:

$$\sigma_k = \frac{\sin k\pi/n}{k\pi/n} \, , \quad 1 \leq k \leq n \, . \tag{46}$$

We see that we may easily obtain the Fourier Series approximation to the smoothed version of a function (which should, therefore, be a "better" approximation in the sense of approximating the smoother function) from the Fourier Series approximation to the function itself.

The result extends nicely to any number of dimensions so that in two dimensions, we have the following sequence of equations.

Let $f(x,y)$ be given and

$$f(x,y) \sim \sum_{-n}^{n} \sum_{-m}^{m} C_{kj} \exp(ikx) \cdot \exp(ijy) \, , \tag{47}$$

then

$$f^*(x,y) = \frac{nm}{4\pi^2} \int_{-\pi/n}^{\pi/n} \int_{-\pi/m}^{\pi/m} f(x+t, \, y+s) \, dtds \, , \tag{48}$$

has a Fourier Series approximation given by

$$f^*(x,y) \sim \sum_{-n}^{n} \sum_{-m}^{m} C_{kj} \, \sigma_k^{(n)} \, \sigma_j^{(m)} \exp(ikx) \exp(ijy), \tag{49}$$

with the sigma factors defined as appropriate by (46).

It is very important to note that any degree of smoothing is obtainable by this technique by simply reapplying the technique to the smoothed versions of the function. In this way, one obtains σ^2, σ^3, etc., factors. The details of this technique and a comment on weighted smoothing will appear elsewhere.

For our purposes, now, we should begin our filter design process with some degree of smoothing applied to the desired filter amplitude spectrum. This should be done carefully, weighing the cost of loss of resolution against expected improvement in the filters' characteristics and the gain in economy of coefficients.

The technique may be used, in addition, to perform the smoothing necessary in a discontinuous amplitude response specification in order to avoid ringing.

Figures 6 through 14 illustrate the use of the Lanczos technique in one and two dimensions to smooth Fourier Series approximations of various orders for functions involving discontinuities.

C. Zero Phase.

Our discussion of filter design principles has so far ignored the question of phase, our attention being directed only to the amplitude spectrum. This is not to say that phase is not important, for indeed it is. Huang [28] has pointed out the great importance of phase considerations in image processing. In geophysics, however, phase lags introduced by a typical two-dimensional digital filter on seismic data would have the effect of shifting the events which we are seeking to new positions in time or space. This kind of distortion may not only change the apparent depth of an event in the earth, but perhaps more important, it may shift it laterally from its correct position with respect to the surface. This kind of distortion could cause us to miss a potential hydrocarbon trap when the well is drilled, even though the trap may well exist in depth at some lateral distance from the well sight.

As a result, we generally seek filters which will operate on the amplitude spectrum of our data in a fixed way but which will leave the phase spectrum unchanged. Such filters are called zero phase filters. Our filter design considerations so far have made no mention of phase.

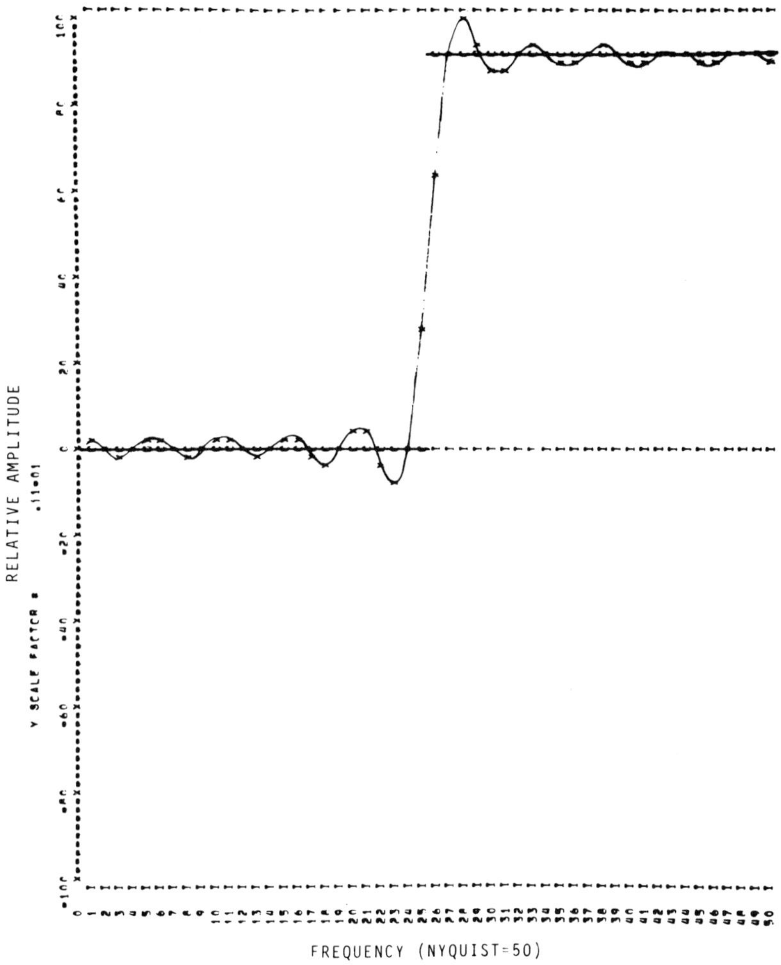

Fig. 6. Fourier Series with 20 coefficients illustrating Gibbs phenomenon.

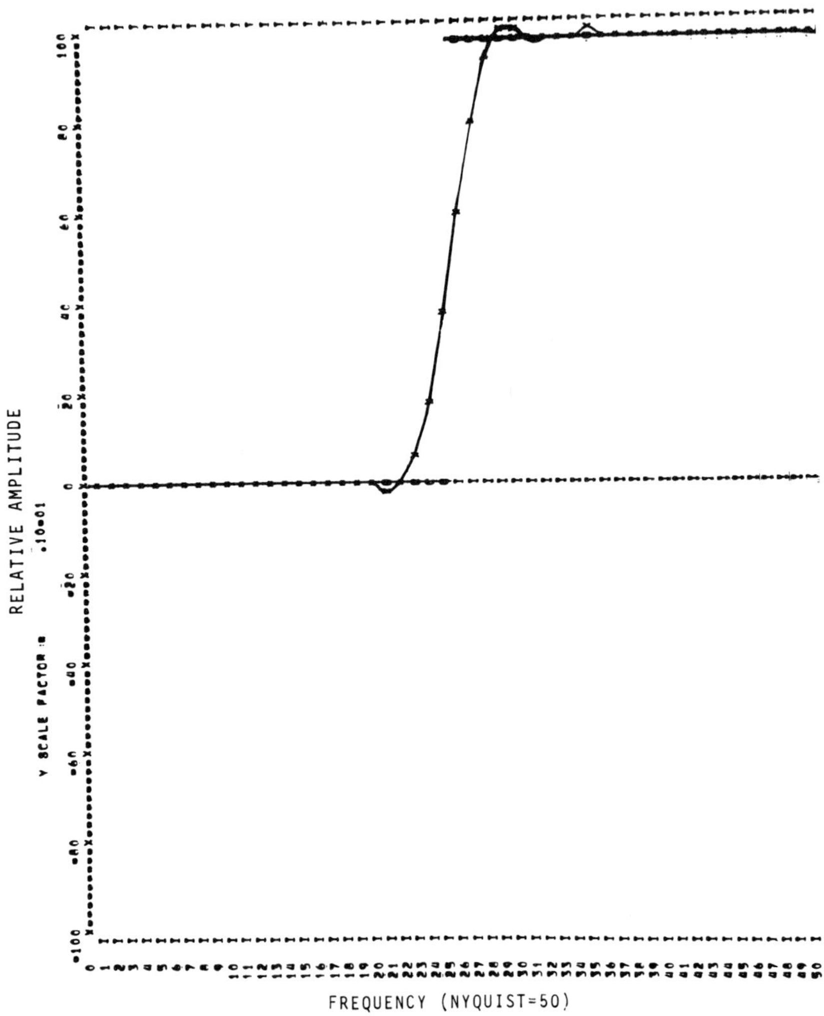

Fig. 7. Same as Fig. 6 smoothed with Lanczos σ factors.

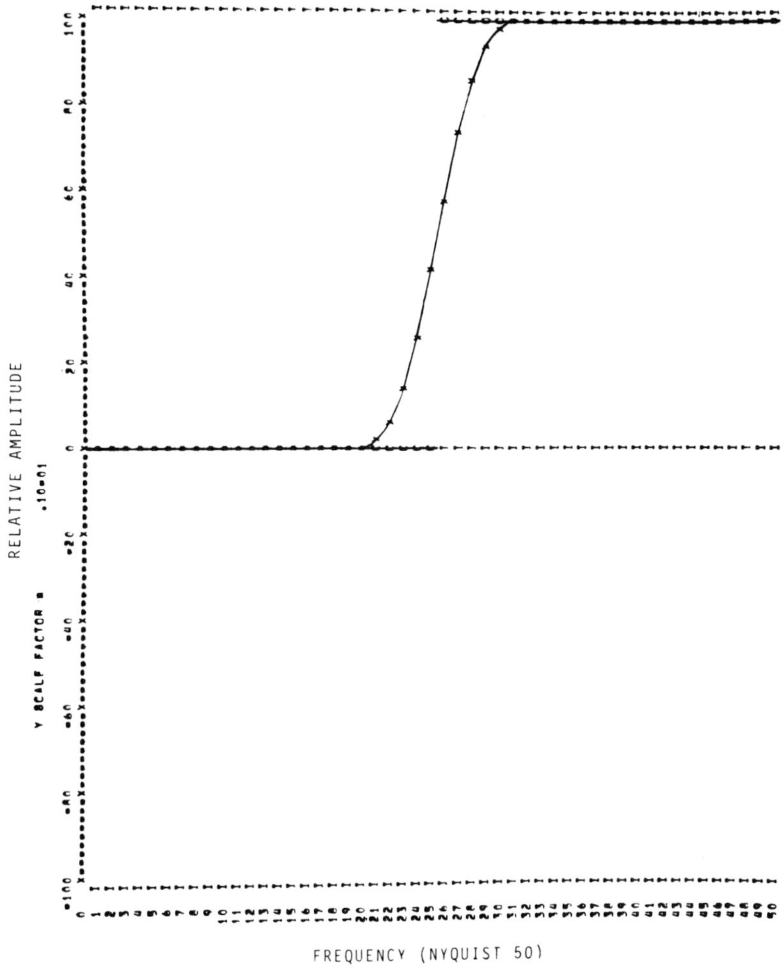

Fig. 8. Same as Fig. 6 smoothed with Lanczos σ^2 factors.

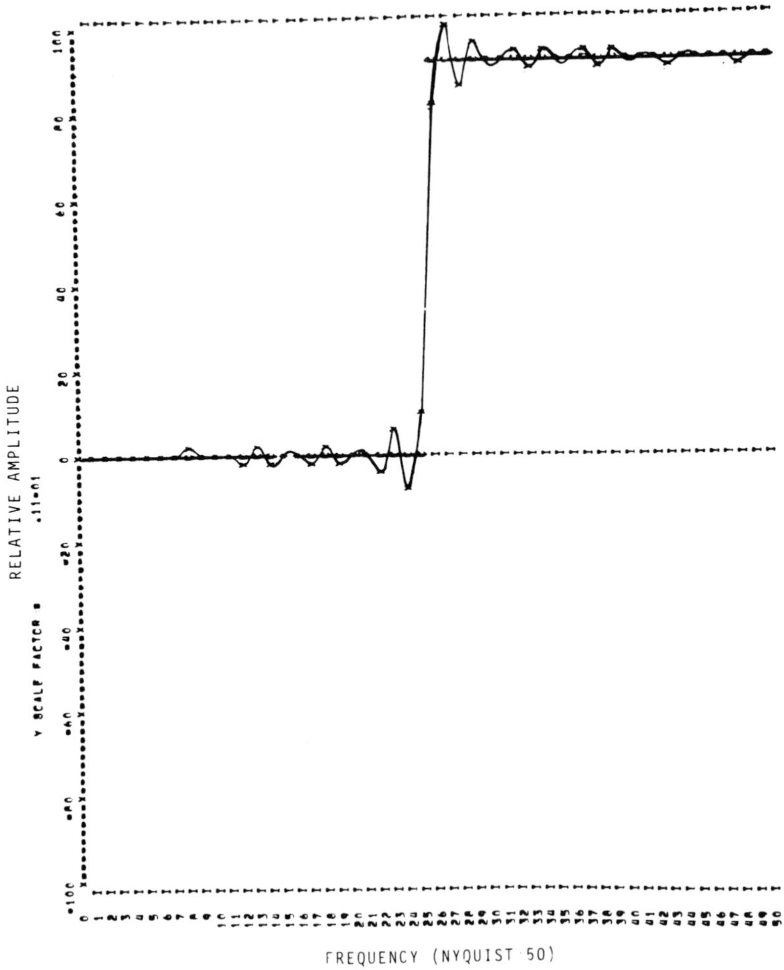

Fig. 9. Fourier Series with 40 coefficients illustrating Gibbs phenomenon.

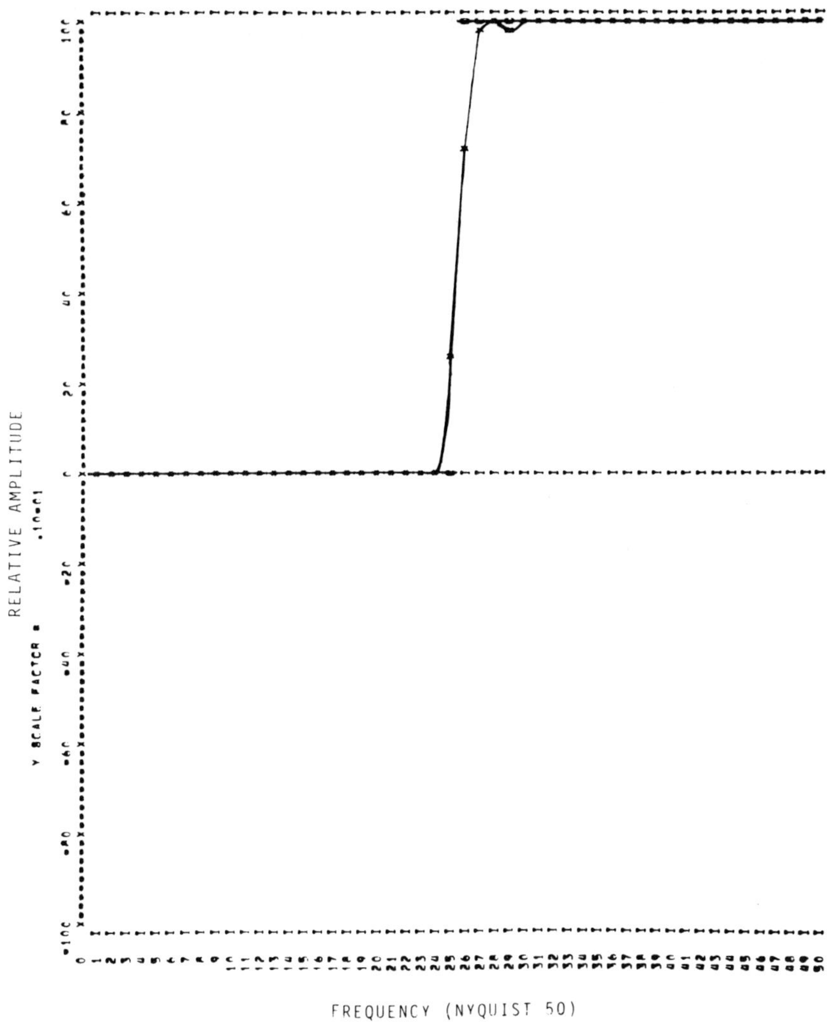

Fig. 10. Same as Fig. 9 smoothed with Lanczos σ factors.

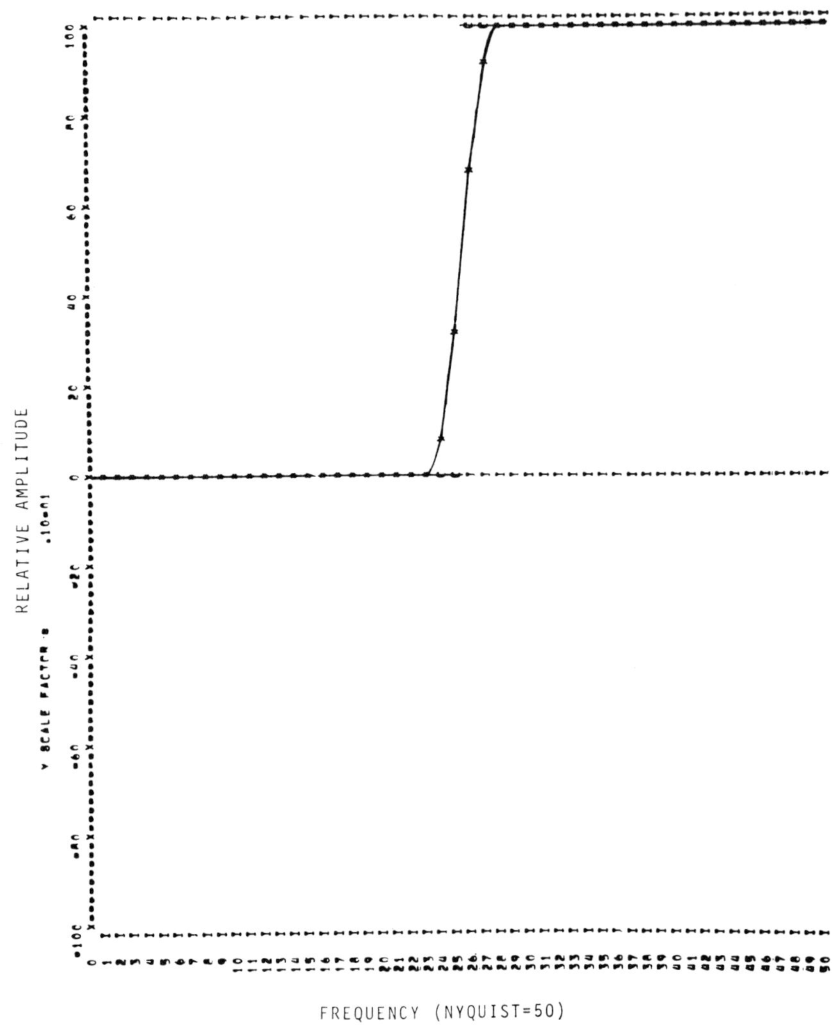

Fig. 11. Same as Fig. 9 smoothed with Lanczos σ^2 factors.

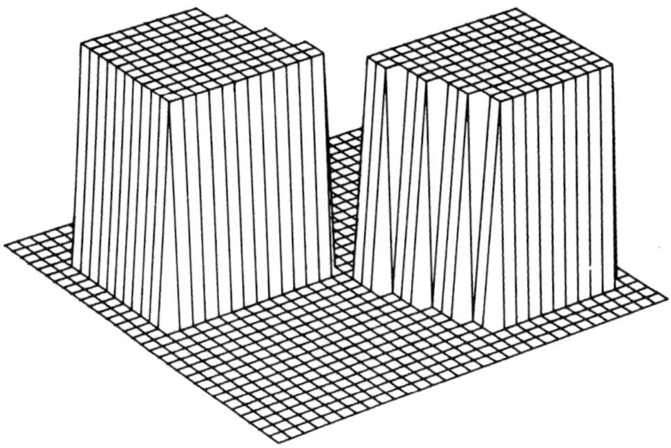

Fig. 12. Surface to be fitted with Fourier Series.

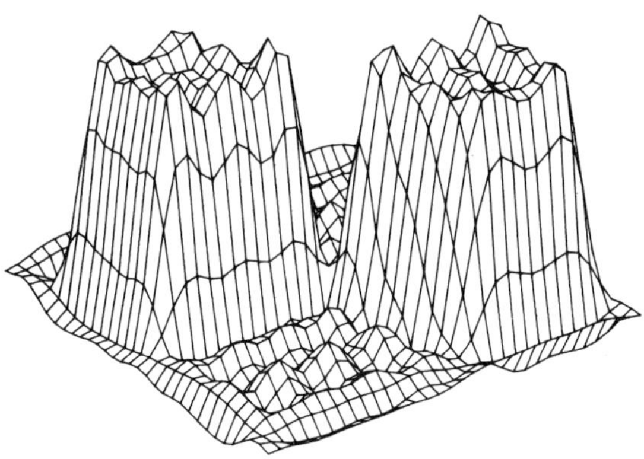

Fig. 13. 8 × 8 partial sum of Fourier Series fit to Fig. 12.

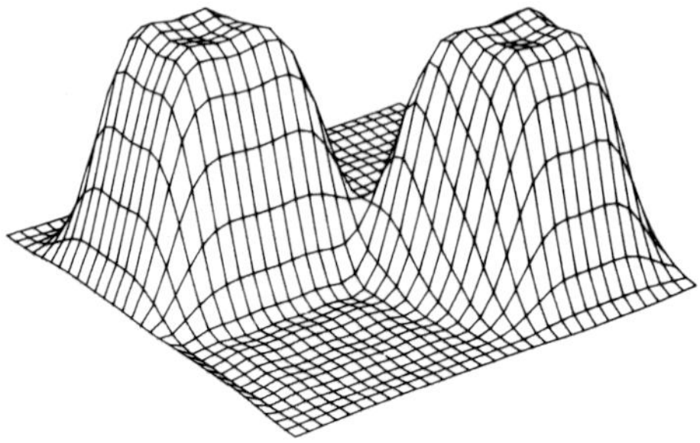

Fig. 14. Fig. 13 smoothed with Lanczos σ factors.

The reason is that the kind of phase control which we seek, i.e., zero phase distortion, is easily obtained when the filter is applied to the data.

The technique commonly used utilizes the concept of the involution of a function. If a function f is given, we define its involution $\tilde{f}$ by the equation

$$\tilde{f}(x) = \bar{f}(-x) , \tag{50}$$

(bar denotes complex conjugation if appropriate).

The technique of obtaining zero-phase filtering relies on the easily checked fact that the Fourier transform of $\tilde{f}$ is the complex conjugate of the Fourier transform of f. As a result, if we filter by f and then by $\tilde{f}$, we will have obtained the result of filtering with a filter whose transform is real, and which, therefore, introduces no translations into the data.

In practice, this procedure is sometimes carried out as follows. Let the input sequence be denoted by a and the filter sequence by f.

First, filter a using f:

$$a \to a*f .$$

Take the involution of a*f and pass the result through the filter f (this is done by filtering a*f backwards),

$$(f*a)^{\sim} \rightarrow f*\,(f*a)^{\sim} \quad .$$

The involution of the result gives the required filter sequence.

$$f*\,(f*a)^{\sim} \rightarrow (f*\,(f*a)^{\sim})^{\sim} = (\tilde{f}*f)*a \ ,$$

where

$$\mathcal{F}(\tilde{f}*f*a) = |\mathcal{F}(f)|^{2} \cdot \mathcal{F}(a) \quad .$$

($\mathcal{F}$ denotes Fourier transform.)

If carried out properly, the final result will be in place at the end of the second pass in which we filter f*a backward through the filter f.

It is most important, now, to notice that the zero-phase distortion filtered output is the result of applying the square of the amplitude spectrum of f to the input rather than the amplitude spectrum of f as desired.

As a result, if we know that we are going to use the zero-phase filtering technique outlined here (at an approximate doubling of our filter time and cost) then the filter design should be based on the square root of the desired filter amplitude response rather than on the desired response itself.

D. Examples.

We have finished our survey of techniques which have been found to be useful in two-dimensional filter design. This would seem to be an appropriate time to illustrate the kind of results obtainable using these procedures.

Figure 4 shows the region of the complex z-w plane used to specify the amplitude spectrum of two-dimensional filters with real coefficients. The missing two quadrants contain reflections through the origin of the spectrum given on the two quadrants shown. If, in addition, a two-dimensional filter exhibits symmetry through the axis separating the quadrants, it is called a symmetric filter. We shall consider some symmetric designs first. Since we really only need one quadrant to indicate a symmetric design, we shall exhibit only one quadrant in these examples. The quadrant to be used is the one shown in Figure 15, viz., the fourth quadrant. The other quadrants are obtained by reflection through the bounding axes and the origin. The origin in these plots is in the upper left-hand corner.

The designs in the examples were not selected for any specific purpose, they are just examples. Figures 20 and 21 are to be compared. One simply has a higher degree of smoothing than the other as may be detected by examining the "pinched" area in the hourglass shape.

Figure 22 shows all four quadrants of a symmetric filter design.

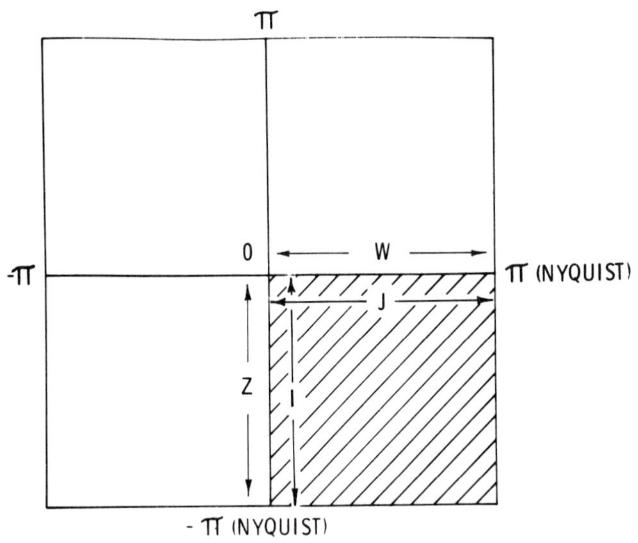

Fig. 15. Frequency domain showing quadrant illustrated in Examples 16 through 21.

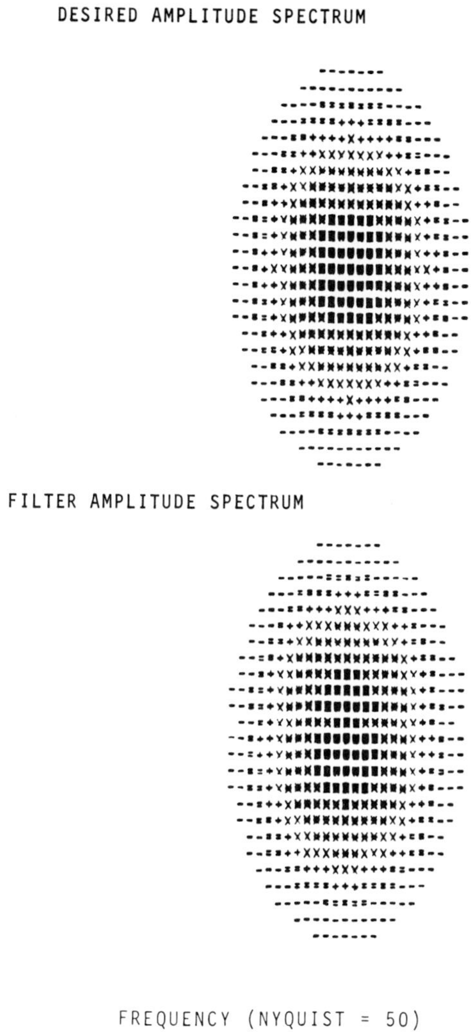

Fig. 16. Desired and actual two-dimensional filter amplitude spectra (fourth quadrant).

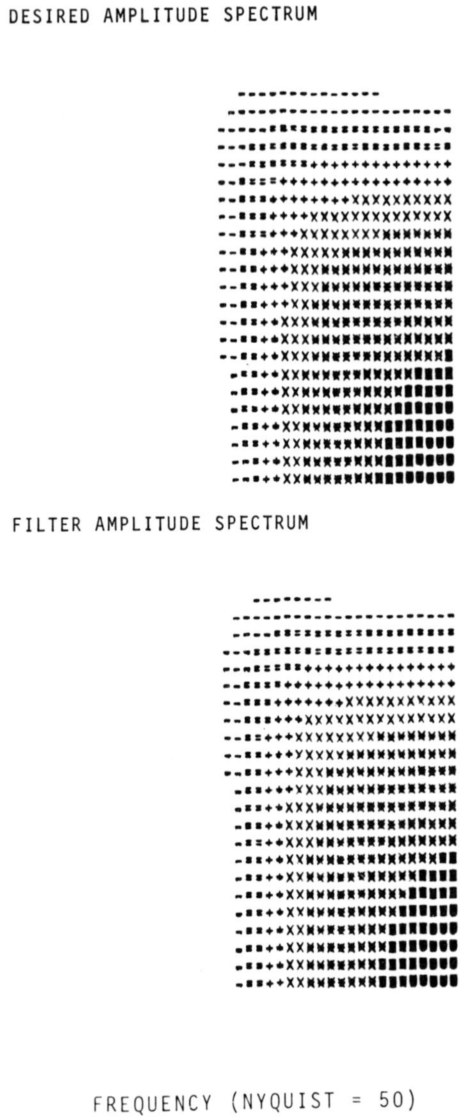

Fig. 17. Desired and actual two-dimensional filter amplitude spectra (fourth quadrant).

DESIRED AMPLITUDE SPECTRUM

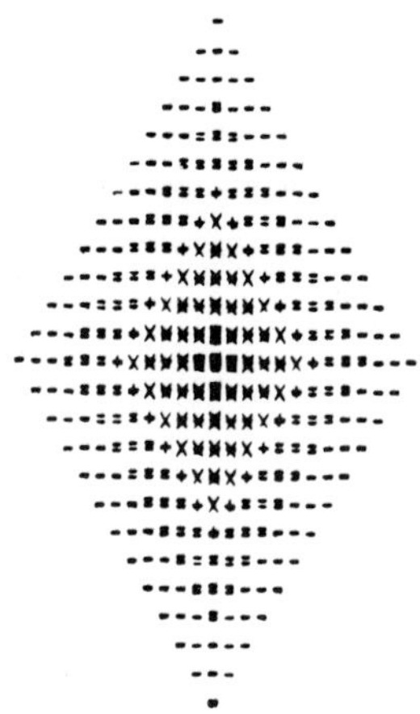

FILTER AMPLITUDE SPECTRUM

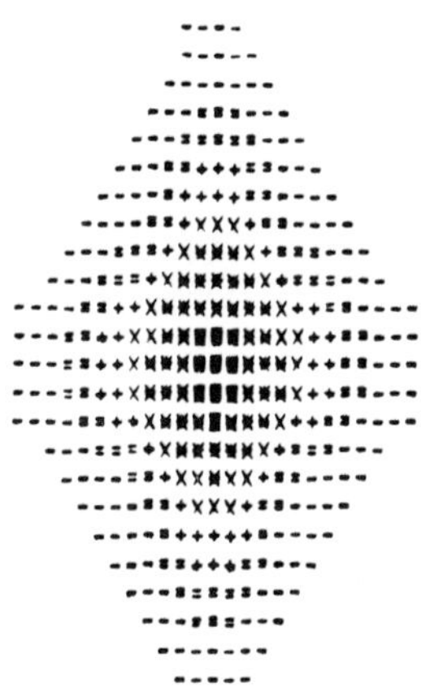

FREQUENCY (NYQUIST = 50)

Fig. 18. Desired and actual two-dimensional filter amplitude spectra (fourth quadrant).

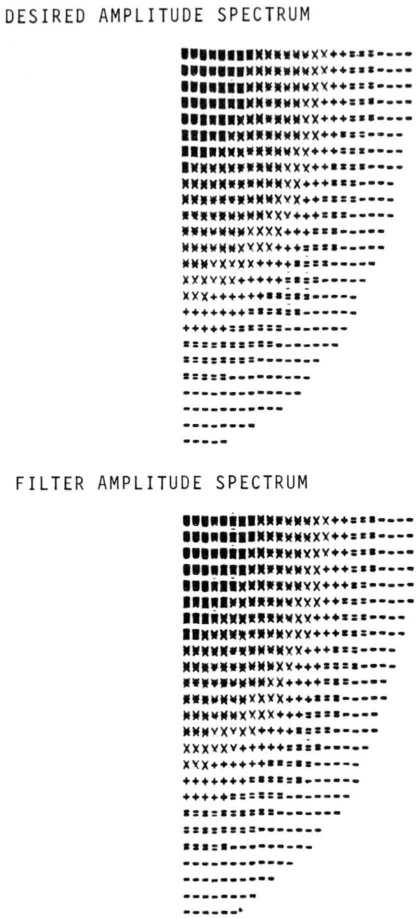

Fig. 19. Desired and actual two-dimensional filter amplitude spectra (fourth quadrant).

DESIRED AMPLITUDE SPECTRUM WITH SMOOTHING

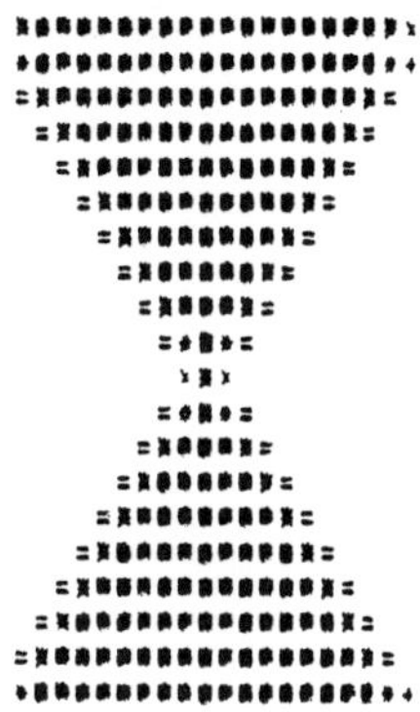

FILTER AMPLITUDE SPECTRUM

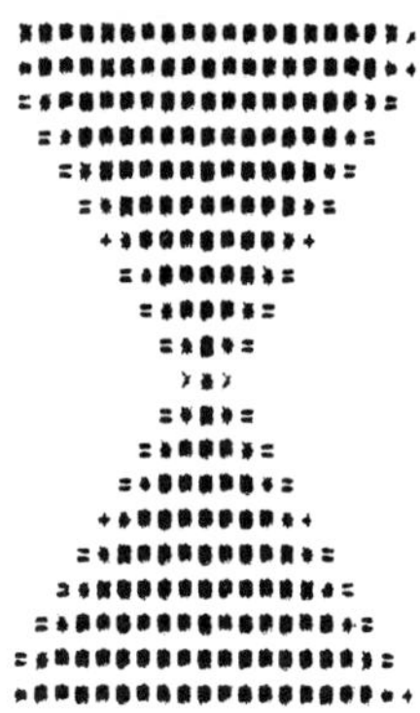

Fig. 20. Desired and actual two-dimensional filter amplitude spectra (fourth quadrant).

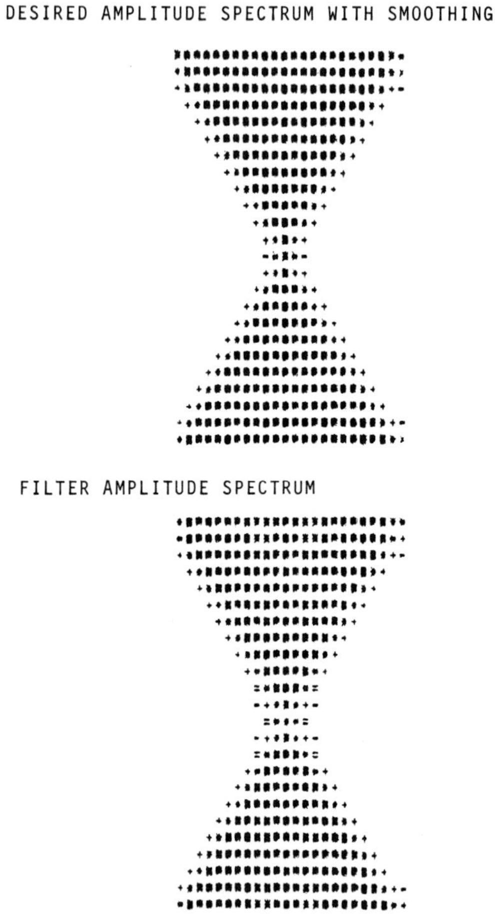

Fig. 21. Desired and actual two-dimensional filter amplitude spectra (fourth quadrant).

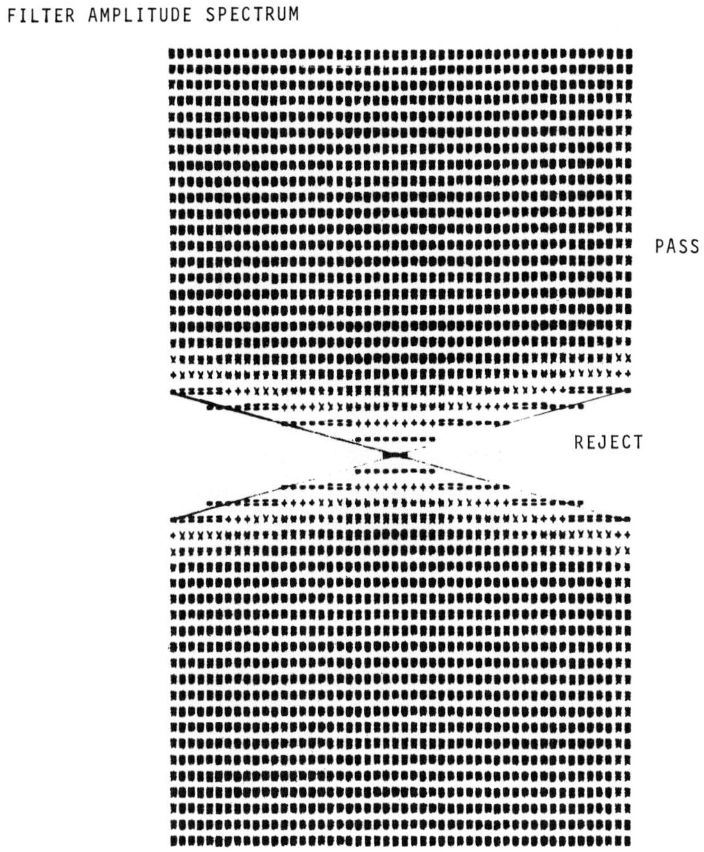

Fig. 22. Symmetric fan reject filter (four quadrants shown).

E. Spatial Aliasing and Nonsymmetric Filters.

At this point, one may well wonder what Figure 23 has to do with digital filtering. Surely, we have all had, at one time or another in our lives, the experience of attending a Western movie. As the intensity of the plot heightens our fears for the well being of the stagecoach and its contents, the inevitable occurs and Indians sweep down upon it from the rear and flanks. The driver whips the team into action and just as escape seems assured, the wheels on the coach appear to reverse and rotate in the wrong direction! Somehow, the

Fig. 23. Example of spatial aliasing in the movies. Photo courtesy of Wells Fargo & Company.

coach always seemed to move ahead anyway, but perhaps we simply conditioned ourselves not to notice what those wheels appeared to be doing. However, the experience is a good example of what we call aliasing, and is the result of a sampling process. The motion picture only samples the scene being portrayed at a rate of perhaps 24 frames per second. If the speed of the wheel is such that it almost, but not quite completes a revolution in one frame, we see what appears to be the spokes on the wheel progressing backward in time, and thus our illusion is created. In fact, there are several speeds at which the new position of the spokes in a frame will be exactly superimposed on the position of the spokes in the preceding frame and the wheel will appear not to rotate at all! At any rate, all of this is an illustration of what is called aliasing and the physical manifestation of it in our example is that our moving object appeared to go the wrong way.

In geophysics, a similar phenomenon may occur if our geophones are placed too far apart to capture correctly the spatial frequencies of our signal. When this happens, our signal may appear indeed to go the wrong way. That is, a wave traveling from the shot across the geophone spread may actually appear to travel across the spread toward the shot! Such information is fallacious and in most cases useless and interferes with the good part of the signal. A signal which contains a broad band of spatial frequencies and which is aliased and moving with some fixed apparent velocity will have a transform with a strong contribution along the dotted line shown at the top of Figure 24. The seismic record will contain contributions from the two points indicated as shown in the lower illustration in Figure 24. The events sloping to the lower right correspond to the point in the right-hand quadrant, and the events sloping the "wrong way" toward the upper right correspond to the aliased part of the signal in the left-hand quadrant in the top illustration.

Even though our signal contains false components introduced by inadequate sampling, we are fortunate in that the transform of the aliased part is isolated from the transform of the unaliased part. As a result, that part of the signal whose transform lies in the left-hand quadrant could be removed by suppressing this part with a two-dimensional filter. Such a (nonsymmetric) filter is illustrated (all four quadrants are shown) in Figure 25.

My point in presenting this discussion is that most papers on two-dimensional filtering seem to be concerned with symmetric filters; whereas, experience has shown that in many cases, nonsymmetric designs are the only good choices.

An example of an optical transform of aliased two-dimensional data is given in Figure 26.

SPATIAL
ALIASING

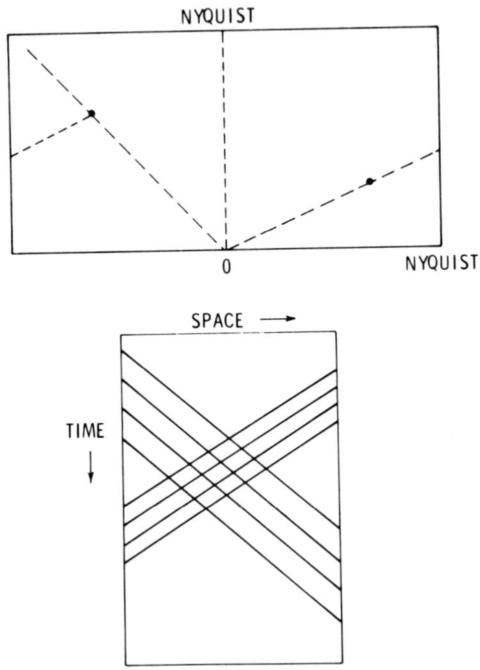

FIGURE 24

Fig. 24. Appearance of aliased and non-aliased events in seismic recording.

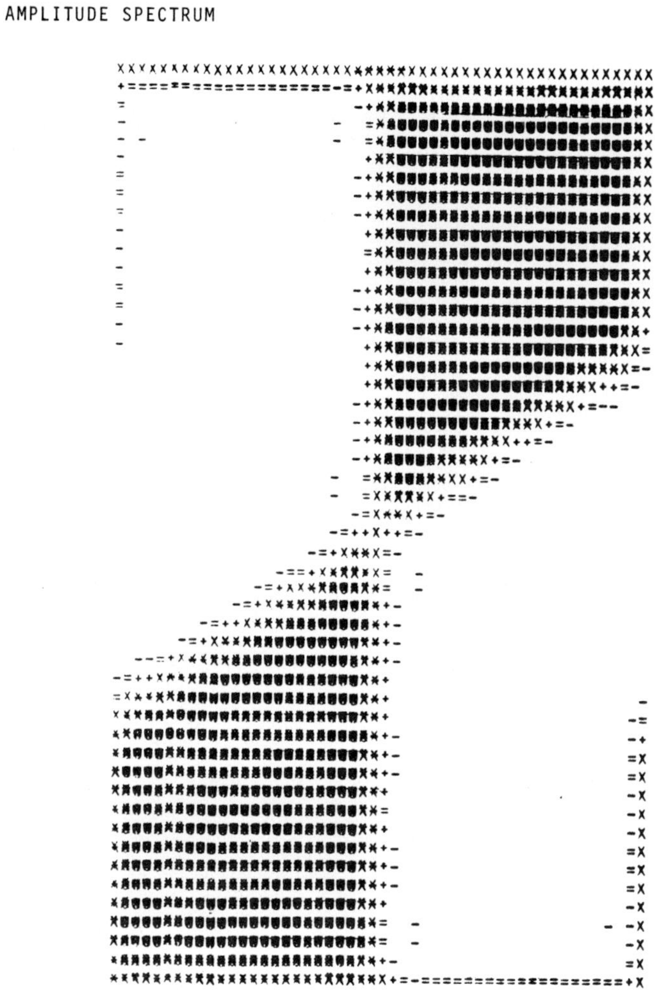

Fig. 25. An example of a nonsymmetric filter (four quadrants shown).

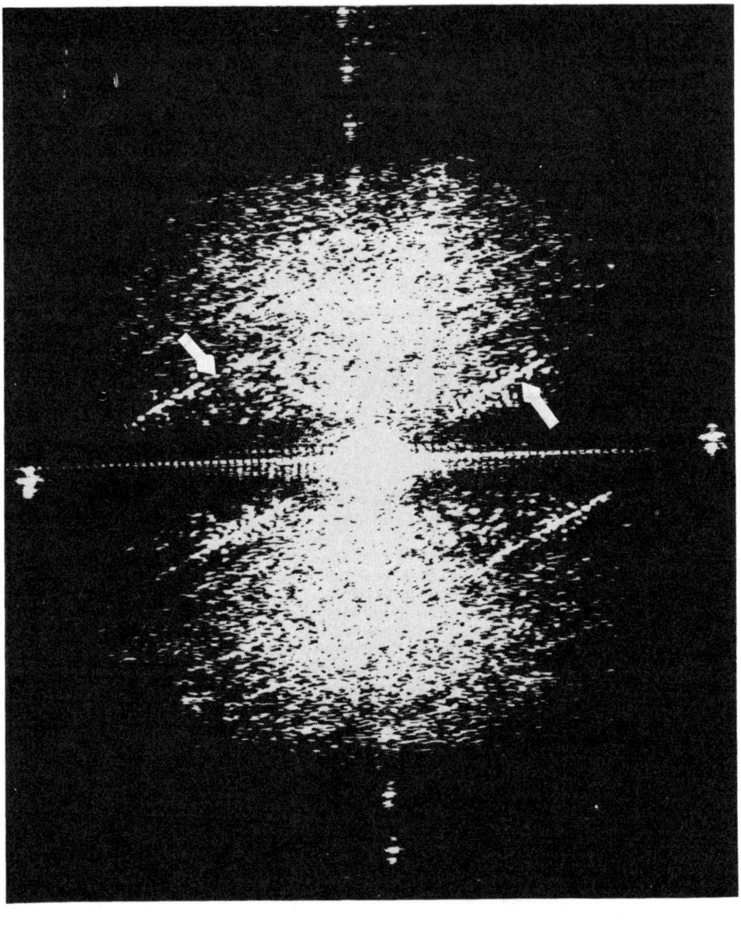

Fig. 26. An optical transform of seismic data exhibiting spatial aliasing (four quadrants shown).

III. Time-Variant Recursive Filters from Two-Dimensional Filter Design.

We come now to an interesting and potentially very useful application of two-dimensional filter design. The idea relates to the fact that we are under no obligation to use two-dimensional filters for two-dimensional filtering. Indeed, if we impose suitable symmetry conditions and proceed in the proper way, it is possible to convert two-dimensional symmetric filters into one-dimensional filters with a parameter.

The parameter may then be used to represent time (or anything else that we might wish to use it for). Figure 27 illustrates the change appropriate to viewing the standard two-dimensional filter transform domain as the transform domain corresponding to a time-variant filter. To interpret a spectrum correctly, we take cross sections through it for fixed values of time. The resulting one-dimensional spectrum is the spectrum of the corresponding one-dimensional filter at that point in time.

To illustrate the use of this technique, a filter was designed which had a very narrow peaked pass band lying along diagonal lines through the origin of the transform domain (Figure 29).

As we apply the filter, progressing in time from zero to some predetermined value, T (20 seconds for the actual example presented at the Symposium), the center frequency of the pass-band slides from 0 Hz to the Nyquist frequency (determined by the sampling rate). The filter was first demonstrated using a noise input which results in a tone which glissandos from 0 Hz to Nyquist in the 20-second interval.

As an interesting application of this "harmonic analyzing filter", a tone with rather discrete harmonics was filtered as indicated in Figure 28. We may imagine that the spectrum of the tone translates upward across the filter spectrum in time (or vice-versa) and the resulting tone is constructed from the intersection of the original tones' spectrum and the filters' pass band. The result, as was demonstrated, is that the individual harmonics comprising the tone are heard

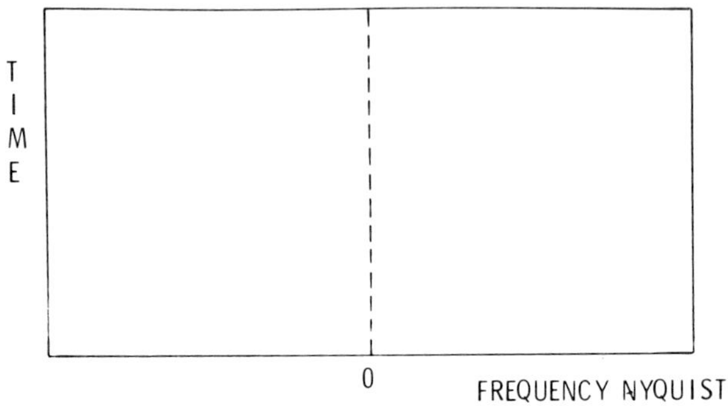

Fig. 27. Domain for designing time-variant filters.

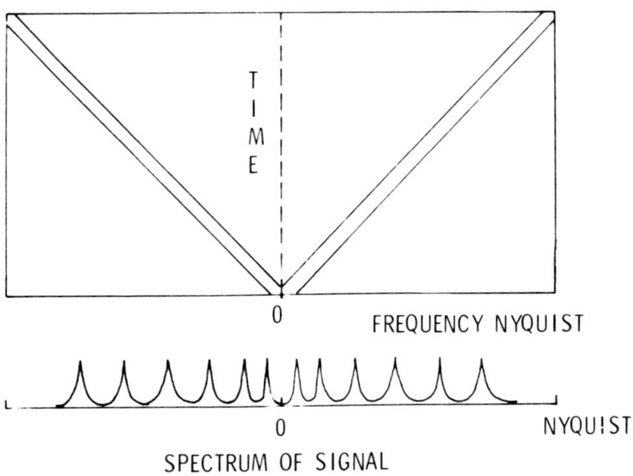

Fig. 28. An example of a time-variant filter design and the spectrum of a signal to be filtered.

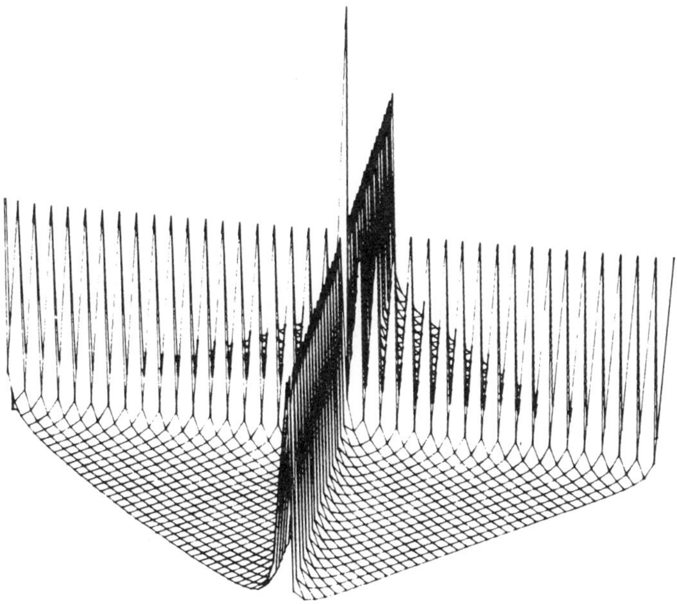

FOUR QUADRANTS

Fig. 29. Three-dimensional plot of a time-variant filter response (axes are time and frequency).

successively in increasing order over the 20-second interval.

An interesting thought now presents itself. We may control the "time" parameter in any way we would like. As a result, a myriad of time-variant filters is obtainable from each design.

As one further illustration, another (single quadrant) time-variant filter spectrum is shown in Figure 30. The pass band is the hatchured region. When used with noise input, the filter produces a sound like surf. To exhibit control of the time function, we simply cycle through the time axis several times as time progresses, producing the sound of successive breakers crashing into the shore.

It is my hope that some use will be made of this technique in seismic work, as I believe it is fraught with possibilities.

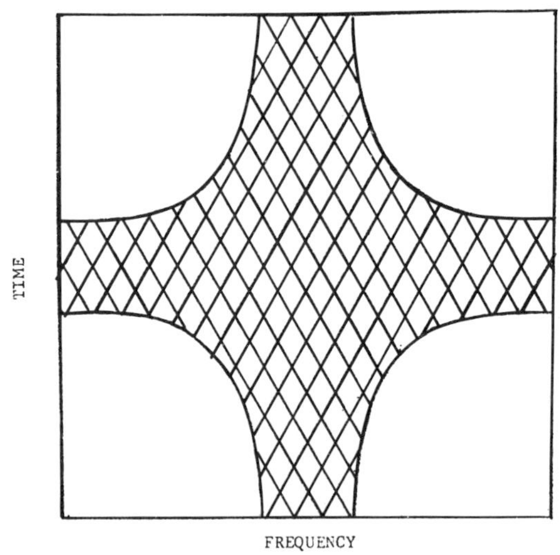

Fig. 30. A time-variant filter example (hatchured region is pass band).

IV. Audio Seismogram.

I have worked for several years in the relatively new field of computer music with my friend and associate, Bing Vassaur. Several years ago, Bing asked why we do not listen to seismograms since the ear can be quite a sensitive detecting device and spans many more octaves and a greater dynamic range than the eye. At the time, I gave many good reasons for not carrying the idea further. However, what better chance could we find to illustrate the idea than a geophysics-related symposium.

Since we are wandering in relatively virgin territory, it seemed prudent to stick with an example whose interpretation would be known.

As a result, we generated a synthetic seismogram from the model shown in Figure 31. We then stretched the time interval and produced an audio seismogram on which we could hear successive arrivals.

The pattern of arrivals is shown below the seismogram. This is the "earth filter" referred to earlier. The numbers refer to the reflecting horizon; the amplitude indicates the number of the multiple from that horizon. The horizontal scale is time. The seismogram may be imagined as the convolution of an audible signal with this "earth filter".

The initial example was generated with a slowly decaying audible signal illustrating the lack of clarity. However, even with a compressed pulse, it was difficult to tell what was going on. However, we observed that we would never really interpret a single trace anyway, and so we selected 12 traces at increasing distance from the shot point and gave each a different pitch in order to distinguish them.

The experiment was then repeated. With a slowly decaying audio wavelet as input, the result was harplike and not very clear. Pulse compression (an example of Wiener filtering) helped greatly in the perception of individual events. An arrival could be heard quite clearly sweeping down the line of geophones. In fact, with a little practice, it became quite easy to distinguish those events which were very "flat" from those which were more curved as illustrated in Figure 32.

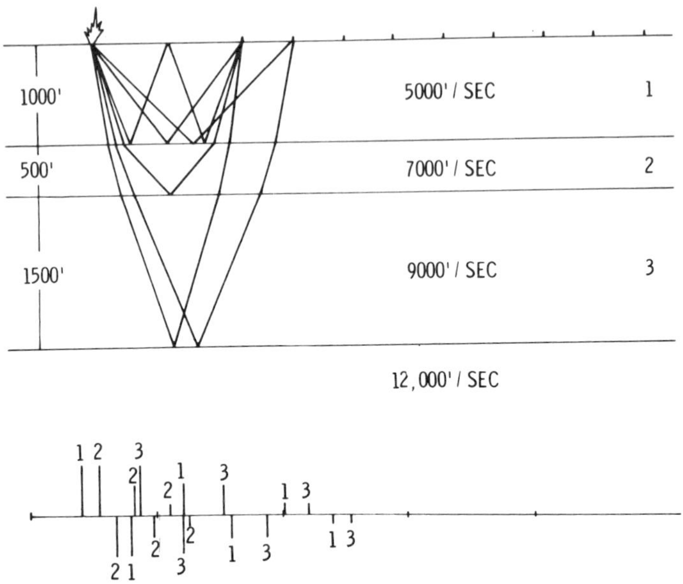

Fig. 31. Synthetic seismogram model.

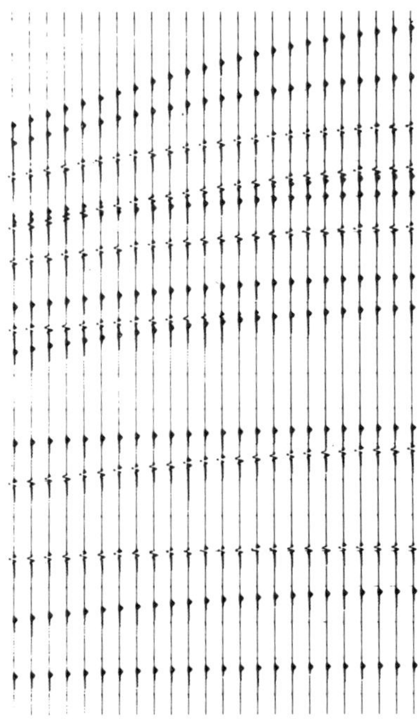

Fig. 32. Record produced by the synthetic seismogram (Fig. 31).

As a final touch in illustrating what filtering is all about, the diagonal band pass filter shown in Figure 29 was preset with the parameter set equal to a constant, selected so that the center frequency of the pass band would coincide exactly with the pitch which had been assigned to one of the geophones. The audio seismogram was filtered and the result was that the selected geophone could be heard clearly above the remaining geophones.

Will the day ever come when we may extract useful information by listening to seismic data? I am not sure, but I am not going to list, either, all those reasons I once gave for not doing it.

Acknowledgment.

The author gratefully acknowledges the support of Amoco Production Company for this work.

Bibliography.

1. Aguilera, S., Debremaecker, J. C., and Hernandez, S., "Design of Recursive Filters," *Geophysics 35*, no. 2, 247-253.
2. Anderson, B. D. O., and Jury, E. I., "Proof of a Special Case of Shanks' Conjecture," *IEEE Trans. Acoust. Speech & Signal Process.*, December, 1976, 574-575.
3. Anderson, B. D. O., Bose, N. K., and Jury, E. I., "A Simple Test for Zeros of a Complex Polynomial in a Sector," *IEEE Trans. Auto. Control AC-19*, 1974, 437-438.
4. Anderson, B. D. O., and Jury, E. I., "Stability Test for Two-Dimensional Filters," *IEEE Trans. Audio & Electroacoustics AU-21*, no. 4, 1973, 366-372.
5. Anderson, B. D. O., and Jury, I. E., "Stability of Multidimensional Filters," *IEEE Trans. Circuits & Systems CAS-21*, 1974, 300-304.
6. Bhattacharyya, B. K., "Recursion Filters for Digital Processing of Potential Field Data," *Geophysics 41*, no. 4, 1976, 712-726.

7. Black, D. I., and Scollar, I., "Spatial Filtering in the Wave-Vector Domain," *Geophysics 34*, no. 6, 1969, 916-923.

8. Bose, N. K., "Implementation of a New Stability Test for Two-Dimensional Filters," *IEEE Trans. Acoust. Speech & Signal Process. ASSP-25*, no. 2, 1977, 117-120.

9. Bose, N. K., "A Criterion to Determine if Two Multivariable Polynomials are Relatively Prime," *Proc. IEEE 60*, no. 1, 1972, 134-135.

10. Bose, N. K., and Newcomb, R. W., "Tellegen's Theorem and Multivariable Realizibility Theory," *Int. J. Electronics 36*, 1974, 417-425.

11. Bose, N. K., and Kamat, P. S., "Algorithm for Stability Test of Multidimensional Filters," *IEEE Trans. Acoust. Speech & Signal Process. ASSP-22*, 1974, 307-314.

12. Bose, N. K., and Jury, E. I., "Positivity and Stability for Multidimensional Filters (Discrete-Continuous)," *IEEE Trans. Acoust. Speech & Signal Process. ASSP-22*, 1974, 174-180.

13. Brophy, F., and Salazar, A. C., "Considerations of the Padé Approximation Technique in the Synthesis of Recursive Digital Filters," *IEEE Trans. Audio & Electroacoustics AU-21*, no. 6, 1973, 500-505.

14. Cheney, E. W., *Introduction to Approximation Theory*, McGraw-Hill Book Company, New York, 1966.

15. Claerbout, J. F., *Fundamentals of Geophysical Data Processing with Applications to Petroleum Prospecting*, McGraw-Hill Book Company, New York, 1976.

16. Clement, W. G., "Basic Principles of Two-Dimensional Digital Filtering," *Geophys. Prospecting 21*, 125-145.

17. Costa, J. M., and Venetsanopoulos, A. N., "A Group of Linear Spectral Transformations for Two-Dimensional Digital Filters," *IEEE Trans. Acoust. Speech & Signal Process. ASSP-24*, 1976, 424-425.

18. Darby, E. K., and Davies, E. B., "The Analysis and Design of Two-Dimensional Filters for Two-Dimensional Data," *Geophys. Prospecting 15*, 383-406.

19. Dudgeon, D. E., "Two-Dimensional Recursive Filtering," Ph.D. Diss., Dept. of Elec. Eng., MIT, Cambridge, Mass., 1974.

20. Ekstrom, M. P., and Woods, J. M., "Two-Dimensional Spectral Factorization with Applications in Recursive Digital Filtering," *IEEE Trans. Acoust. Speech & Signal Process. ASSP-24*, no. 2, 1976, 115-128.

21. Ekstrom, M. P., and Woods, J. M., "Correction to 'Two-Dimensional Spectral Factorization with Applications to Recursive Digital Filtering,'" *IEEE Trans. Acoust. Speech & Signal Process. ASSP-24*, no. 6, 1976, 585.

22. Embree, P., Burg, J. P., and Backus, M. M., "Wide Band Velocity Filtering - The Pie-Slice Process," *Geophysics 28*, no. 6, 1963, 948-974.

23. Fail, J. P., and Grau, G., "Fan Filters," *Geophys. Prospecting 11*, 1963, 131-163.

24. Fail, J. P., Grau, G., and Layotte, P. C., "Improvement of the Signal/Noise Ratio by Means of Fan Filtering," *Geophys. Prospecting 12*, 1964, 258-282.

25. Genin, Y., and Kamp, Y., "Counterexample in the Least-Square Inverse Stabilization of 2D-Recursive Filters," *Elec. Lett. 11*, 1975, 330-331.

26. Genin, Y., and Kamp, Y., "Two-Dimensional Stability and Orthogonal Polynomials on the Hypercircle," Special Issue of *IEEE Proc. Multidimensional Syst.*, June, 1977.

27. Huang, T. S., "Stability of Two-Dimensional Recursive Filters," *IEEE Trans. Audio & Electroacoustics AU-20*, no. 2, 1972, 158-163.

28. Huang, T. S., Burnett, J. W., and Deczky, A. G., "The Importance of Phase in Image Processing Filters," *IEEE Trans. Acoust. Speech & Signal Process. ASSP-23*, no. 6, 1975, 529-542.

29. Huang, T. S., "Two-Dimensional Windows," *IEEE Trans. Audio & Electroacoustics AU-20*, no. 1, 1972, 88-89.

30. Hu, J. V., and Rabiner, L. R., "Design Technique for Two-Dimensional Digital Filters," *IEEE Trans. Audio & Electroacoustics AU-20*, no. 4, 1972, 249-257.

31. Hubral, P., "Stacking Filters and Their Characterization in the (F-K) Domain," *Geophys. Prospecting 22*, 1974, 722-735.

32. Justice, J. H., "Frequency Domain Design of Recursive Filters using Padé Approximation Techniques," Amoco

Technical Report, 1976.

33. Justice, J. H., "A Levinson-Type Algorithm for Two-Dimensional Wiener Filtering," Special Issue of *IEEE Proc. Multidimensional Syst.*, June, 1977.

34. Justice, J. H., "Frequency Domain Design of Two-Dimensional Recursive Filters using Padé Approximation Techniques," Amoco Technical Report, 1976.

35. Justice, J. H., and Shanks, J. L., "Stability Criterion for N-Dimensional Digital Filters," *IEEE Trans. Auto. Control AC-18*, no. 3, 1973, 284-286.

36. Lanczos, C., *Discourse on Fourier Series*, Hafner Publishing Company, Princeton, New Jersey, 1961.

37. Lanczos, C., *Applied Analysis*, Prentice-Hall, Incorporated, Englewood Cliffs, New Jersey, 1956.

38. Lavin, P. M., and Devane, J. F., "Direct Design of Two-Dimensional Digital Wavenumber Filters," *Geophysics 35*, no. 6, 1970, 1073-1078.

39. Maria, G. A., and Fahmy, M. A., "On the Stability of Two-Dimensional Digital Filters," *IEEE Trans. Audio & Electro-acoustics AU-21*, 1973, 470-472.

40. Maria, G. A., and Fahmy, M. A., "An ℓ_p-Design Technique for Two-Dimensional Digital Recursive Filters," *IEEE Trans. Acoust. Speech & Signal Process. ASSP-22*, 1974, 15-21.

41. Mersereau, R. M., Mecklenbrauker, W. F. G., and Quatjeri, T. F., Jr., "McClellan Transformations for Two-Dimensional Digital Filtering - Part I: Design," *IEEE Trans. Circuits & Syst. CAS-23*, 1976, 405-414.

42. Mitra, S. K., Sagar, A. D., and Pendergrass, N. A., "Realizations of Two-Dimensional Recursive Digital Filters."

43. Mersereau, R. M., and Dudgeon, D. E., "Two-Dimensional Digital Filtering," *Proc. IEEE 63*, no. 4, 1975, 610-623.

44. Nikolic, Z. J., "A Recursive Time-Varying Band Pass Filter," *Geophysics 40*, no. 3, 1975, 520-526.

45. Oppenheim, A. V., and Schafer, R. W., *Digital Signal Processing*, Prentice-Hall, Incorporated, Englewood Cliffs, New Jersey, 1975.

46. Peacock, K. L., and Treitel, S., "Predictive Deconvolution - Theory and Practice," *Geophysics 34*, 155-169.
47. Pendergrass, N. A., Mitra, S. K., and Jury, E. I., "Spectral Transformations for Two-Dimensional Digital Filters," *IEEE Trans. Circuits & Syst. CA-23*, 1976, 26-35.
48. Pistor, P., "Stability Criterion for Recursive Filters," *IBM J. Res. Develop.*, 1974, 59-71.
49. Rader, C. M., and Gold, B., "Digital Filter Design Techniques in the Frequency Domain," *Proc. IEEE 55*, 1967, 149-171.
50. Read, R. R., Shanks, J. L., and Treitel, S., "Two-Dimensional Recursive Filtering," *Topics in Applied Physics* Springer-Verlag, New York, v. 6, 131-176.
51. Read, R. R., and Treitel, S., "The Stabilization of Two-Dimensional Recursive Filters via the Discrete Hilbert Transform," *IEEE Trans. Geosci. Elec. GE-11*, 1973, 153-160, 205-207.
52. Robinson, E. A., "Predictive Decomposition of Time Series with Applications to Seismic Exploration," *Geophysics 42*, 1967, 418-484.
53. Robinson-Treitel Reader, Seismograph Services Corporation, Tulsa, Oklahoma, 1969.
54. Robinson, E. A., *Multichannel Time Series Analysis with Digital Computer Programs* Holden-Day, San Francisco, California, 1967.
55. Sengbush, R. L., and Foster, M. R., "Optimum Multichannel Velocity Filters," *Geophysics 33*, no. 1, 1968, 11-35.
56. Shanks, J. L., "Two-Dimensional Recursive Filters," *SWIEECO Record 21*, 1969, 19E1-19E8.
57. Shanks, J. L., "Recursion Filters for Digital Processing," *Geophysics 32*, no. 1, 1967, 33-51.
58. Shanks, J. L., Treitel, S., and Justice, J. H., "Stability and Synthesis of Two-Dimensional Recursive Filters," *IEEE Trans. Audio & Electroacoustics AU-20*, no. 2, 1972, 115-128.
59. Suresh, B. R., and Shenoi, B. A., "Exact Realization of 2-Dimensional Digital Filters by Separable Filters," *Elec. Lett. 12*, no. 10, 1976.

60. Treitel, S., and Robinson, E. A., "The Design of High-Resolution Digital Filters," *IEEE Trans. Geosci. Elec. GE-4*, no. 1, 1966, 25-38.

61. Treitel, S., Shanks, J. L., and Frasier, C. W., "Some Aspects of Fan Filtering," *Geophysics 32*, no. 5, 1967, 789-800.

62. Treitel, S., and Shanks, J. L., "The Design of Multistage Separable Planar Filters," *IEEE Trans. Geosci. Elec. GE-9*, no. 1, 1971, 10-27.

63. Twogood, R. E., and Mitra, S. K., "Computer-Aided Design of Separable Two-Dimensional Digital Filters," *IEEE Trans. Acoust. Speech & Signal Process. ASSP-25*, no. 2, 1977, 165-169.

64. Wiggins, R. A., "On Factoring the Correlations of Discrete Multi-Variable Stochastic Processes," Ph.D. Diss., Dept. of Geology and Geophysics, MIT, Cambridge, Mass., 1965.

65. Wiggins, R. A., and Robinson, E. A., "Recursive Solution to the Multichannel Filtering Problem," *J. Geophys. Res. 70*, no. 8, 1965, 1885-1891.

66. Wiggins, R. A., "ω-K Filter Design," *Geophys. Prospecting 14*, 427-440.

67. Wood, L., and Treitel, S., "Seismic Signal Processing," *Proc. IEEE 63*, no. 4, 1975, 649-661.

68. Zurflueh, E. G., "Applications of Two-Dimensional Linear Wavelength Filtering," *Geophysics 32*, 1967, 1015-1035.

ADAPTIVE PROCESSING OF SEISMIC DATA

Stanley J. Laster
The University of Tulsa

Introduction.

Most of us when in doubt tend to lick our finger and hold it up to sense the prevailing nature of the wind. Subsequent action usually depends on the result. In a nutshell that is adaptive processing. This paper is a somewhat tutorial review of what can be done and what has been done in adaptive processing of seismic data. If one has a long period of observation prior to making a decision, the results of any action taken should almost be optimum. Fixed filters designed using past observations of the data lead to Wiener filtering and Maximum Likelihood filtering. An enormous body of literature deals with these processes. We will ignore them here in favor of filter systems which change continually in time. These fall under the category of Stochastic Approximation (Albert and Gardner, 1967, and Widrow, 1966).

In dealing with seismic data, one attempts to determine the structure of the geological discontinuities in the earth by observing reflections of elastic waves. These waves travel downward to the discontinuity and back to the receivers in a direction almost perpendicular to the earth's surface. Unfortunately, the earth also supports many other types of natural vibration, most of which have their origin near the surface. We call these disturbances noise. Our attempt to probe the earth by means of seismic waves also generates undesirable vibrations analogous to multiple reflections in two parallel mirrors. We also describe these as noise, although they are source generated. The purpose of processing seismic data is to enhance the fundamental (primary) reflections from the subsurface, while attenuating the undesired

ISBN 0-12-257250-5

ambient and source-generated seismic noise. The processing algorithm requires a description of both the signal (desired output) and the noise. The signal and noise must differ in some respect if we are to succeed. We find this to be a convenient dividing point, depending on whether the signal and noise differ in their time attributes or their space attributes. When the signal and noise have different time characteristics, we can achieve our goal by operating on the output of a single receiver. We refer to these as single-channel processes, and they are the simplest to implement. Spatial differences require the output of many receivers, and are computationally more difficult. The discussion below will be organized along these lines.

Single-Channel Adaptive Processing.

Most single-channel processing of seismic data, with the exception of obvious things such as scaling and automatic gain control, deals with pulse shaping or "wavelet processing." In particular, the user wants to remove the effects of his recording system and imperfections in the earth, such as scattering and attenuation. The resultant process is usually referred to as deconvolution.

Deconvolution is a catch-all title that covers two processes. In the first place, we wish that our source and receiving system could generate and record all frequencies. A system that performs equally well at all frequencies is referred to as "white," borrowing a term from optics. Unfortunately, real systems are not white, failing most significantly at high frequencies. Pulses which should be sharp are spread out in time. To some extent this effect can be overcome by the deconvolution process termed "Whitening Deconvolution" or "Spiking Deconvolution." Historically, this was the first type of deconvolution offered and referred simply to sharpening of pulses. The success of this method depends on the validity of the random reflection model (Robinson and Silvia, 1978), in which primary seismic reflections are assumed to arrive at the receiver randomly in time.

On the other hand, even with very sharp source pulses, the data may be contaminated with undesired reflections (the source-generated noise mentioned above). The undesired elements are separated in time from the primary reflection, and recur periodically. This behavior is most striking in marine work, where the bulk of the energy is trapped in the water layer. This is shown schematically in Figure 1. It should be noted that the multiple reflections are strictly periodic only if the waves travel vertically through the water layer. In deep water, the delay between the primary

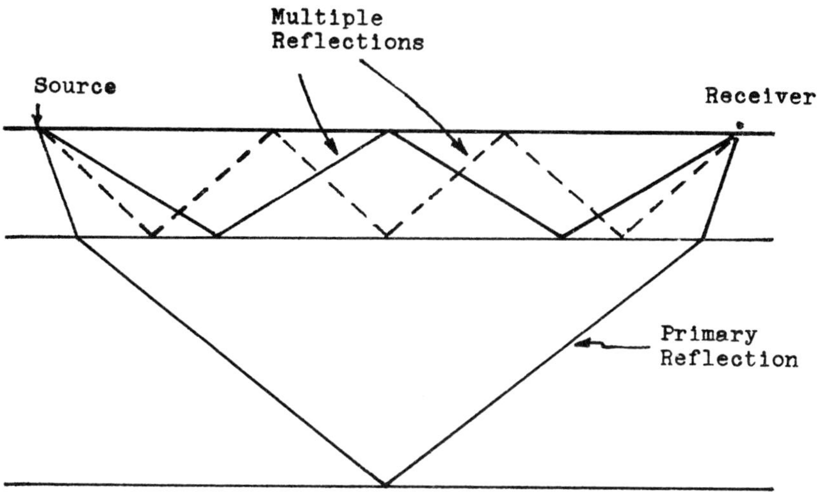

Fig. 1. Schematic representation for primary reflection and multiple reflections. Delay times between successive multiples are equal only for vertical incidence.

reflection and subsequent multiples can be quite large. Deconvolution to eliminate multiples is referred to as "Predictive Deconvolution" (Peacock and Treitel, 1969).

Although the two types of deconvolution discussed have different purposes, they are mathematically very similar. The desired filter operator is characterized as a prediction error operator. Past values of the observed data are used to predict the next data sample to be observed. The predicted value will usually not be exactly correct, and the error between the observed and predicted value is the prediction error. If we let the past data be represented by the vector

$$X_n = (x_{n-q+1}, x_{n-q}, \ldots, x_{n-q-p+1}) \ ,$$

then the filter application equation has the form

$$Y_{n+1} = x_{n+1} - F_n{}^T X_n$$

where Y_{n+1} is the output, x_{n+1} is the value to be predicted, and F_n is a vector of filter weights. The parameter q is the prediction distance, and p is the number of past data points to be used in the prediction. For whitening deconvolution, q is usually unity, while for predictive deconvolution, it is approximately equal to the difference in arrival times of the successive multiples (expressed as the number of time samples). Thus, in the latter case, a short filter operator can be used even though the time delay is large. The filter coefficients are determined by means of minimizing the mean squared prediction error,

$$\overline{Y_{n+1}^2} = F^T \overline{XX^T} F - 2F^T \overline{X_n x_{n+1}} + \overline{x_{n+1}^2} \ .$$

The derivatives of $\overline{Y_{n+1}^2}$ with respect to the components of F are set to zero, yielding a set of linear equations. If the correlation matrix $\overline{XX^T}$ and the cross-correlation vector $\overline{X_n x_{n+1}}$ are known, these are just the Wiener equations. In the adaptive case, the minimization is achieved by iterative use of the Steepest Descent Method, which says (Widrow, 1966) (a) move (in the space of filter weights) opposite to the direction of the gradient, and (b) move a distance proportional to the gradient. This rule gives

$$F_{n+1} = F_n - k \nabla_F \overline{Y_{n+1}^2} \quad .$$

The gradient is not known, so we use as an estimate $\nabla_F Y_{n+1}^2$ (no averaging). This leads to the filter update equation

$$F_{n+1} = F_n + 2k Y_{n+1} X_n \quad .$$

Thus, after each application of the filter operator, a new improved filter can be computed. The parameter k is a convergence parameter which determines how quickly the process can change, and how stable it will be. The choice of k will often have profound effects on the process. For input data that have been normalized to unit power, a value of about .001 appears to be suitable in many cases.

The use of adaptive deconvolution, since it is a time-varying process, should allow for some failures in the conceptual model of the deconvolution process. Particularly, in the predictive case, it should allow for changes in pulse shape between the various orders of multiple events, and for uncertainties in the delay time between events. Examples of this are discussed by Wang and Treitel (1971) and more recently by Griffiths, Smolka, and Trembly (1977).

A somewhat different approach has been used by Burg (1967). While he also uses a prediction error filter, he requires that the filter work equally well in either the forward or backward time direction (assuming time stationarity of the data). Minimization of the combined forward and backward error leads to equations for the filter coefficients, except that the coefficients are generated by means of the Levinson algorithm (Wiener, 1948). This procedure generates a filter of length n+1 coefficients from a filter of length n, and a single number determined from a minimization problem.

The Levinson algorithm is of the form

$$\begin{pmatrix} 1 \\ g_1^{n+1} \\ g_2^{n+1} \\ \vdots \\ \\ g_{n+1}^{n+1} \end{pmatrix} = \begin{pmatrix} 1 \\ g_1^n \\ g_2^n \\ \vdots \\ g_n^n \\ 0 \end{pmatrix} + a_{n+1} \begin{pmatrix} 0 \\ g_n^n \\ \\ \vdots \\ g_1^n \\ 1 \end{pmatrix}$$

where g_n^m is the mth coefficient of the filter of length n, and a_{n+1} is the minimization parameter (often called the "reflection coefficient"). Obviously, g_{n+1}^{n+1} is just a_{n+1}. The minimization problem is well known and the reader is referred to Wiener's book for details. The beauty of this method is that it guarantees minimum phase filters. In connection with this fact, it is interesting to note that the Burg type filter will generally not allow a long delay between the predicted point and the past data. This is easily seen by considering a three-point filter constructed from a two-point filter whose first coefficient is zero,

$$\begin{pmatrix} 1 \\ a_3 g_2 \\ g_2 \\ a_3 \end{pmatrix} = \begin{pmatrix} 1 \\ 0 \\ g_2 \\ 0 \end{pmatrix} + a_3 \begin{pmatrix} 0 \\ g_2 \\ 0 \\ 1 \end{pmatrix} .$$

Thus, the three-point filter will have all non-zero coefficients except in the unlikely case that a_3 is also zero.

Burg's technique when applied to a fixed length of data can be used to produce a fixed filter analogous to the Wiener filter. In fact, this can provide an alternative definition of the autocorrelation function of the process. Another application is the estimation of the frequency spectrum of the process, yielding the "Maximum Entropy" estimate. The expression for this estimate is

$$P(f) = \frac{\Delta t \, P_N}{\left|1 + \sum_{m=1}^{N} g_m Z^m\right|^2}$$

where

$$Z = \exp(-\, 2\pi i f \Delta t)$$

and P_N is the power not predictable with an N point filter. This estimator is useful for data that contain narrow spectral lines, since the dominant contributions come from the neighborhood of the zeros of the denominator.

Burg has extended this technique to make use of the adaptive procedure (Riley and Burg, 1972). The Burg procedure can be recast as a ladder network (Figure 2). The input comes in at one end, and the forward and backward output for filters of various length appears at the lateral taps. The circle containing Z is a unit delay operator. The coefficients a_n are the same that we have discussed above, but now are computed iteratively according to the formula as follows:

$$a_m^i = -\, N_m^i / D_m^i$$

$$N_m^i = (1 - \alpha)\, N_m^{i-1} + 2\alpha\, f_{m-1}^i\, b_{m-1}^i$$

$$D_m^i = (1 - \alpha)\, D_m^{i-1} + \alpha[(f_{m-1}^i)^2 + (b_{m-1}^i)^2] \quad .$$

The parameter α is a memory parameter analogous to the convergence parameter k. It can be shown that this adaptive process yields essentially the same result as the fixed Burg

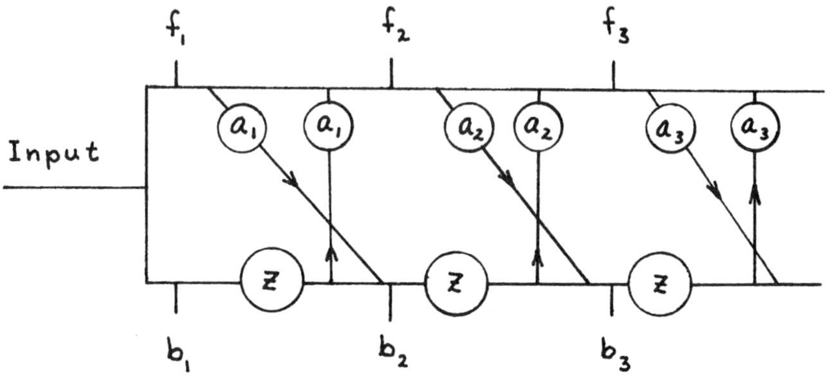

Fig. 2. Ladder network representation of the Burg algorithm. The tapped outputs f_n and b_n are the forward and backward outputs for an n-point filter.

algorithm if the following restrictions are observed: (a) process through the data allowing only the first coefficient to vary; fix this coefficient at its final value; and (b) repeat using the second coefficient, etc. This suggests that in application of this technique, we begin adapting only a small number of coefficients, and increasing the number as time goes on.

Despite the fact that the Burg procedure began by making use of strict time stationarity, it has been shown that this adaptive process works very well in practice and is quite stable.

Multichannel Adaptive Processes.

In this instance, we are dealing with the output of two or more sensors. In principle, these can be any type of sensors and may be different, so long as the desired signals on the two outputs have some coherent relationship. However, in actual practice, the channels dealt with are the output of several spatially separated identical receivers, usually sensing the velocity of ground motion. Thus, we are dealing with a two-dimensional process, although the two dimensions--space and time--are not treated equally. There are usually many more time samples than space samples.

The use of adaptive processing requires an adequate description of the signal, the simplest of which is that the signal appears to be the same on every input channel. Since we are considering the propagation of elastic waves, this requires that the desired wave arrive at the same time on every receiver, and have the same pulse shape. The pulse shape requirement, with some minor exceptions, is usually satisfied. However, the time of arrival depends on the source receiver distance and the type of wave. Thus, the data will require an initial time alignment before any filtering is done.

The simplest situation occurs for critically refracted seismic waves. These events are observed beyond some critical distance (Figure 3), and travel at constant velocity along the surface of a flat refractor. The time shift required is a constant for each receiver, and depends only on the source receiver distance. Adaptive processing offers the opportunity to attenuate strong coherent arrivals of different velocity from the desired arrival. No particular advantage is gained against random noise, and the desired signal can actually be degraded if the refractor deviates in velocity or if its surface is not plane. Nevertheless, the process is so simple and fits the problem so well that it would seem worthy of testing. Fixed (Wiener) multichannel filters have been applied to multichannel data and were shown to be useful (Laster and Linville, 1968), but this author is not aware of any significant test of the adaptive procedure. For practical purposes, this may not be important since the seismic refraction method is no longer widely used.

The seismic reflection method has proved to be much more effective. Although many geometric configurations have been

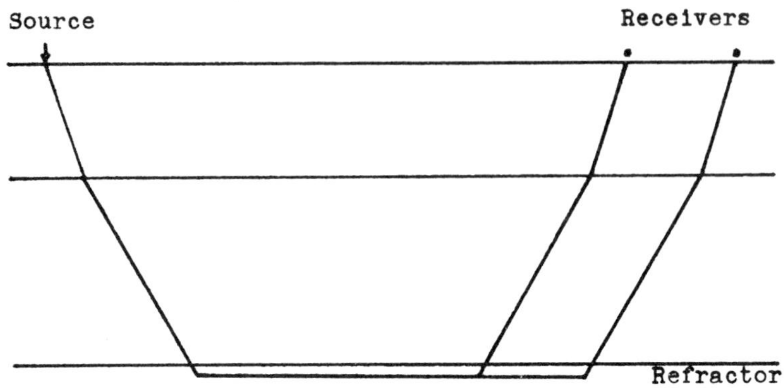

Fig. 3. Schematic representation for critically refracted event. Travel time vs. distance curve is a straight line.

used, the standard of the industry is the Common Depth Point Method introduced in the early 1960s. This ranks with Digital recording as a revolution in exploration for petroleum. Using this method, data from several source-receiver pairs (Figure 4) are combined to enhance the desired reflection while attenuating the undesired noise (usually multiply reflected waves). The strategy is to align the desired signal while hopefully misaligning the noise. The same strategy is used in adaptive processing.

In order to evaluate adaptive processing of seismic data, we must carefully describe those processes which would be

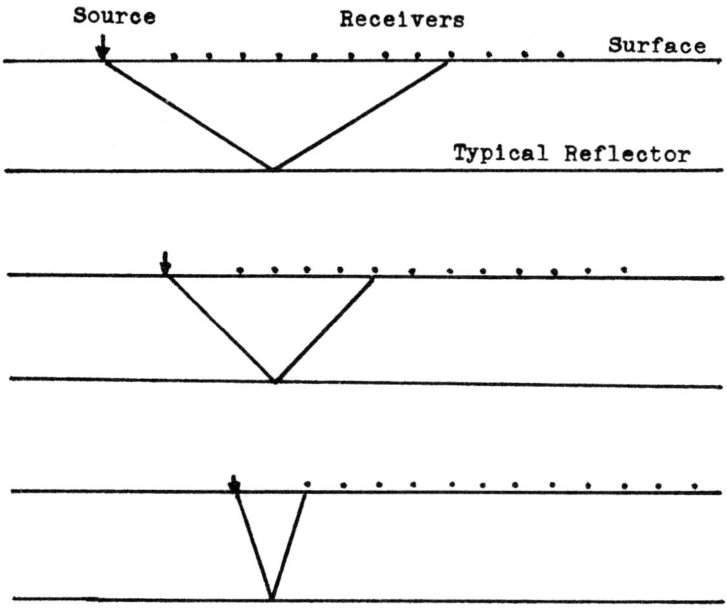

Fig. 4. Schematic diagram for Common Depth Point geometry, twelve receivers with two receiver group move-up. This gives 300% coverage.

performed using the standard techniques, those which are common to the two procedures, and those which are peculiar to either techniques. Also, some of the common procedures will be more effective in one environment than the other.

Reflected waves behave in an entirely different manner from refracted waves. The apparent velocity depends on source-receiver distance (offset) and also on the depth of the reflecting interface. Thus, the time shift required to align the signal varies with receiver location and expected arrival time--a time variant time shift. This is referred to as the Normal Moveout Correction (NMO). It is required whether we use the standard summation process or the adaptive process. Since most of the problems occur here, we will discuss this process in detail.

The reflected arrival time at a source receiver separation x is given by

$$T^2(x) = T_o^2 + x^2/V^2(T_o)$$

where T_o is arrival time at zero separation, and $V(T_o)$ is the Root Mean Square velocity. Note that V depends on T_o. This is valid if the earth is composed of plane layers in which the true (interval) velocity depends only on the depth. When this assumption is valid, it is desirable to choose these layers (artificially, if need be) so as to have the same thickness, as measured in terms of two-way travel time. We will always take this two-way time to be equal to the sample interval. Thus, in principle, the seismogram for zero source-receiver distance should consist of a string of impulses, each being the reflection from a different interface. This picture is clearly too simple, first because there are multiple reflections, and second because the signal pulse shape is of finite length, not an impulse.

The NMO correction process as defined in the travel-time equation above is a mapping of data on an observed output at offset x and arrival time T(x) to a corrected data trace simulating zero offset and arrival time T_o. We can think of T_o as being given, in integral multiples of Δt, and T(x) being computed as the location to obtain the data sample. T(x) will usually not be an integral multiple of Δt so some type of interpolation is required. Except for this interpolation, the NMO correction does not alter the amplitude of the data. It can easily be seen from the equation that for large T_o or small x, T(x) and T_o are not very much different. Thus, the pulse shape of the signal is very little altered. On the other hand, for large x and small values of T_o, the correction is large, and changes rapidly with T_o. This leads to serious distortion of the pulse through stretching. It is also obvious that multiply reflected events will not be corrected properly since they satisfy a different travel-time equation. An excellent summary of the properties of the normal moveout correction is given by Dunkin and Levin (1973).

Having applied the NMO correction, and being aware of its limitations, we can next sum the data to enhance the signals and attenuate noise. It has been assumed that the signal is identical on all data channels after correction, but this is known to be only an approximation, particularly for the shallow reflections recorded on receivers with large offset. In fact, in many cases, this part of the data will not be included in the sum, a process called muting. However, for the data which are used, the samples are summed across all channels (source-receiver pairs) for each time point. The summing also is referred to as "Stacking," and provides significant improvement if the noise is random across channels. Note that no other specific knowledge of the noise has been used.

The common depth point summation method can be thought of in terms of a filtering operation where a one-point filter is applied to each data channel and then the results summed. In this sense, it is a multichannel filter system. By increasing the length of the filter operator applied to each channel, it is possible to obtain some additional improvement, particularly if the noise exhibits coherent properties from channel to channel. Such filters have been designed using the Wiener equations, but their application is expensive and they are not robust against changes in the signal and noise characteristics. By designing the filter operators adaptively, it is possible that the latter difficulty could be overcome, although the resulting process would still be more expensive to apply that ordinary CDP summation.

The adaptive multichannel filter system can be implemented in a number of ways. The method chosen here is the Full Gradient Projection Method (Burg, Booker, and Holyer, 1967). In this case, the mean data channel (i.e., the CDP summation trace) is subtracted from each input data channel (these are the NMO corrected channels). After subtraction, the resulting channels should contain only noise. These noise-only channels are then used to predict the remaining noise on the CDP summation trace. The predicted noise is subtracted, leaving an output which should contain less noise.

Mathematically, this is written as

$$\overline{x_t} = \frac{1}{N} \sum_{i=1}^{N} x_{i,t}$$

$$Y_{t-s} = \overline{x}_{t-s} - \sum_{i,j} f_{i,j}(\overline{x}_{t-j} - x_{i,t-j})$$

where $\overline{x_t}$ is the CDP summation, Y_{t-s} is the final output, and $f_{i,j}$ is the jth filter coefficient for channel i. In order that the filter system actually preserve signals, it is necessary that the filter coefficients satisfy a constraint,

$$\sum_i f_{i,j} = \delta_{js} , \quad \text{Kronecker's delta} .$$

This simply requires that events which arrive simultaneously on all channels be passed without degradation. The filters used are two-sided in time, and the subscript s refers to the output point.

The actual determination of the filter coefficients is based on the minimum mean square error criterion, and proceeds exactly like the single-channel case discussed previously in this paper. The iterative, or update, equation for the filter coefficients is as follows:

$$f_{i,j}(t + 1) = f_{i,j}(t) + 2k\, Y_{t-s}(\overline{x}_{t-j} - x_{i,t-j}) .$$

As before, k is a convergence parameter, which for data with power normalized to unity should be about .001. By summing the above equation over i (the channel counter), it can be seen that $f_{i,j}(t + 1)$ will satisfy the constraint equation, if $f_{i,j}(t)$ satisfies it. Thus, it is only necessary that the initial set of filter coefficients be chosen to satisfy the constraint. A convenient choice is

$$f_{i,j} = \frac{1}{N} \delta_{js}$$

which are just the ordinary CDP summation operators. When performing computations using limited arithmetic precision, the filters will sometimes stray from the constraint due to round-off error. When this happens, the simplest procedure is to divide the constraint error for each time point by the number of channels, and subtract from the coefficients. Usually, this leads to little loss in performance of the filter system.

A computer program using the Full Gradient Projection Method is given in the Appendix. This program expects as input data traces that are NMO corrected. It appears that there are no good examples of multichannel adaptive processing of seismic reflection data in the literature. Sherwood and Poe (1972) mention a process which is presumably the same as that described here, and show a synthetic example. The results were very good. Booker, Ong, Rekeita, and the present author did extensive processing on a limited set of marine data at Texas Instruments in 1972. The results were very promising, although not spectacular. Unfortunately, these results were not published. This author is of the opinion that the adaptive multichannel process may be more effective on land-recorded data, where the primary noise is seismic surface waves. These tend to be large in relation to reflections and are often reasonably coherent.

Several generalizations of the multichannel adaptive method can be devised. Booker and Ong (1971) have shown how the system can be modified to satisfy several constraints simultaneously. Unfortunately, it is not clear that this is a useful modification for geophysical purposes. Also, the addition of additional hard constraints generally degrades the noise rejection performance of the system. Other generalizations combine more than one adaptive process. For example, one might apply NMO to align the multiple reflections, apply predictive deconvolution, remove the NMO appropriate to the multiples, apply NMO to align the primary reflections, and sum or adaptively filter. There are many processing systems of this kind in existence, and some of them perform very well. However, most of them are proprietary, and it is difficult to obtain information as to how they actually work.

Some General Comments.

Despite this author's enthusiasm for adaptive processes, it should be evident that they are not a panacea. We cannot expect to achieve spectacular results without knowing a great deal about the signal and noise. In the following paragraphs, a few specific problems will be discussed, and a few practical suggestions given.

Perhaps the biggest problem that occurs is that the actual signal present in the data does not match the signal model exactly. Experience has shown that filter systems which offer the highest improvement are most sensitive to this problem. Simply put, the signal looks like noise to the processor, and the problem becomes more severe with increasing signal-to-noise ratio. The causes for the problem are numerous--for example, NMO distortion, variations in the recording system from channel to channel, and errors in recording geometry. A particularly pathological example occurs when one data channel goes dead in a multichannel system. After a relatively short time, the adaptive equations have shifted all the non-zero filter weights to the dead channel. We have the ideal situation in which the filter weights satisfy the constraint, the output power is minimum (in fact, zero!). Unfortunately, there is also no signal. Thus, one must be on the alert for dead channels in using multichannel processes, particularly in real time.

Another common problem is too large a convergence parameter. Since the input data are presumed to have some random component, the output of the process will fluctuate, even with a fixed filter system. However, in adaptive systems, the filter weights themselves also fluctuate leading to "Misadjustment Noise." This tends to increase with increasing convergence parameter. In the extreme case, the filters may not converge at all. The system is unstable and both the output and the filter weights increase without bound.

A more subtle case occurs when the input data are severely oversampled. To the filter coefficients, the data appear to be very low frequency and change slowly in time. On a single-channel basis, it is clear that new data points will be highly predictable from past data points even though the time interval spanned by the filter has very little information content. This usually leads to severe signal degradation. Besides, processing of data that are severely oversampled is computationally inefficient. Such data should be resampled, first applying an anti-alias filter if necessary.

Finally, adaptive filter systems prefer data that are approximately stationary in time. Simple data conditioning that leads to this situation can often greatly improve performance. For example, adaptive systems are particularly sensitive to transients. For this reason, it is often desirable to begin processing a seismic trace at the wrong end (in time) and proceed back up the trace to earlier arrival times. In this way, we minimize the trauma of the first arrivals. Another useful trick is to apply a time-varying scaling to the data so that the average amplitude in any gate of a given length is constant. This is very similar to automatic gain control, and will sometimes lead to significant improvement. The scaling can be removed after adaptive filtering, if the gain function is recorded.

Summary.

Adaptive filter systems appear to offer some advantages in geophysical data processing. In particular, the algorithms are simple and require a minimum description of the signal to be extracted. The computations required are primarily multiply and add operations which are easy to implement in special digital hardware. The single-channel processes have been widely studied and are used for elimination of multiples in marine work. The multichannel processes have not been so widely studied, at least not in reflection applications, but would seem to offer some possible advantages over the usual Common Depth Point summation method.

There are a number of pitfalls that can occur, such as imperfect signal models. However, the ultimate usefulness of these methods can only be evaluated after more results have been published.

References.

Albert, A. E., and Gardner, L. A., Jr., *Stochastic Approximation and Nonlinear Regression*, The M.I.T. Press, Cambridge, Mass., 1967.

Burg, J. P., "A New Analysis Technique for Time Series Data," Presented at the 37th Annual Meeting of the Society of Exploration Geophysicists, Oklahoma City, 1967.

Burg, J. P., Booker, A., and Holyer, R., "Adaptive Filtering of Seismic Array Data," Advanced Array Research, Special Report No. 1, Texas Instruments, Inc., 1967.

Dunkin, J., and Levin, F., "Effect of Normal Moveout on the Seismic Pulse," *Geophysics 38*, no. 4, p. 635, 1973.

Griffiths, L. J., Smolka, F. R., and Trembly, L. D., "Adaptive Deconvolution: A New Technique for Processing Time Varying Seismic Data," *Geophysics 42*, no. 4, p. 742, 1977.

Laster, S., and Linville, A. F., "Preferential Excitation of Refracting Interfaces by Use of a Source Array," *Geophysics 33*, no. 1, p. 49, 1968.

Peacock, K., and Treitel, S., "Predictive Deconvolution - Theory and Practice," *Geophysics 34*, no. 2, p. 155, 1969.

Riley, D., and Burg, J. P., "Time and Space Adaptive Deconvolution Filters," Presented at the 42nd Annual Meeting of the Society of Exploration Geophysicists, Anaheim, Calif., 1972.

Robinson, E. A., and Silvia, M. T., *Digital Signal Processing and Time Series Analysis*, Holden-Day Inc., San Francisco, 1978.

Sherwood, J., and Poe, P., "Continuous Velocity Estimation and Seismic Wavelet Processing," *Geophysics 37*, p. 769, 1972.

Wang, R., and Treitel, S., "Adaptive Signal Processing through Stochastic Approximation," *Geophys. Prospecting 19*, p. 718-727, 1971.

Wiener, N., *Extrapolation, Interpolation and Smoothing of Stationary Time Series*, The M.I.T. Press, Cambridge, Mass., 1949.

Widrow, B., "Adaptive Filters I: Fundamentals," Technical Report No. 6764-6, Systems Theory Laboratory, Stanford University, 1966.

Appendix.

```
C       MULTICHANNEL ADAPTIVE FILTER
C
C       NC = NUMBER OF CHANNELS  (MAX 24)
C       NF = NUMBER OF FILTER POINTS PER CHANNEL (MAX 11)
C       NP = NUMBER OF DATA POINTS PER CHANNEL (MAX 1000)
C       NDP = NUMBER OF DEPTH POINTS TO PROCESS
C       NBAK = 0 PROCESS IN FORWARD TIME DIRECTION
C              1 PROCESS BACKWARD IN TIME STARTING AT LAST DATA
C                POINT
C       H = NORMALIZED ADAPTIVE CONVERGENCE PARAMETER
C           (USUALLY ABOUT .001)
C       SIGNAL IS ASSUMED TO BE TIME ALIGNED ON THE INPUT DATA
C
        DIMENSION X(1000,24),F(11,24),B(1000),Y(1000)
        READ (5,100) NC,NF,NP,NDP,NBAK
    100 FORMAT(16I5)
        READ (5,102) H
    102 FORMAT(8E10.0)
        IDP=0
    232 DO 1 I=1,NC
        READ (14) (X(J,I),J=1,NP)
      1 CONTINUE
C
C       COMPUTE ACTUAL CONVERGENCE PARAMETER
C
        SUM=0.
        DO 34 I=1,NP
        DO 34 J=1,NC
     34 SUM=SUM+X(I,J)**2
         G=H*NP*NC/SUM
         A=1./NC
         MF=1+NF/2
C
C       INITIALIZE
C
C            COMPUTE CDP STRAIGHT STACK (SIGNAL ESTIMATE)
C
         DO 2 I=1,NP
        B(I)=0.
        DO 3 J=1,NC
      3 B(I)=B(I)+X(I,J)*A
         DO 4 J=1,NC
      4 X(I,J)=B(I)-X(I,J)
      2 CONTINUE


C
C            INITIALIZE FILTERS
C
        DO 5 I=1,NC
        DO 6 J=1,NF
      6 F(I,J)=0.
      5 F(MF,I)=A
```

```
C     FILTER APPLICATION
C
          DO 9 I=1,MF
      9 Y(I)=B(I)
         DO 7 II=1,NP
         I=II
         IF (NBAK.GT.0) I=NP-II+1
          M=I-MF
          IF (M.LT.1) GO TO 7
          Y(M)=B(M)
          DO 8 J=1,NC
         DO 8 K=1,NF
          L=I-K
         IF (L.LT.1) GO TO 8
          Y(M)=Y(M)-F(K,J)*X(L,J)
      8 CONTINUE
C
C        FILTER UPDATE
C
         DO 10 K=1,NF
         DO 10 J=1,NC
         L=I-K
         IF (L.LT.1) GO TO 10
         F(K,J)=F(K,J)+2.*G*Y(M)*X(L,J)
     10 CONTINUE
      7 CONTINUE
C
C        WRITE NEAR TRACE
C
         WRITE (15) (X(I,1),I=1,NP)
C
C        WRITE CDP STRAIGHT STACK
C
         WRITE (15) (B(I),I=1,NP)
C
C        WRITE ADAPTIVE STACK
C
         WRITE (15) (Y(I),I=1,NP)
C


C        WRITE FILTER WEIGHTS
C
         WRITE (6,231)((F(I,J),I=1,NF),J=1,NC)
    231 FORMAT(1H1/(1P5E20.8))
         IDP=IDP+1
         IF (IDP.LT.NDP) GO TO 232
         STOP
         END
```

A COMPARATIVE CASE STUDY OF SEVERAL SPECTRAL ESTIMATORS

Don McIntire*
Anderson Clayton Foods
Richardson, Texas

Introduction.

In [3] the authors present a technique for spectral estimation of discrete and continuous time series. This procedure, labeled G-SPEC in [3], is a competitor for such alternatives as ARMA type estimators, ARSPEC, and MEM, presented in [3]. Devoted completely to the analysis of the time series of chemical process concentrations presented as Series A in Box and Jenkins [2], this paper compares several spectral techniques with that of G-SPEC and in addition considers the applicability of an "ARMA" type spectral estimator as suggested in [3]. This analysis, using various spectral estimation methods, should provide additional understanding of the G-SPEC method and ARMA type estimator.

Series A.

A plot of the actual data set appears in Figure 1. The goal of this paper is not to model Series A as an ARMA process, but to provide an adequate estimator of the spectrum. Indeed this series does not obviously lend itself to ARMA modeling. However, several models for the series were suggested in [2] and their two final models could provide spectral estimators of the ARMA type for comparison. Box and Jenkins consider both a stationary and a nonstationary model. These models

*This work was supported by ONR Contract No. N00014-75-C-0439.

ISBN 0-12-257250-5

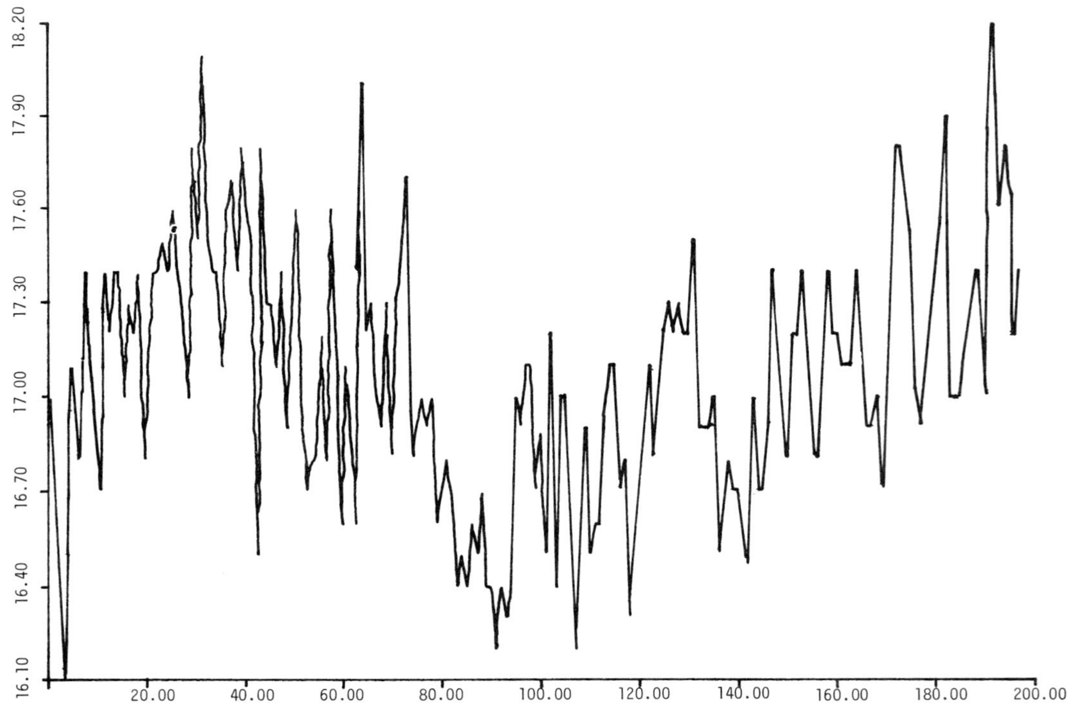

Fig. 1. Series A chemical process concentration, two-hourly readings.

are

$$(1 - 0.92B)(X_t - \bar{X}) = (1 - 0.58B)\ a_t \tag{1}$$

and

$$(1 - B)\ X_t = (1 - 0.70B)\ a_t \tag{2}$$

respectively where the operator B is defined by $BX_t = X_{t-1}$. For forecasting the nonstationary model is preferred since both (1) and (2) are quite similar and the nonstationary model does not assume a fixed level for the process as does the stationary model. However, for spectral analysis, the stationary model is the more acceptable selection, and hence will be the one adopted here. Therefore, in this paper, comparisons will be given to stationary models only.

Spectral Estimators.

The spectrum under model (1) which we will denote as $S_{ARMA}(\cdot)$ is given by

$$S_{ARMA}(\omega) = 2\frac{|1 - 0.58e^{-2\pi i\omega}|^2}{|1 - 0.92e^{-2\pi i\omega}|^2}, \quad 0 \le \omega \le 0.5,$$

where $i = \sqrt{-1}$. This spectrum is plotted in Figure 2. Consider also the estimation scheme denoted ARSPEC in [3]. This too is a model-building technique adapted to spectral estimation. According to the FPE criterion of Akaike [1], a seventh order autoregression is selected,

$$(1 - .37B - .20B^2 - .02B^3 - .01B^4 + .01B^5 - \\ - .06B^6 - .16B^7)(X_t - \bar{X}) = a_t \quad . \tag{3}$$

Figure 3 is a plot of the spectrum of this process which shall

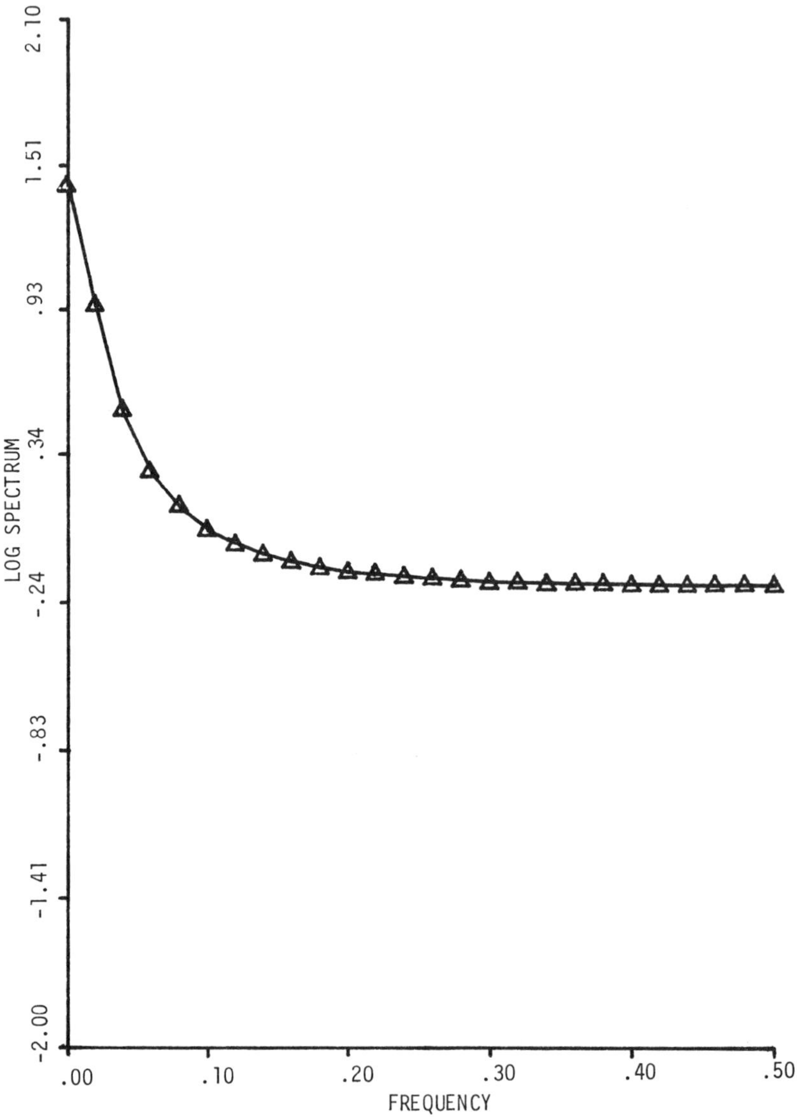

Fig. 2. Plot of the ARMA(1,1) spectral density $S_{ARMA}(\cdot)$.

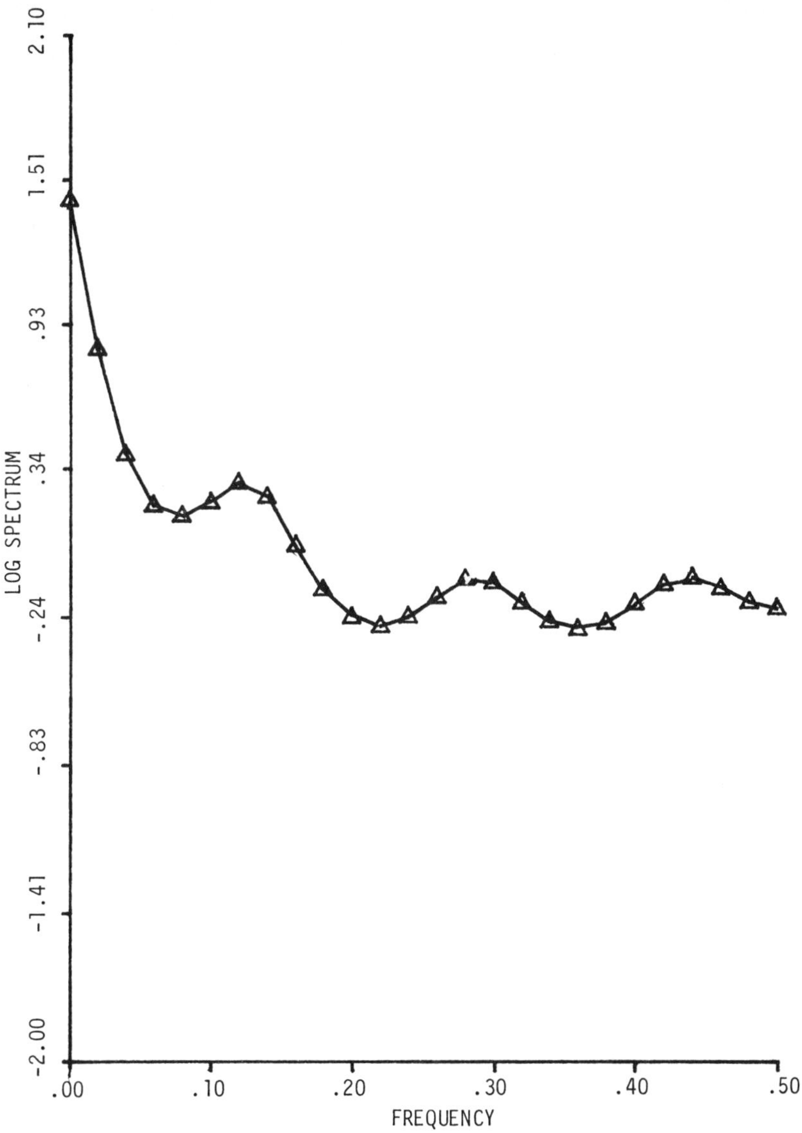

Fig. 3. Plot of the AR(7) spectral density S_{ARSPEC}.

be referred to as $S_{ARSPEC}(\cdot)$, i.e.,

$$S_{ARSPEC}(\omega) = 2\{|1 - .37e^{-2\pi i\omega} - .20e^{-4\pi i\omega} - .02e^{-6\pi i\omega} - .01e^{-8\pi i\omega} + .01e^{-10\pi i\omega} - .06e^{-12\pi i\omega} - .16e^{-14\pi i\omega}|^2\}^{-1} .$$

The maximum entropy method (MEM) described in [3] also was employed to obtain a spectral estimator. Again, the FPE criterion selects a seventh-order autoregression. The prediction error filter obtained corresponds to the AR(7) model

$$(1 - .36B - .19B^2 - .02B^3 - .02B^4 + .02B^5 - .07B^6 - .19B^7)(X_t - \bar{X}) = a_t ,$$

and hence the method produces essentially the same estimator as ARSPEC. Figure 4 displays the Parzen window estimate with bandwidth 0.04. The Rectangular, Bartlett, and Tukey windows give similar estimates. Finally, Figure 5 shows the spectral estimator for the automatic G-spectral estimation scheme, G-SPEC. In Table 1, the orders of the G-transform selected by G-SPEC at various frequencies are shown. Notice that at low frequencies relatively high-order transforms are selected, whereas at high frequencies low orders are chosen. This corresponds to selecting high-order autoregressions for estimates at low frequencies and low-order autoregressions to estimate the spectrum at high frequencies. In particular observe that $G_7^{(1)}$ is selected at the zero frequency. This is essentially equivalent to the selection of a seventh-order autoregression, which in this instance is the same result obtained by FPE. That is, at $\omega = 0$ G-SPEC, ARSPEC, and MEM are very nearly equivalent, while the tendency to select lower order transformations at the high frequencies suggests an affinity toward the

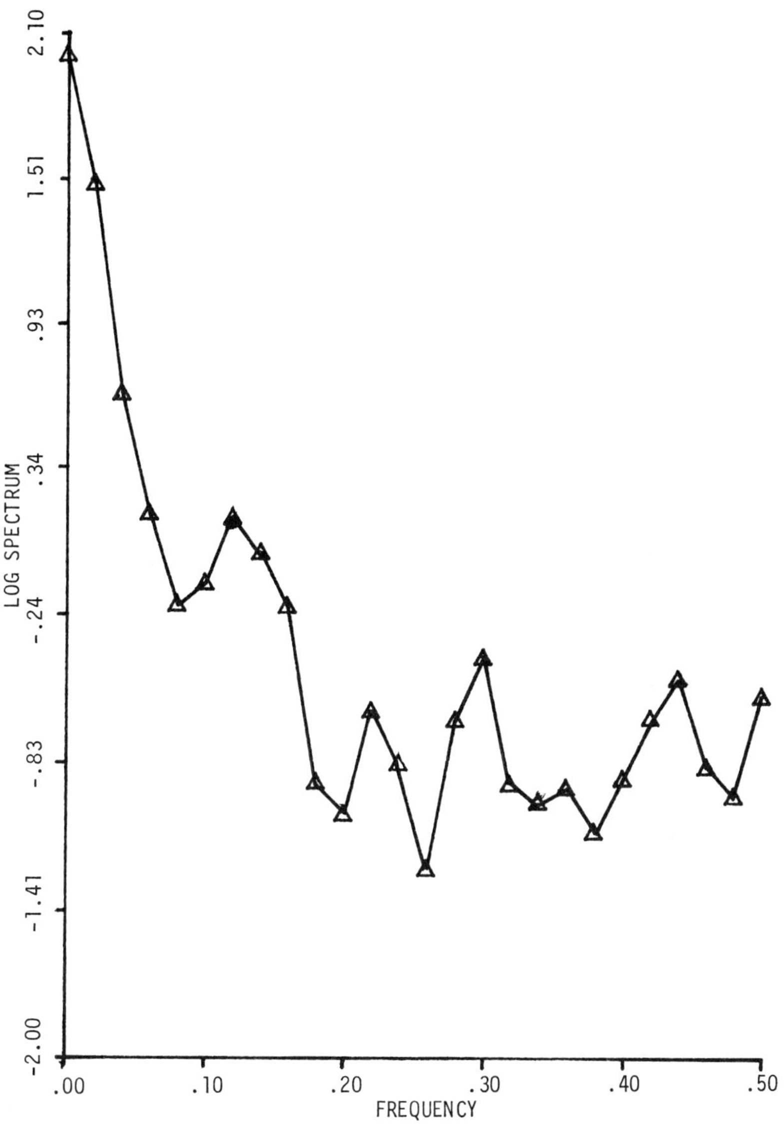

Fig. 4. Parzen window estimate of the spectrum with bandwidth 0.04.

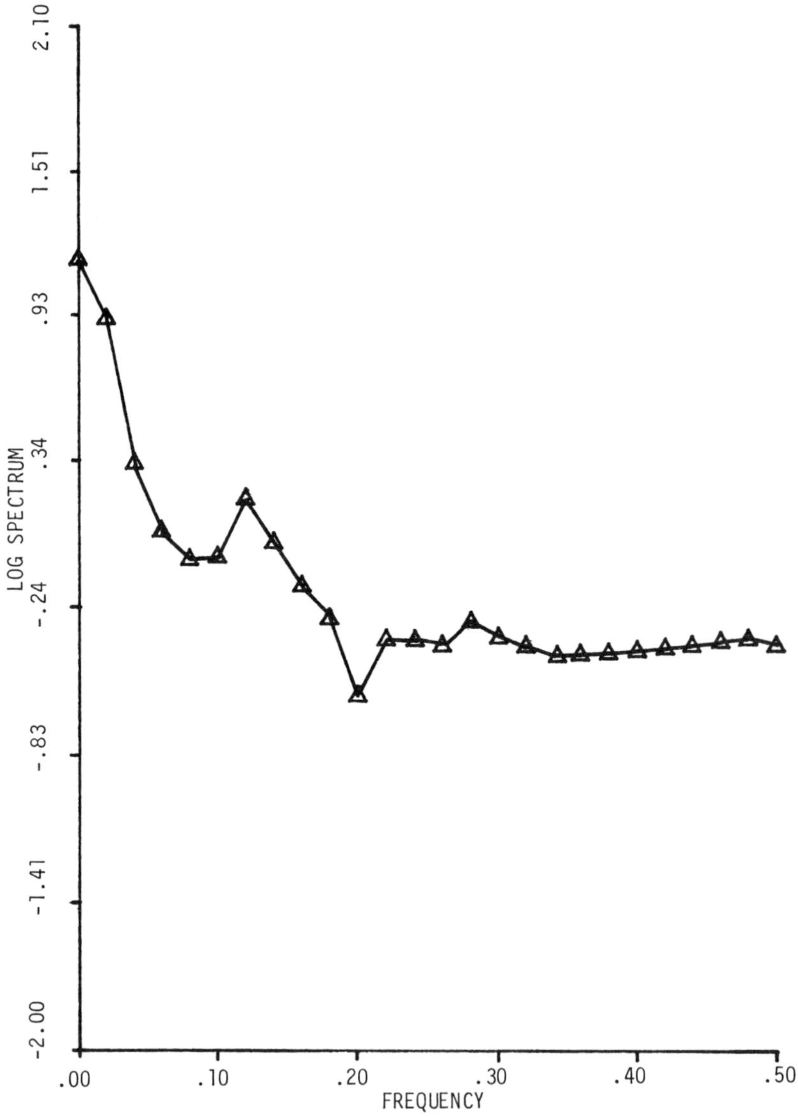

Fig. 5. G-SPEC estimate of the spectrum.

TABLE 1

Orders of the G-transform selected by G-SPEC at the Indicated Frequencies

$(0, G_7^{(1)})$	$(.02, G_5^{(1)})$	$(.04, G_{13}^{(1)})$	$(.06, G_{12}^{(1)})$	$(.08, G_6^{(1)})$
$(.1, G_6^{(1)})$	$(.12, G_{10}^{(1)})$	$(.14, G_{12}^{(1)})$	$(.16, G_9^{(1)})$	$(.18, G_8^{(1)})$
$(.2, G_{13}^{(1)})$	$(.22, G_8^{(1)})$	$(.24, G_2^{(1)})$	$(.26, G_2^{(1)})$	$(.28, G_8^{(1)})$
$(.3, G_5^{(1)})$	$(.32, G_{10}^{(1)})$	$(.34, G_{14}^{(1)})$	$(.36, G_3^{(1)})$	$(.38, G_3^{(1)})$
$(.4, G_2^{(1)})$	$(.42, G_3^{(1)})$	$(.44, G_3^{(1)})$	$(.46, G_3^{(1)})$	$(.48, G_3^{(1)})$

Box-Jenkins models, (1) and (2) near $\omega = 0.5$. In the G-matrix at 0.5 frequency, Table 2, it should be noticed that due to lack of power at this frequency, most orders of the transform give about the same result. Then, since both high- and low-order autoregressions yield essentially the same value for the spectrum at high frequencies, G-SPEC reduced to the simpler model. Close inspection of the S- and R-matrices (see [4] and [5]) shows the stationary model to be more correctly modeled as the ARMA(7,1),

$$(1 - .42B - .17B^2 - .26B^7)(X_t - \bar{X}) =$$

$$(1 + .26B)\, a_t \quad .$$

Actually, the preferred model for forecasting purposes is the nonstationary model,

TABLE 2

G-matrix for the Highest Frequency, ω = 0.50

N / M	0	1	2	3	4	5	6	7
0	0.000	1.428	0.290	0.600	0.761	0.453	1.502	0.343
1	1.176	1.108	0.467	0.526	0.491	1.867	1.855	0.380
2	0.952	-2.731	0.545	0.549	0.412	1.499	0.294	0.362
3	0.229	0.071	0.522	0.445	0.467	0.365	0.334	0.365
4	0.017	-0.204	0.390	0.428	0.309	0.334	0.334	0.353
5	0.454	1.104	0.475	0.336	0.330	0.334	0.330	0.341
6	0.889	0.836	0.330	0.331	0.331	0.330	0.330	0.345
7	0.680	-2.533	0.337	0.330	0.334	0.330	0.343	0.343
8	0.103	-0.019	0.315	0.335	0.335	0.357	0.344	0.346
9	-0.063	-0.267	0.368	0.334	0.372	0.393	0.339	0.275
10	0.286	0.691	0.244	0.356	0.362	0.354	0.415	
11	0.599	0.576	0.409	0.364	0.364	0.978		
12	0.525	2.035	0.327	0.362	0.309			
13	0.235	0.177	0.449	0.333				
14	0.151	-0.080	0.255					
15	0.364	0.689						
16	0.560							

N / M	8	9	10	11	12	13	14
0	0.366	0.366	0.366	0.333	0.260	0.197	0.264
1	0.364	0.367	0.471	0.297	0.353	0.259	0.259
2	0.362	0.359	0.321	0.221	0.041	0.264	0.325
3	0.378	0.341	0.924	-0.022	-0.028	0.331	
4	0.343	0.344	0.251	0.043	0.300		
5	0.342	0.340	0.415	0.293			
6	0.353	0.335	0.305				
7	0.342	0.310					
8	0.305						

$$(1 - B)(1 + .68B + .50B^2 + .50B^3 + .50B^4 + $$

$$+ .50B^5 + .50B^6)\ X_t = a_t \quad .$$

As has been indicated by the number of models suggested, there is some question as to what is the most appropriate ARMA model here, but the G-transform seems to adjust with the frequency to accommodate differing ARMA models. Hence, it should be clear, contrary to this author's first impression, that the G-spectral estimator is in no way tied to a fixed ARMA model. Notice that all spectral estimators presented here other than the windows and G-SPEC are not frequency dependent.

Table 3 is the G-matrix for frequency 0.20. In Figure 6, the spectrum at this frequency appears out of character with the other frequencies. Indeed, inspection of the G-matrix indicates that the difficulty was with the automatic selection of the order and not with the G-spectral estimate. Examination of Table 3 suggests that $G_8^{(1)}$ should have been selected. Figure 6 presents G-SPEC with this adjustment.

For completeness, Figure 7 is included. This is the plot of the spectrum of the ARMA(7,1) model suggested earlier,

$$(1 - .42B - .17B^2 - .26B^7)(X_t - \bar{X}) = (1 + .26B)\ a_t \quad .$$

Note that all spectral estimates other than the ARMA, model (1), present a peak near 0.12 frequency. Possibly, some "window carpentry" could smooth the window estimates somewhat, but the appropriate bandwidth is not known, especially prior to analyzing the data. The ARSPEC and MEM techniques find their power in autoregressive type models. They are also frequency independent in the sense that the selection of the model does not rely on the frequencies at which the power is to be determined. However, the G-SPEC theoretically covers a much broader class of models than even the ARMA class and is dependent on ω for order selection.

TABLE 3

Matrix for the Frequency $\omega = 0.20$

N / M	0	1	2	3	4	5	6	7
0	0.000	0.274	0.411	0.429	0.321	0.432	0.410	-0.154
1	0.430	0.390	0.273	0.445	0.449	0.415	0.417	0.393
2	0.355	0.409	0.425	0.747	0.428	0.410	0.554	0.396
3	0.452	0.430	0.429	0.433	0.444	0.377	0.336	0.389
4	0.410	0.425	0.443	0.422	0.367	0.369	0.360	0.393
5	0.438	0.449	0.451	0.310	0.382	0.356	0.354	1.709
6	0.461	0.441	0.340	0.334	0.351	0.135	0.348	0.389
7	0.419	0.380	0.285	0.357	0.354	0.352	0.361	0.603
8	0.349	0.358	0.348	0.353	0.352	0.352	0.573	0.572
9	0.366	0.339	-0.750	0.354	0.352	0.125	0.601	-0.693
10	0.317	0.355	0.360	0.357	0.096	0.098	-0.092	
11	0.384	0.370	0.346	0.375	0.125	0.714		
12	0.359	0.344	0.344	0.428	0.978			
13	0.327	0.346	1.377	0.487				
14	0.369	0.429	0.499					
15	0.464	0.482						
16	0.504							

N / M	8	9	10	11	12	13	14
0	0.396	0.392	0.395	0.131	0.384	0.253	0.254
1	0.394	0.393	0.393	0.369	0.374	0.268	0.443
2	0.391	0.395	0.553	0.382	0.442	0.600	0.478
3	0.389	0.419	0.402	0.637	0.512	0.532	
4	0.431	0.426	0.333	0.414	0.577		
5	0.413	0.510	0.147	0.940			
6	0.365	-1.196	-0.273				
7	0.260	-0.013					
8	0.118						

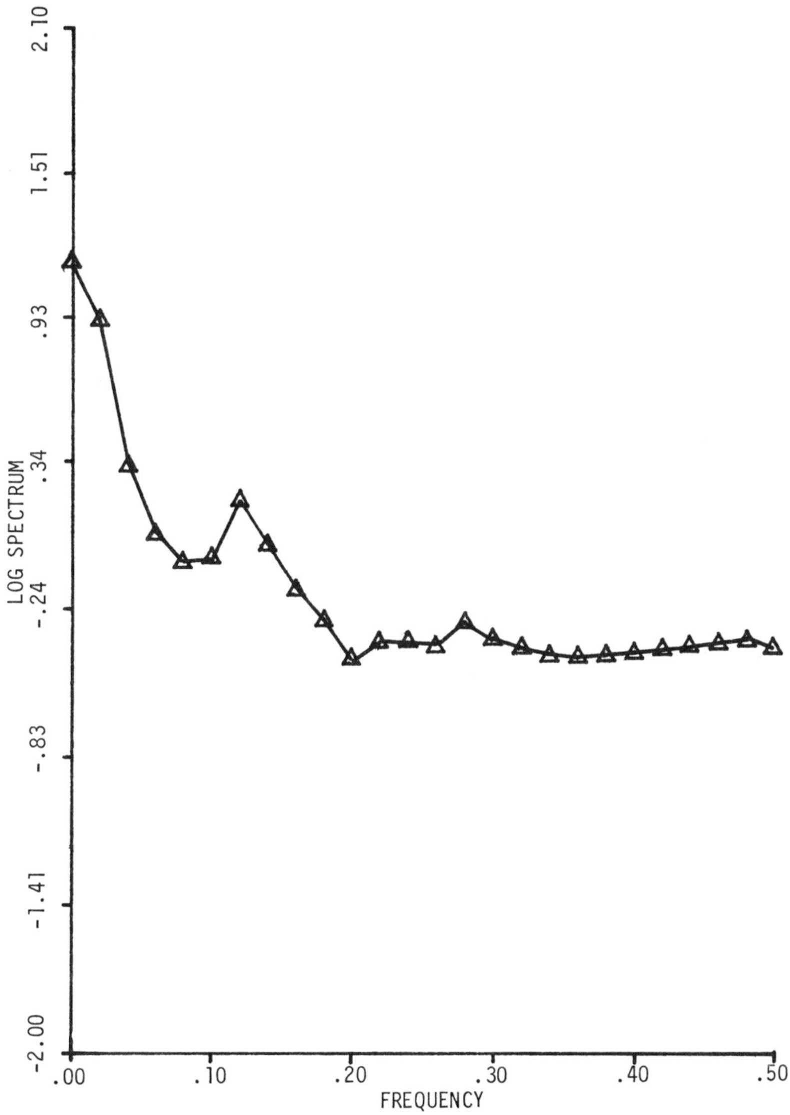

Fig. 6. G-SPEC estimate with frequency $\omega = 0.20$ *adjusted.*

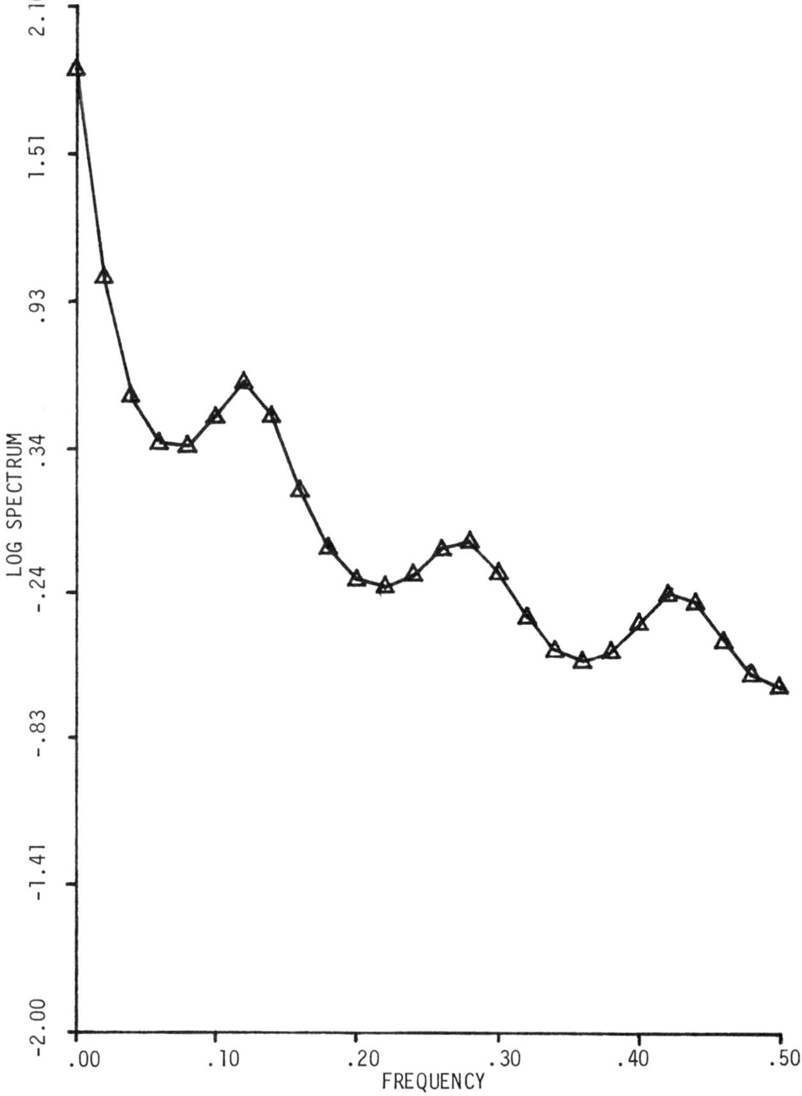

Fig. 7. Plot of the spectral density of fitted ARMA(7,1) model.

Conclusion.

Various spectral estimators have been presented for the time series known as Series A in Box and Jenkins [2]. The MEM, ARSPEC, and GSPEC all seem to work well. The Parzen Window also gave a satisfactory result but to a lesser degree than the others. The more interesting finding was that the ARMA spectral estimator suggested by Model 1 did not perform as well. Examination of Figure 7 indicates that the problem may lie in the selection of the model used. If this is the case, this would suggest that finding an appropriate ARMA model for producing $S_{ARMA}(\cdot)$ may be a more difficult problem than the primary problem indicated here, i.e., finding an adequate spectral estimator.

Acknowledgment.

The author would like to express his appreciation to H. L. Gray for his aid and encouragement in preparing this paper, and also to D. F. Findley for his helpful suggestions.

Bibliography.

[1] Akaike, H. (1969) "Fitting Autoregressive Models for Prediction." *Ann. Inst. Statist. Math.* *21*, 243-247.

[2] Box, G. E. P., and Jenkins, G. M. (1970) *Time Series Analysis: Forecasting and Control*. Holden-Day, San Francisco, California.

[3] Gray, H. L., Houston, A. G., and Morgan, F. W. "On G-spectral Estimation." *Applied Time Series Analysis*. Academic Press, New York, to appear in 1978.

[4] Kelley, G. D. "A New Approach to ARMA Modeling - The Nonstationary Case." Ph.D. Dissertation, Southern Methodist University, 1977.

[5] McIntire, D. D. "A New Approach to ARMA Modeling." Ph.D. Dissertation, Southern Methodist University, 1977.

APPLICATION OF HOMOMORPHIC FILTERING TO SEISMIC DATA PROCESSING*

A. Oppenheim and J. Tribolet
Massachusetts Institute of Technology

I. Introduction.

The field of seismic data processing has traditionally been an important area for the exploration, development, and refinement of new signal processing ideas. Since the raw data is relatively expensive to acquire, most of the signal processing is carried out by computer off-line, there are no real-time constraints and the potential benefits make practical the use of large computers with complicated algorithms. The detailed representation of the seismic signal involves a relatively complicated model and the overall processing often requires the use of a set of techniques, each perhaps, based on different simplifications of the model. One class of techniques is based on a representation of the signal as a convolution of components, with the basic signal processing task being deconvolution.

More specifically, this simplified model is illustrated in Figure 1. Ideally what is desired is a measurement of the impulse response of the earth. Toward this end, a seismic source consisting perhaps of a dynamite charge or for marine seismic exploration, an air gun, is used as an excitation. Reflections received from the subsurface structure are then measured with a receiver. If the source were, in fact, an impulse, then the received signal would consist of

*This work was supported in part by the Advanced Research Projects Agency monitored by ONR under Contract N00014-75-C-0951-NR 049-308 and in part by the National Science Foundation under Grant ENG71-02319-A02.

ISBN 0-12-257250-5

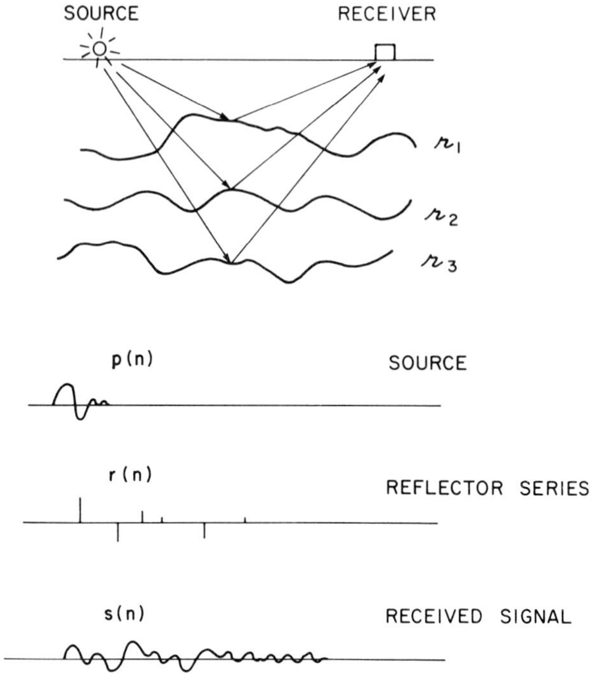

Fig. 1. Simplified model representing the seismic signal as a convolution of components.

a train of impulses of unequal amplitude corresponding to each of the subsurface reflectors. What is in fact received, however, is the convolution of the source pulse with the reflector series and if the duration of the source pulse is not short compared with the spacing between the reflectors (as is typically the case), then it is generally difficult to detect in the received signal the individual reflectors. Consequently, the basic problem is to deconvolve the effect of the source from the received signal. There are a number of approaches typically available for carrying out a deconvolution. One of the most common is the use of inverse filtering, that is processing the composite signal s(n) by a filter whose frequency response is the reciprocal of the Fourier transform of the source pulse. Obviously, in order to use inverse filtering, the source pulse (or its Fourier transform) must be known or estimated. Furthermore, if only a causal inverse filter is used, then the source pulse must be minimum phase. Another approach to recovering the reflector series from the composite signal s(n) is to assume that the reflector series is a white random process. Using the techniques of Wiener filtering, an optimum filter is then designed to whiten the sequence s(n) to obtain an approximation to the reflector series. In the most straightforward application of this approach, the source pulse is assumed to be minimum phase. If it is in fact a mixed phase pulse, the deviation from minimum phase manifests itself in a dispersion of the approximation to the reflector series. Consequently, it is sometimes necessary to use a formulation of the Wiener inverse filter problem in which a mixed phase inverse filter is obtained[(R1)].

Homomorphic filtering represents an alternative approach to the deconvolution of the seismic signal. This class of filtering techniques represents a generalization of linear filtering to encompass a variety of non-linear filtering problems. One particular class of non-linear filtering problems to which it has been successfully applied in a number of applications is that of deconvolution. Within the context of seismic signal processing, a number of approaches have been explored for the application of this class of non-linear filtering techniques. One consists of applying the method directly to extract the reflector series from the received signal. A second corresponds to using homomorphic filtering to estimate the source pulse. From this estimate, an inverse filter is then designed using one of several possible techniques including Wiener inverse filter design.

The objective of this paper is to review the various ways in which homomorphic filtering has been explored for use with seismic data processing. Before doing so, however, we discuss in the next section the basic principles of homomorphic filtering.

II. Homomorphic Filtering.

When we are faced with the problem of filtering signals which have been added, we very often use a linear filter. A principal reason is that linear filters are analytically convenient for dealing with signals that have been added. There are many examples for which signals have been added and for which a non-linear filter is superior to a linear filter. The principal advantage with linear filtering applied to added signals, however, is that if the behavior of the filter for each of the components separately is known, then the behavior for the sum is the sum of the responses, as a consequence of the principle of superposition. The notion of homomorphic deconvolution is to apply the same basic reasoning to deconvolution. Toward this end, we consider a class of systems matched to convolution in the same way that linear systems are matched to addition. Thus, the class of systems that we consider have the superposition property illustrated in Figure 2. It can be shown that this class of systems can be represented by the cascade of three systems, as shown in Figure 3. The first system with transformation $D_*[\cdot]$ maps a convolution of components to a sum. The system $L[\cdot]$ is a linear system and the system $D_*^{-1}[\cdot]$ is the inverse of the system $D_*[\cdot]$ and, consequently, maps a sum of components to a convolution. The system $D_*[\cdot]$ and its inverse are specified by the class of homomorphic systems with which we are dealing. Thus, all of the flexibility within this class lies in the linear system $L[\cdot]$. The transformation $D_*[\cdot]$ can be defined more specifically by the property that the z-transform (or Fourier transform) of its output is equal to the complex logarithm of the z-transform (or Fourier transform) of its input, i.e., if

$$\hat{x}(n) = D_*[x(n)] \tag{1a}$$

then

$$\hat{X}(z) = \log X(z) \ . \tag{1b}$$

Since the z-transform is a complex valued function, its logarithm must be carefully defined. In particular, since the transformation $D_*[\cdot]$ maps convolution to addition, the transformation of Eq. 1b is required to map multiplication to addition and this imposes a particular interpretation on the complex logarithm. We will return to a further

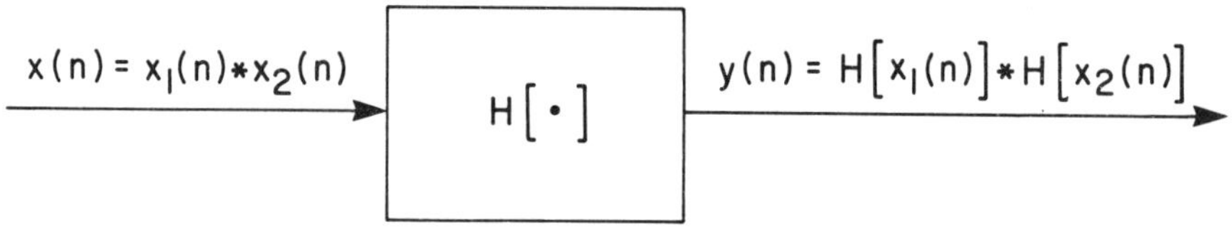

Fig. 2. System satisfying superposition under convolution.

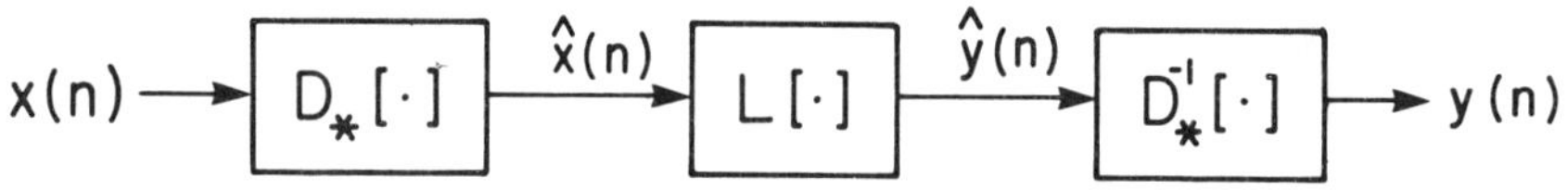

Fig. 3. Canonic decomposition of a homomorphic system.

consideration of this when we discuss some of the computational considerations in the use of homomorphic deconvolution.

The output $\hat{x}(n)$ of the system $D_*[\cdot]$ is referred to as the complex cepstrum of the input $x(n)$. This terminology is motivated by the relationship between this transformation and the cepstrum as proposed by Bogert, Healy, and Tukey for the detection of echoes. Specifically, the cepstrum of a signal was defined as the power spectrum of the logarithm of the power spectrum. Since the cepstrum was directed toward echo detection rather than a deconvolution, retention of phase information was not important. Thus, it does not utilize phase information and involves the logarithm of real, positive values. By contrast, the output of the system $D_*[\cdot]$ is referred to as the complex cepstrum in reference to the fact that it requires the use of the complex logarithm. However, it is important to note that the complex cepstrum is a real valued sequence. It should be noted also that the cepstrum as defined by Bogert, Healy, and Tukey is proportional to the even part of the complex cepstrum.

III. Properties of the Complex Cepstrum.

In carrying out a deconvolution using homomorphic filtering, an appropriate choice for the linear system $L[\cdot]$ must be made. This choice is intimately connected to a number of properties of the complex cepstrum as summarized below:

Property 1. The complex cepstrum of a convolution of two (or more) signals is the sum of their individual complex cepstra. This, of course, is simply a restatement of the fact that the system $D_*[\cdot]$ maps convolution to addition.

Property 2. The complex cepstrum $\hat{x}(n)$ of a signal $x(n)$ whose spectrum is smooth tends to be concentrated around low-time values. For example, the complex cepstrum of an exponential $(\alpha^n)u(n)$ (where $u(n)$ is the unit step sequence) is given by $1/n\ \alpha^n\ u(n-1)$.

Property 3. The complex cepstrum of a minimum phase signal is zero for $n < 0$ and the complex cepstrum of a maximum phase signal is zero for $n > 0$. In general, a signal x(n) can be decomposed into a convolution of a minimum phase component and a maximum phase component. Thus, the complex cepstrum of these two components occupy disjoint time intervals and, consequently, factorization of a signal into its minimum phase and maximum phase components is easily carried out through the use of the complex cepstrum.

Property 4. Consider a reflector series which is minimum phase and for which the time of arrival of the first reflection is N, so that

$$x(0) \neq 0$$

$$x(n) = 0 \qquad 0 < n < N \quad .$$

Then, the complex cepstrum $\hat{x}(n)$ is zero for $0 < n < N$.

All of these properties follow in a relatively straightforward way from the definition of the transformation $D_*[\cdot]$. Specifically, since

$$\hat{X}(z) = \log X(z) \tag{2}$$

it follows by differentiating that

$$\frac{d\hat{X}(z)}{dz} = \frac{1}{X(z)} \frac{dX(z)}{dz} \quad . \tag{3}$$

Let us consider X(z) to be a rational function of the form

$$X(z) = \frac{N(z)}{D(z)} \quad . \tag{4}$$

Then, from Eq. 3,

$$\frac{d\hat{X}(z)}{dz} = \frac{N'(z)D(z) - D'(z)N(z)}{ND' - DN'} \tag{5}$$

where prime denotes differentiation with respect to z. Since the z-transform of $n\hat{x}(n)$ is $-zd\hat{X}(z)/dz$, we see that the poles of the z-transform of $n\hat{x}(n)$ occur at the poles and zeroes of the z-transform of x(n). Thus, if all poles and zeroes of X(z) lie inside the unit circle, corresponding to

$x(n)$ being minimum phase, then all poles of $-z d\hat{X}(z)/dz$ lie inside the unit circle so that $n\hat{x}(n)$ and, consequently, also $\hat{x}(n)$ is zero for $n < 0$. A similar type of argument can be constructed for a maximum phase signal. Also, from Eq. (5) it follows that for $X(z)$ of the form of Eq. (4), $n\hat{x}(n)$ will consist of a sum of exponentials with exponential frequencies corresponding to each of the poles and zeroes associated with $x(n)$.

Equation (3) can be rephrased in the time domain. Specifically,

$$-z\, X(z) \frac{d\hat{X}(z)}{dz} = -z \frac{dX(z)}{dz} \tag{6}$$

so

$$nx(n) = x(n) * [n\hat{x}(n)] \tag{7}$$

or

$$\sum_{k=-\infty}^{+\infty} k\hat{x}(k)x(n-k) = nx(n) \ . \tag{8}$$

If $x(n) = 0$ for $n < 0$ and $\hat{x}(n) = 0$ for $n < 0$ so that $x(n)$ is minimum phase, then Eq. (8) can be used to express $\hat{x}(n)$ explicitly in terms of $x(n)$ as:

$$\hat{x}(n) = \frac{1}{x(0)} [x(n) - \sum_{k=0}^{n-1} \left(\frac{k}{n}\right) \hat{x}(k)x(n-k)] \quad n > 0 \ . \tag{9}$$

From this, it follows that if $x(n)$ is zero for $0 < n < N$, then $\hat{x}(n)$ is also zero in that interval.

IV. Computational Considerations in the Use of Homomorphic Deconvolution.

The principal computational step in using homomorphic deconvolution is the computation of the complex cepstrum. For minimum phase data, Eq. (9) could in principle, be used to compute the complex cepstrum. However, if Eq. (9) is applied to mixed phase data, even if it is close to minimum phase, the computation is numerically unstable. Furthermore, the use of Eq. (9) is generally inefficient.

The procedure for determination of the complex cepstrum generally relies on an explicit computation of the inverse Fourier transform of the logarithm of the Fourier transform of the data. This is most efficiently carried out using the Fast Fourier Transform algorithm to compute samples of the Fourier transform, equally spaced in frequency. In using this procedure, there are a number of computational considerations that arise.

Phase Unwrapping. Since the Fourier transform is a complex-valued function, its logarithm must be appropriately defined. More specifically, the logarithm of a complex number has as its real part the logarithm of the magnitude and as its imaginary part, the phase. The angle of a complex number is only unique to within an integer multiple of 2π and this ambiguity must be resolved in computing the phase of the Fourier transform. In order that the logarithmic transformation map multiplication to addition as required, it can be shown that the phase must be generated as a continuous function of frequency. Thus, if the phase is computed modulo 2π, as is typically the case, it must be "unwrapped" to obtain a continuous phase curve.

A simple algorithm based on examination of the sampled phase to locate jumps of 2π is generally unsatisfactory. A more successful algorithm has been developed by Tribolet[R2].

As part of the phase unwrapping procedure, a linear ramp is removed from the phase to make the resulting phase curve continuous at $\omega = 0$ and $\omega = \pi$. Since we assume that the data is causal, any maximum phase component must correspond only to zeroes in the z-transform, as discussed by Buhl et al.[R3]. The number of zeroes, i.e., the length of the maximum phase component, corresponds to the slope of the linear ramp in the phase.

Exponential Weighting. As discussed in the previous section, the complex cepstra of the minimum and maximum phase components of a signal occupy different time regions. Furthermore, from Property 4 we will see that it is particularly advantageous for the reflector series to be minimum phase. If it is not, exponential weighting can be applied to the data. This has the effect of radially scaling the poles and zeroes of the z-transform. Thus, zeroes outside the unit circle in the z-plane can be moved inside the unit circle by means of exponential weighting.

Cepstral Aliasing. Since the complex cepstrum is computed from samples of its Fourier transform, time aliasing of the complex cepstrum results. This can be reduced by appending zeroes to the data before computing the Fourier transform, corresponding to the use of finer spectral sampling. The use of exponential weighting will generally also help if enough is applied to make the reflector series minimum phase. In general, the problem with the use of exponential weighting to control aliasing is that if it is applied to reduce the duration of (and hence the aliasing in) the positive time component (corresponding to the minimum phase component of the original data), then it will increase the duration of (and hence aliasing in) the negative time component.

Oversampling. Oversampling of the original data or, more generally, bands of frequency with low amplitude is generally undesirable because of the fact that the complex cepstrum utilizes the log of the Fourier transform. Thus, it may be necessary to implement appropriate interpolation and resampling prior to computing the cepstrum.

Effects of Additive Noise in the Data. The detailed effects are not well understood. In general, for data without sharp resonances or antiresonances, if the signal-to-noise ratio is large, additive noise does not seem harmful. It has been suggested[R4] that noise in the phase curve tends to influence the long-time portion of the cepstrum.

In the presence of a high Q pole or zero, additive noise could potentially cause such a pole or zero to cross the unit circle, thus changing its contribution in the complex cepstrum to negative time rather than positive time. Whether or not this is an effect which is harmful in the overall processing remains to be resolved.

Postfiltering of the Data. As exploited by Buhl et al.[(R3)], postfiltering of the data can generally be accomplished by addition of the appropriate sequence in the cepstral domain.

V. Homomorphic Deconvolution by Time Windowing the Cepstrum.

In Section III, we saw that the structure of the complex cepstrum is such that the complex cepstrum of the source pulse and that of the reflector series may tend to occupy disjoint time intervals. For example, if the source pulse has a relatively smooth spectrum, its complex cepstrum will tend to be concentrated around the time origin. If the reflector series is minimum phase, or has been made minimum phase by exponential weighting, then it contributes to the complex cepstrum only for times greater than or equal to the time of the first reflection. Furthermore, the complex cepstrum naturally factors a signal into its minimum phase and maximum phase components with the minimum phase part contributing to the complex cepstrum only for positive time and the maximum phase part contributing only for negative time. This tendency for different components to occupy different time intervals in the cepstral domain suggests the possibility of separating these components by gating (i.e., windowing) the cepstrum. A window around the time origin for example would tend to retain the contribution from the source pulse; whereas, a window that retained only the long-time portion of the complex cepstrum would be used to recover a minimum phase reflector series. This time window then represents the linear system in the canonic form of Figure 3. The use of a low-time window on the complex cepstrum corresponds to smoothing, i.e., low-pass filtering the log spectrum. Similarly, the use of a long-time window on the complex cepstrum corresponds to retaining the rpaidly varying components, i.e., high-pass filtering the log spectrum.

Cepstral gating (or "liftering" as it is sometimes referred to) was first explored for seismic data by Ulrych[(R4)], and was directed toward the deconvolution of teleseismic events. Figure 4 shows the results obtained by Ulrych for a synthetic example. In Figure 4a is shown the impulse

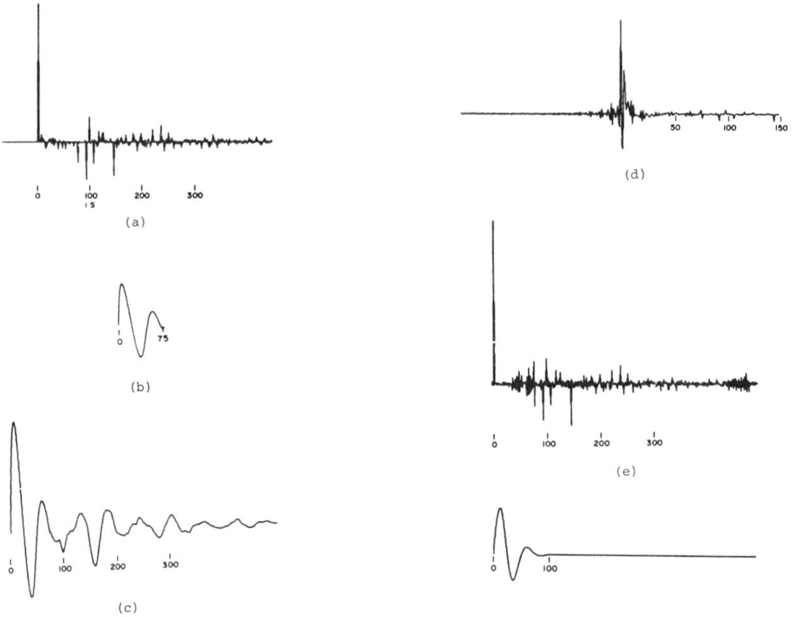

Fig. 4. Homomorphic deconvolution of a synthetic teleseismic effect (after Ulrych) including (a) theoretical impulse response of crust near Leduc, Alberta (after O. Jensen); (b) an assumed seismic wavelet; (c) synthetic seismogram; (d) complex cepstrum of the trace of Figure 4c exponentially weighted with α = 0.985; (e) high-pass output; and (f) low-pass output.

response of the earth's crust, 4b the assumed seismic wavelet, and 4c the synthetic seismogram corresponding to the convolution of 4a and 4b. The complex cepstrum was computed with an exponential weighting of 0.985, and is shown in Figure 4d. Finally, Figures 4e and 4f show the results of high-time and low-time gating of the complex cepstrum. The cutoff time used as 25 samples. In Figure 5 is shown a similar example with teleseismic data. In Figure 5a is shown the original event, after low-pass filtering with a cutoff frequency of three Hertz to reduce the effects of additive noise. Figures 5b and 5c show the recovered wavelet obtained from the low-time portion of the complex cepstrum and the deconvolved crustal impulse response resulting from the long-time portion of the complex cepstrum. The cutoff time used was 16 data samples (at a sampling rate of 20 samples per second) and an exponential weighting factor of 0.965 was applied to the original data. In Figure 5d is

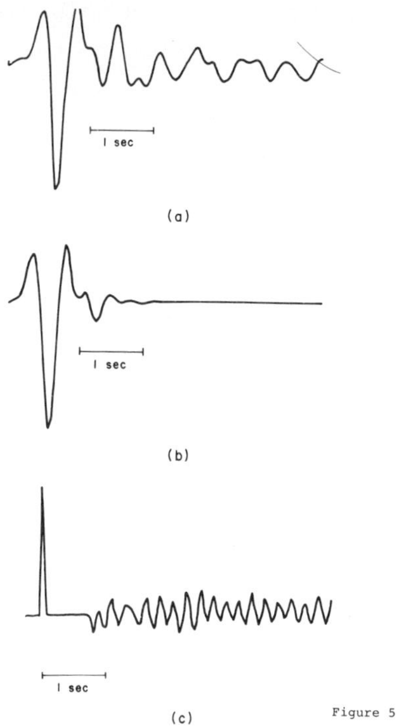

Fig. 5. Homomorphic deconvolution of a real teleseismic event (after Ulrych) including (a) original event after low-pass filtering; (b) recovered wavelet; and (c) recovered crustal impulse response.

shown the convolution of the recovered wavelet with the impulse response which had been computed for the site based on P wave velocities and formation thicknesses from well logs and crustal study data.

The use of homomorphic deconvolution by cepstral gating also has been applied to seismic data by Buhl, Stoffa, and Bryan[(R3)]. Their work focused on shallow-water marine seismic data, obtained from a single free-firing air gun and a single channel detector streamer. For data of this type, the source pulse corresponding to the air gun signature exhibits a bubble pulse periodicity. This source pulse is then convolved with the impulse response of the water column which consists of equally spaced impulses and then finally with the reflector series characterizing the structure beneath the water column. Thus, in this case, the components to be removed in the deconvolution are the air gun pulse with its associated bubble pulse periodicity, convolved with the impulse response of the water column.

Both the water column reverberation and the bubble pulse periodicities each generate an infinite train of decaying impulses in the complex cepstrum. As shown by Buhl et al., long-time gating the complex cepstrum of such a signal with a cutoff time greater than the first N complex cepstrum contributions will eliminate the first N multiples in the original reverberation pattern and will reduce the remaining multiples to at most 1/(N+1) of their original value. This then provides guidance in choosing an appropriate value for the cutoff time. Thus, for example, with a water depth of 60 meters, the water column has its first two complex cepstrum locations at .08 and .16 seconds. Additionally, with an air gun whose spectrum peaks at 18 Hz, the period of the air gun is .055 seconds so that the first three complex cepstrum contributions of the bubble pulse occur at .055, .11, and .165 seconds. By choosing the cutoff time of the long-time cepstral gate to be greater than .2, the first two water column multiples are eliminated and the remainder are reduced to at most one-third of their original amplitude. Also, the first three bubble pulse oscillations are eliminated and the remainder are attenuated by a factor of four. From the above discussion, we see that as the water depth increases, the effect of the water column reverberations becomes more significant, and if the water column is too deep, the cepstral cutoff time required may be so large as to adversely affect the desired reflector series.

For the data examined by Buhl et al., the water depth was 30 meters, and consequently the cepstral cutoff time was dictated primarily by the bubble periodicities. The bubble pulse oscillation for their data had a period of 0.055 seconds and, consequently, they chose a cepstral cutoff time of 0.176 seconds, corresponding to 44 points. In their processing, they also applied bandpass filtering to the output trace. This filtering was in fact accomplished by <u>adding</u> to the gated complex cepstrum, the cepstrum of the desired bandpass filter. Since addition in the cepstral domain gets transformed to convolution, in obtaining the output trace, this provides a convenient way of implementing the filtering. In their work, an exponential weighting of .998 was used. Their results are illustrated in Figure 6, with 10 unprocessed traces on the left-hand side and 10 processed traces on the right-hand side. The processing includes deconvolution and bandpass filtering using the

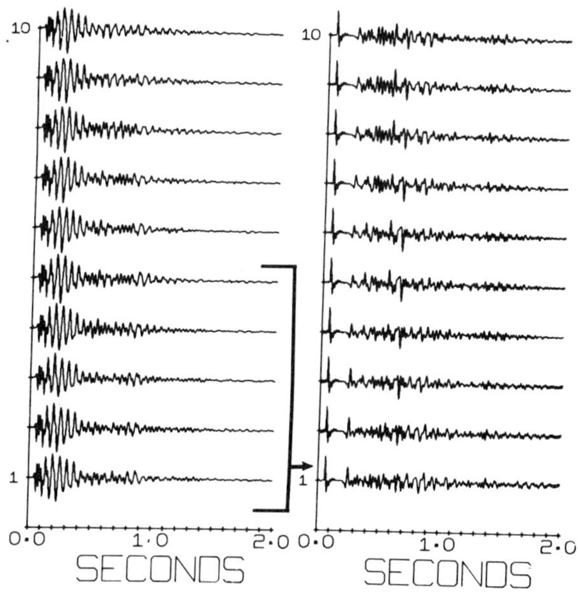

Fig. 6. Homomorphic deconvolution of marine seismic data (after Buhl et al.). On the left are the unprocessed traces and on the right the processed traces.

complex cepstrum, programmed gain control, and mixing. The programmed gain control is a linear increase in gain of 1 to 10 from 0 to 1 second, and a constant gain of 10 thereafter. They used an equal weight summing mix of five processed traces as indicated in the figure. In Figure 7 are shown 125 original traces and 120 processed traces in profile fashion.

Gallemore[(R5)] has attempted to evaluate the use of homomorphic deconvolution of marine seismic data and to compare its performance to the use of an optimal tapped delay line

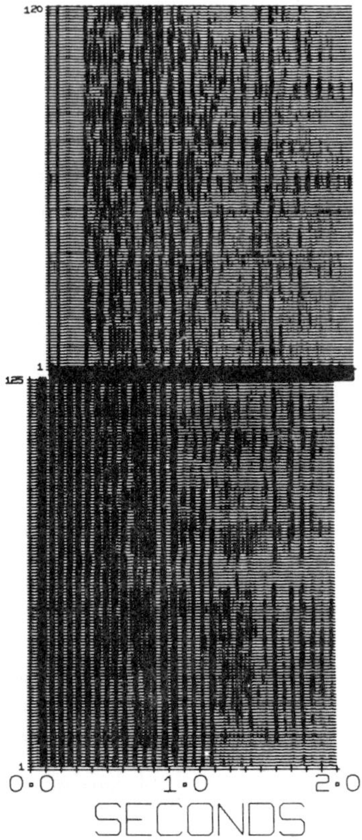

Fig. 7. Results of homomorphic deconvolution of marine seismic data displayed in profile fashion (after Buhl et al.).

filter. The criteria for evaluation were the percent of multiple energy removed, percent of reflector distortion, and visual improvement of the data when both methods were applied to synthetic seismic data. The indications from the study were that homomorphic deconvolution appears to have greater potential in shallow-water applications while the tapped delay line filter appears to be more efficient for deep-water data.

VI. Homomorphic Filtering and the Design of Linear Inverse Filters.

In the above discussion, the approach has been to use homomorphic deconvolution directly, to obtain the reflector series. Such procedure applies whenever the cepstra of the signal components occupy, approximately, disjoint time intervals. In some cases, however, this assumption is not justified. An example is shown in Figure 8 in which the average amplitude apectra of all readings on an event from Western Australia is shown. As discussed by Clayton and Wiggins[(R6)], it is clear from this data that the source pulse has a sharp antiresonance corresponding to a zero near the unit circle. Due to this zero, the complex cepstrum of the source will have a long duration and, in particular, will tend to strongly overlap with the complex cepstrum of the reflector series.

A second example in which the complex cepstra of the source pulse and reflector series will tend to overlap has been reported by Otis and Smith[(R7)], in attempting to enhance the resolution of a suite of reflections corresponding to a given depth interval. In their example, the reflector series is mixed phase and, consequently, its complex cepstrum may exhibit both low-time and high-time components. Obviously, examples may occur in which there are both effects, i.e., a tendency for the complex cepstrum of the source pulse to be long due to sharp resonances or antiresonances, and a tendency for the complex cepstrum of the reflector series to exhibit low-time components.

One approach which can be used in the first case, where the complex cepstra of the two components overlap, but the reflector series contributes only high-time components, is to use low-time gating of the complex cepstrum to obtain an initial estimate of the source pulse. This estimate can then be used to design a Wiener inverse filter with which the input data can then be filtered. This approach as been suggested by Treitel and Wang[(R8)]. In Figure 9 is shown a comparison between this approach and the use of a unit prediction distance Wiener deconvolution of the input data, followed by digital bandpass filtering in the range of 10 to 50 Hz. For the results obtained using homomorphic filtering followed by Wiener filtering, a cepstral cutoff time $n_c = 30$ msec was used. The initial source pulse estimate obtained by low-time cepstral gating also can be used in the design of

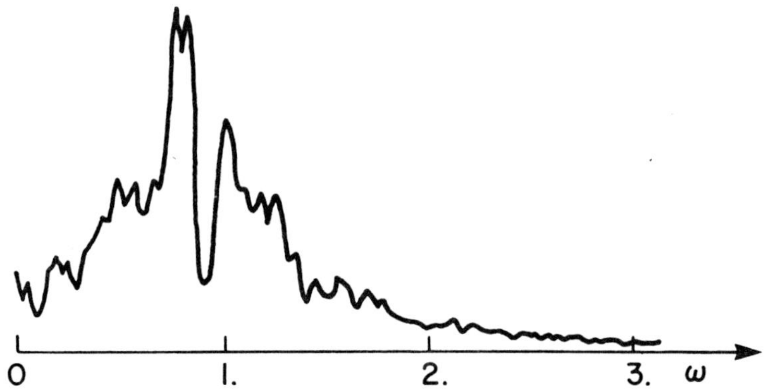

Fig. 8. Average amplitude spectrum of all readings on an event from Western Australia (after Clayton and Wiggins).

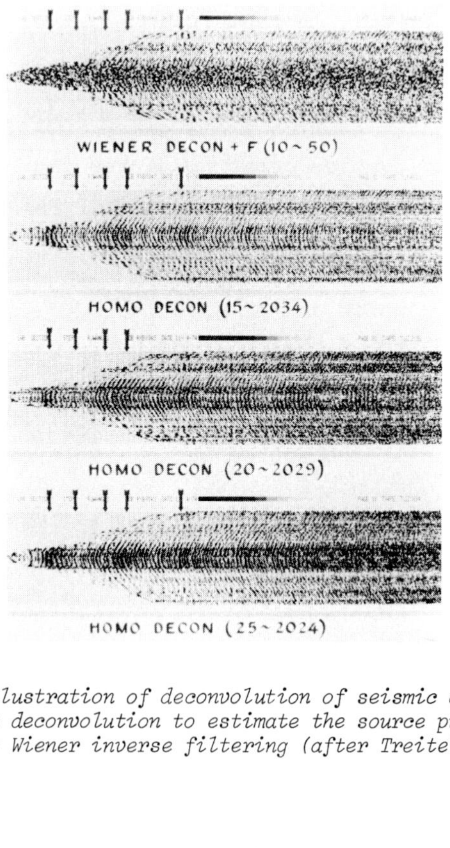

Fig. 9. Illustration of deconvolution of seismic data using homomorphic deconvolution to estimate the source pulse, followed by Wiener inverse filtering (after Treitel and Wang).

other types of filters. The use of homomorphic predictive filters is one alternate approach presently being investigated[R9].

Another approach which appears to be effective more generally when the complex cepstra of the two components overlap is a procedure referred to as cepstral stacking. This method assumes that we have available a suite of traces, all with the same source pulse but different reflector series. The kth seismic trace $s_k[n]$ is given by

$$s_k[n] = p[n] * r_k[n] \qquad k = 1,\ldots,N$$

where $p[n]$ is the source pulse and $r_k[n]$ is the reflector series corresponding to the kth trace. Consequently, the complex cepstrum of the kth trace is given by

$$\hat{s}_k[n] = \hat{p}[n] + \hat{r}_k[n] \ .$$

Cepstral averaging yields the following:

$$\langle\hat{s}[n]\rangle = \frac{1}{N}\sum_{k=1}^{N} \hat{s}_k[n] = \hat{p}[n] + \frac{1}{N}\sum_{k} \hat{r}_k[n] \ .$$

The successful recovery of the cepstrum of the source pulse depends upon whether the second term on the right-hand side of the above equation, that is $\langle\hat{r}[n]\rangle$, tends to be zero. Noting that a sequence consisting of impulses with random arrival times has a complex cepstrum consisting also of impulses with random arrival times, one is tempted to conclude that given a set of random sequences, their cepstral average will tend to be zero as the number of sequences increases.

The statistics of such averaging procedures have been established only for the real cepstrum[R10]. However, experiments with real and synthetic data suggest that the above conjecture also is true for the complex cepstrum. From the point of view of seismic data processing, the requirement that the sequences $\hat{r}_k[n]$ be random amounts to selecting different suites of reflections for each channel.

To illustrate cepstral averaging, a set of cepstra arising from random reflector series has been averaged and is depicted in Figure 10a. In Figure 10b, a set of cepstra arising from a plane layered structure is averaged and for this situation the average does not, obviously, tend to be zero.

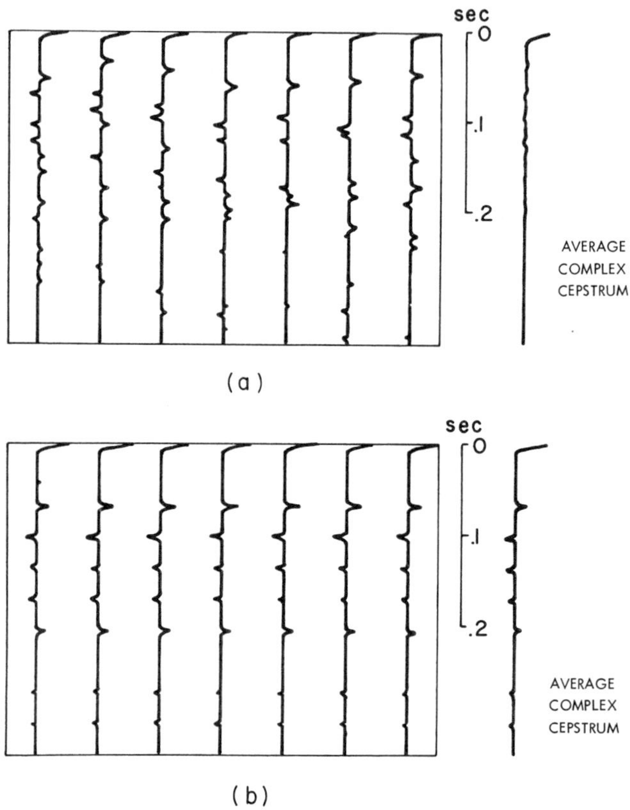

Fig. 10. Complex cepstra of idealized impulse responses including (a) in a geologically complex earth; and (b) in a flat-layered earth (after Otis and Smith).

Conclusions.

In the above discussions, we have tried to outline what appear to be the principal considerations in the use of homomorphic filtering for deconvolution of seismic data. Preliminary results indicate that this approach may have considerable potential, particularly when used in conjunction with other methods, such as the Wiener inverse filter design, etc. However, there are many issues which remain to be resolved and many refinements yet to be explored. Many of these are suggested by the way in which homomorphic filtering has been applied in other areas. In the attached bibliography, the application of homomorphic filtering to other areas is included, and the interested reader is encouraged to explore these.

References

R1 Treitel, S., and E. A. Robinson, "The Design of High-Resolution Digital Filters," *IEEE Trans. Geoscience Electronics GE-4*, 1, June, 1966.

R2 Tribolet, J. M., "A New Phase Unwrapping Algorithm" to be published in the *IEEE Trans. Acoustics, Speech and Signal Processing*.

R3 Stoffa, P. L., P. Buhl, and G. M. Bryan, "The Application of Homomorphic Deconvolution to Shallow-Water Marine Seismology. Part I: Models. Part II: Real Data," *Geophysics 39*, 4, 401-426, August, 1974.

R4 Ulrych, T., "Application of Homomorphic Deconvolution to Seismology," *Geophysics 36*, 4, 650-660, August, 1971.

R5 Gallemore, J. B., "A Comparative Evaluation of Two Acoustic Signal Dereverberation Techniques," S.M. and E.E. Thesis, Massachusetts Institute of Technology, June, 1976.

R6 Clayton, R. W., and R. A. Wiggins, "Source Shape Estimation and Deconvolution of Teleseismic Bodywaves," submitted to *Geophysics*.

R7 Smith, R. B., and R. M. Otis, "Homomorphic Deconvolution by Log Spectral Averaging," submitted to *Geophysics*.

R8 Treitel, S., private communication on homomorphic deconvolution.

R9 Tribolet, J. M., A. V. Oppenheim, and G. E. Kopec, "Deconvolution by Homomorphic Prediction," presented at the 45th Annual International Meeting of the Society of Exploration Geophysicists, Denver, Colorado, October 12-16, 1975, and at the 29th Annual Midwestern Exploration Meeting, Dallas, Texas, March 7-9, 1976.

R10 Stockham, T. G., Jr., T. M. Cannon, and R. B. Ingebretsen, "Blind Deconvolution through Digital Signal Processing," *IEEE Proc. 63*, 4, 678-692, April, 1975.

Bibliography

General Theory

1. Bogert, B. P., M. J. Healy, and J. W. Tukey, "The Quefrency Analysis of Time Series for Echoes: Cepstrum, Pseudo-Autocovariance, Cross-Cepstrum, and Saphe Cracking," in M. Rosenblatt (Ed.), *Time Series Analysis*, John Wiley and Sons, Incorporated, New York, 209-243, 1963.
2. Oppenheim, A. V., "Nonlinear Filtering of Convolved Signals," *Quarterly Progress Report* 54, Massachusetts Institute of Technology, 168-175, January, 1965.
3. Oppenheim, A. V., "Superposition in a Class of Nonlinear Systems," *Technical Report* 432, Massachusetts Institute of Technology, March, 1965.
4. Oppenheim, A. V., "Optimum Homomorphic Filters," *Quarterly Progress Report* 77, Massachusetts Institute of Technology, 248-260, April 15, 1965.
5. Bogert, B. P., and J. F. Ossanna, "The Heuristics of Cepstrum Analysis of a Stationary Complex Echoed Gaussian Signal in Stationary Gaussian Noise," *IEEE Trans. Inform. Theory IT-12*, 3, 373-380, July, 1966.
6. Oppenheim, A. V., "Generalized Superposition," *Inform. Control 11*, 5-6, 528-536, November-December, 1967.
7. Oppenheim, A. V., R. W. Schafer, and T. G. Stockham, Jr., "Nonlinear Filtering of Multiplied and Convolved Signals," *Proc. IEEE 56*, 8, 1264-1291, August, 1968.
8. Oppenheim, A. V., "Generalized Linear Filtering," *Digital Processing of Signals*, McGraw-Hill Book Company, 1969.
9. Schafer, R. W., "Echo Removal by Discrete Generalized Linear Filtering," *Technical Report* 466, Massachusetts Institute of Technology, February, 1969. Also Ph.D. Thesis, Massachusetts Institute of Technology, February, 1968.

10. Noll, A., "The Cepstrum and Some Close Relatives," *Signal Processing*, Academic Press, London, 11-22, 1973.

11. Stoffa, P. L., P. Buhl, and G. M. Bryan, "Cepstrum Aliasing and the Calculation of the Hilbert Transform," *Geophysics 39*, 4, 543-544, August, 1974.

12. Oppenheim, A. V., G. Kopec, and J. Tribolet, "Signal Analysis by Homomorphic Prediction," *IEEE Trans. Acoustics, Speech and Signal Processing ASSP-24*, 4, 327-332, August, 1976.

13. Skinner, D. P., and D. G. Childers, "Real-Time Composite Signal Decomposition," to be published in the *IEEE Trans. Acoustics, Speech and Signal Processing*.

14. Tribolet, T. M., "A New Phase Unwrapping Algorithm," to be published in the *IEEE Trans. Acoustics, Speech and Signal Processing*.

15. Childers, D. G., and D. P. Skinner, "The Cepstrum: A Guide to Processing," manuscript submitted for publication.

Seismic Processing

16. Cohen, T., "Source-Depth Determinations using Spectral Pseudo-Autocorrelation and Cepstral Analysis," *Geophys. J. Royal Astro. Soc. 20*, 223-231, 1970.

17. Ulrych, T., "Application of Homomorphic Deconvolution to Seismology," *Geophysics 36*, 4, 650-660, August, 1971.

18. Ulrych, T., O. G. Jensen, R. M. Ellis, and P. G. Sommerville, "Homomorphic Deconvolution of Some Teleseismic Events," *Bull. Seismological Soc. Amer. 62*, 5, 1253-1265, March, 1972.

19. Stoffa, P. L., P. Buhl, and G. M. Bryan, "The Application of Homomorphic Deconvolution to Shallow-Water Marine Seismology. Part I: Models," *Geophysics 39*, 4, 401-416, August, 1974.

20. Buhl, P. P. L. Stoffa, and G. M. Bryan, "The Application of Homomorphic Deconvolution to Shallow-Water Marine Seismology. Part II: Real Data," *Geophysics 39*, 4, 417-426, August, 1974.

21. Sommerville, P. G., "Time Domain Determination of Earthquake Fault Parameters from Short-Period P-Waves," Ph.D. Thesis, British Columbia University, December, 1975.

22. Shensa, M. J., "Complex Exponential Weighting applied to Homomorphic Deconvolution," *Geophys. J. Royal Astro. Soc. 44*, 379-387, 1976.

23. Stone, D. G., "Almost Homomorphic Filters," presented at the 29th Annual Midwest Society of Exploration Geophysicists Meeting, Dallas, Texas, March 7-9, 1976.

24. Smith, R. B., and R. M. Otis, "Homomorphic Deconvolution by Log Spectral Averaging," submitted to *Geophysics*.

25. Tribolet, J. M., A. V. Oppenheim, and G. E. Kopec, "Deconvolution by Homomorphic Prediction," presented at the 45th Annual International Meeting of the Society of Exploration Geophysicists, Denver, Colorado, October 12-16, 1975, and at the 29th Annual Midwestern Exploration Meeting, Dallas, Texas, March 7-9, 1976.

26. Clayton, R. W., and R. A. Wiggins, "Source Shape Estimation and Deconvolution of Teleseismic Bodywaves," submitted to *Geophysics*.

27. Treitel, S., private communication on homomorphic deconvolution.

Speech Processing

28. Noll, A., "Short-Time Spectrum and 'Cepstrum' Techniques for Vocal-Pitch Detection," *J. Acoust. Soc. Amer. 36*, 296-302, February, 1964.

29. Noll, A., "Cepstrum Pitch Determination," *J. Acoust. Soc. Amer. 41*, 2, 293-309, February, 1967.

30. Oppenheim, A. V., and R. W. Schafer, "Homomorphic Analysis of Speech," *IEEE Trans. Audio Electroacoust. AU-16*, 2, 221-226, June, 1968.

31. Noll, A., "Clipstrum Pitch Determination," *J. Acoust. Soc. Amer. 44*, 6, 1585-1591, December, 1968.

32. Oppenheim, A. V., "A Speech Analysis-Synthesis System based on Homomorphic Filtering," *J. Acoust. Soc. Amer. 45*, 458-465, February, 1969.

33. Schafer, R. W., and L. R. Rabiner, "System for Automatic Formant Analysis of Voiced Speech," *J. Acoust. Soc. Amer. 47*, 2, Pt. 2, 634-648, February, 1970.

34. Stockham, T. G., Jr., "Restoration of Old Acoustic Recordings by Means of Digital Signal Processing," Preprint, 41st Convention, Audio Engineering Society, New York, October, 1971.

35. Kopec, G. E., A. V. Oppenheim, and J. M. Tribolet, "Speech Analysis by Homomorphic Prediction," to be published in the *IEEE Trans. Acoustics, Speech and Signal Processing*.

Picture Processing

36. Stockham, T. G., Jr., "Image Processing in the Context of a Visual Model," *Proc. IEEE 60*, 7, 828-842, July, 1972.

37. Cole, E. R., "The Removal of Unknown Image Blurs by Homomorphic Filtering," Utah University Computer Science Department, UTEC-CSc-74-029, June, 1973.

38. Cannon, T. M., "Digital Image Deblurring by Nonlinear Homomorphic Filtering," Utah University Computer Science Department, UTEC-CSc-74-091, August, 1974.

39. Ingebretsen, R. B., "Log Spectral Estimation for Stationary and Nonstationary Processes," M. S. Thesis, Utah University, 1975.

40. Stockham, T. G., Jr., T. M. Cannon, R. B. Ingebretsen, "Blind Deconvolution through Digital Signal Processing," *Proc. IEEE* 63, 678-692, April, 1975. (refer to for other related reports and Master's theses)

Other

41. Halpeny, O. S., "Epoch Detection by Cepstrum Analysis," M. S. Thesis, Florida University, 1970.

42. Prabhakar, J., and S. C. Gupta, "Separation of Rayleigh and Poisson Density Functions through Homomorphic Filtering," *Proc. National Elec. Conf.* 605-610, December, 1970.

43. Kemerait, R. C., "Signal Detection and Extraction by Cepstrum Techniques," Ph.D. Dissertation, Florida University, 1971.

44. Senmoto, S., "Adaptive Decomposition of Composite Signals in Noise," Ph.D. Dissertation, Florida University, 1971.

45. Kemerait, R., and D. G. Childers, "Composite Signal Decomposition by Cepstrum Techniques," *IEEE Region 3 Convention Record* K1-1 - K1-4, 1972.

46. Senmoto, S., and D. G. Childers, "Analysis of a Composite Signal by Complex Cepstrum and Adaptive Filter," *Trans. Inst. Elec. Commun. Engrs.* Part A, 9-16, 1972.

47. Senmoto, S., and D. G. Childers, "Adaptive Decomposition of a Composite Signal of Identical Unknown Wavelets in Noise," *IEEE Trans. Sys., Man., and Cyber. SMC-2*, 59-66, January, 1972.

48. Kemerait, R., and D. G. Childers, "Detection of Multiple Echoes Immersed in Noise," *Proc. 15th Midwest Symp. on Circuit Theory* 1-10, May 4-5, 1972.

49. Kemerait, R., and D. G. Childers, "Signal Detection and Extraction by Cepstrum Techniques," *IEEE Trans. Inform. Theory IT-18*, 6, 745-759, November, 1972.

50. Kemerait, R., and D. G. Childers, "Decomposition of Pulse-Type Data by Cepstrum Techniques," *Proc. IEEE Southeast Conf.* F4-1 - F4-4, April 30 - May 2, 1973.

51. Tenold, J., D. H. Crowell, R. H. Jones, T. H. Daniel, D. F. McPherson, and A. N. Popper, "Cepstral and Stationarity Analysis of Full-Term and Premature Infants' Cries," *J. Acoust. Soc. Amer. 56*, 3, 975-980, September, 1974.

WAVES PROPAGATING IN RANDOM MEDIA AS STATISTICAL TIME SERIES

Enders A. Robinson
Consultant for Amoco Production Company
Tulsa, Oklahoma

Abstract.

Wave propagation in an inhomogeneous medium may be studied by the familiar device of replacing a generally continuous function by a step function. In this way, the medium is divided into a set of strata with constant parameters in each stratum. The amplitudes of waves propagating in the stratified medium are normalized so that the square of wave amplitude gives energy. Alternatively, one can consider wave motion as a particle process where square of wave amplitude gives the probability of the particle. The total energy (or probability) injected in the system is unity, and the flow of this energy through time and space is followed as reactive energy is transformed into reflected and transmitted energy. In those cases in which the reflection coefficients of the interfaces are small, the reflected wave is equal to the convolution of a minimum-delay reverberation wavelet with the reflection coefficient sequence. In many physical problems in which large amounts of data are handled, the reflection coefficient sequence may be considered as a realization of a random process, in which case the reflected wave is the corresponding realization of a time series generated by a statistical minimum-delay model. In the seismic exploration for petroleum and natural gas, the reflection coefficient series represents the reflectivity of the earth's crust and so can be used to make contour maps of the earth's structure at depth. Empirically, it is known that sections of the earth's crust have uncorrelated reflection coefficient sequences, and as a result the reflectivity

ISBN 0-12-257250-5

of the deep earth's crust can be obtained by deconvolution of reflection seismograms recorded on the surface.

The Diffusion Equation and the Wave Equation.

In many problems in physics and engineering, traveling waves are studied because it is important to pinpoint the time for a disturbance to propagate from one place to another. Traveling wave phenomena appear in signal and power transmission, mechanics, acoustics, seismology, thermal conduction, diffusion, and in many other areas of science and engineering.

The first-order differential equations for electrical transmission lines are called the telegrapher's equations, which are

$$\frac{\partial v}{\partial x} = - Ri - L \frac{\partial i}{\partial t} , \quad \frac{\partial i}{\partial x} = - Gv - C \frac{\partial v}{\partial t}$$

where v is voltage across the transmission line, i is current in the transmission line, x is distance on the line, R is series resistance per unit length, L is series inductance per unit length, G is shunt conductance per unit length, and C is shunt capacitance per unit length.

Both of these equations contain loss terms, namely Ri and Gv, and energy storage terms, namely $L\, \partial i/\partial t$ and $C\, \partial v/\partial t$. For different circumstances, these equations can be specialized. There are two particularly important cases, namely (1) the diffusion case in which the series inductance and shunt conductance of the electric transmission line are neglected (i.e., L = 0 and G = 0) and (2) the lossless case in which loss on the transmission line is neglected (i.e., R = 0 and G = 0). In the first case, the two telegrapher's equations may be combined to give the second-order partial differential equation, called the diffusion equation,

$$\frac{\partial^2 v}{\partial x^2} = RC \frac{\partial v}{\partial t}$$

where 1/RC is the diffusion constant. This equation represents the case of diffusion waves. In the second case, the two telegrapher's equations may be combined to give the second-order partial differential equation, called the <u>wave equation</u>,

$$\frac{\partial^2 v}{\partial x^2} = LC \frac{\partial^2 v}{\partial t^2}$$

where $1/\sqrt{LC}$ is the velocity of propagation. This equation represents the case of lossless waves. The characteristic impedance $Z = \sqrt{L/C}$ also is defined.

Equations for waves in other branches of science can be obtained by use of analogies. That is, the telegrapher's equations have analogs in nearly all the other wave phenomena commonly studied. For example, one of the telegrapher's equations for heat conduction is Fourier's law of heat conduction, and the other is a continuity equation for heat based upon the principle of storage of heat and heat capacity. These equations for heat conduction correspond to the electric transmission line in the diffusion case, so the resulting equation is the diffusion equation. On the other hand, the telegrapher's equations for various types of acoustic waves are based on Newton's second law and D'Alembert's principle, and on the principle of conservation of mass which results in a continuity equation. These acoustic equations correspond to the electric transmission line in the lossless case, so the resulting equation is the wave equation.

As it turns out, many of the wave phenomena commonly studied fall into one of these two cases, i.e., diffusion (with the diffusion equation) or lossless (with the wave equation). For example, by using the analogy of longitudinal particle velocity and longitudinal stress for voltage and current, we obtain the wave equation for sound waves in solids. Because of such analogies, we will describe our variables in electric terms, and the interested reader may then use the corresponding analogy to convert the variables into the terms of the particular phenomena at hand.

Furthermore, for the rest of this paper, we will be concerned only with lossless waves as described by the wave equation.

Stratified Medium.

In setting up the mathematical model of a physical system, we use the familiar method of replacing an arbitrary function by a step function. Consider the lower half-space where $x = 0$ denotes the surface and positive x is measured vertically down. A function $c(x)$ is specified for each value of x. Instead of the continuous dependence of c on x, we think of the half-space as being divided into a set of horizontal strata, with $c(x)$ having a constant value in each stratum.

We write the wave equation as

$$\frac{\partial^2 v}{\partial x^2} = \frac{1}{c^2} \frac{\partial^2 v}{\partial t^2}$$

where $c = c(x)$ is a positive constant for the stratum in question. The wave equation is one of the few partial differential equations whose general solution can be found explicitly. To do so, we introduce new coordinates of $\eta = x - ct$ and $\xi = x + ct$. Then, by the chain rule, we have

$$\frac{\partial^2 v}{\partial x^2} = \frac{\partial^2 v}{\partial \xi^2} + 2 \frac{\partial^2 v}{\partial \xi \partial \eta} + \frac{\partial^2 v}{\partial \eta^2}$$

$$\frac{\partial^2 v}{\partial t^2} = c^2 \left[\frac{\partial^2 v}{\partial \xi^2} - 2 \frac{\partial^2 v}{\partial \xi \partial \eta} + \frac{\partial^2 v}{\partial \eta^2} \right]$$

so the wave equation becomes

$$\frac{\partial^2 v}{\partial \xi \partial \eta} = 0 \ .$$

This equation is satisfied if and only if

$$v = v_d(\eta) + v_u(\xi)$$

where v_d and v_u are arbitrary waveforms. That is, the solution of the wave equation is of the form

$$v(x, t) = v_d(x - ct) + v_u(x + ct)$$

which is due to D'Alembert in 1747. For constant propagation velocity c, the voltage and current each propagate along the x-axis as a wave. The waveform $v_d(x - ct)$ propagates in the plus x-direction and represents a downgoing wave. The waveform $v_u(x + ct)$ propagates in the minus x-direction and represents an upgoing wave. From the telegrapher's equations, it follows that the voltage for each waveform is proportional to the current of that waveform and the constant of proportionality is just the characteristic impedance Z. Calling i_d and i_u the downgoing and upgoing current waveforms respectively, we have

$$v_d = Zi_d \ , \quad v_u = - Zi_u \ .$$

Thus, voltage and current may be expressed in terms of downgoing and upgoing waves as

$$v = v_d + v_u = Z(i_d - i_u)$$

$$i = i_d + i_u = \frac{1}{Z} (v_d - v_u) \ .$$

Scattering Variables.

The choice of an appropriate formalism is particularly important in studies of time series generated by a stratified system. The scattering quantities represent a set that is particularly appropriate in dealing with a stratified system. While the conventional quantities are voltage and current, the quantities employed in the scattering formalism are linear combinations of voltage and current. More particularly, the scattering quantities are the downgoing and upgoing waves.

However, instead of choosing the scattering quantities in terms of either voltage or current, it is more convenient to make the unit (voltage · current)$^{1/2}$. Thus, we write

$$i\sqrt{Z} = i_d\sqrt{Z} + i_u\sqrt{Z}$$

$$\frac{v}{\sqrt{Z}} = i_d\sqrt{Z} - i_u\sqrt{Z} \quad .$$

If we define normalized voltage $\tilde{v}$ and normalized current $\tilde{i}$ as

$$\tilde{v} = \frac{v}{\sqrt{Z}} \ , \quad \tilde{i} = i\sqrt{Z}$$

and the normalized downgoing waveform d and normalized upgoing waveform u as

$$d = i_d\sqrt{Z} \ , \quad u = i_u\sqrt{Z}$$

we have

$$\tilde{i} = d + u$$

$$\tilde{v} = d - u \quad .$$

The energy is voltage times current, which is

$$\text{Energy} = vi = \tilde{v}\tilde{i} \ .$$

In case of complex quantities, the energy should be written as

$$\text{Energy} = \text{Re}(vi^*) = \text{Re}(v^*i)$$

where Re stands for "real part" and the asterisk denotes complex conjugate.

In certain applications, the wave motion may be considered as representing a particle process where energy may be considered as the probability of the particle.

Reflection and Transmission Coefficients.

The subject of this section is the reflection and transmission of traveling waves at interfaces. We want to emphasize that the amplitude of a reflection is not just a property of the material in the stratum where the reflection occurs, but it is an interface property depending on the material on both sides of the interface. In fact, a thin layer of extraneous material on the interface beween two materials will change the reflection. As a result, in constructing the mathematical model of our physical system, we must make the thicknesses of the discrete strata sufficiently fine in order to account for all significant thin layers.

We depict distance x as increasing downward on a vertical axis and time t as increasing to the right on a horizontal scale. An interface (x = constant) is then represented by a horizontal line. All our wave motion is at normal incidence, that is, all waves move up and down vertically, but in order to include the dimension of time, a wave component at a particular place and time will be indicated by a vector in the x,t plane. The vector points downward and to the right for a downgoing wave and upward and to the right for an upgoing wave.

Let us now consider the case of an incident unit downgoing pulse on an interface. See Figure 1. Throughout this paper, we use the word "pulse" to indicate a spike (i.e., a sharply peaked waveform of very short time duration) which contains energy equal to the square of its amplitude. Thus, a unit pulse contains energy equal to unity squared which is one. We assume the incident pulse is at the interface (x = constant) at a given instant of time that defines the timing line (t = constant). The vector shows the direction of the incident pulse, but the circled ⊠ shows the location. More specifically, we assume that the incident pulse is slightly above the interface and slightly to the left of the timing line, that is, the incident pulse is in the second quadrant defined by the axes of the interface and timing line. This incident pulse of unit amplitude is converted by the

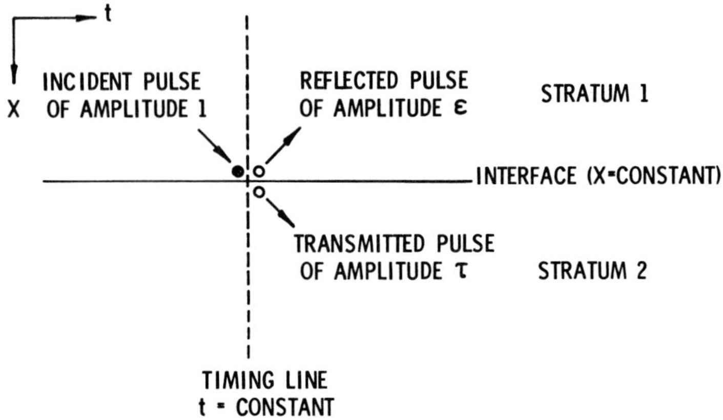

Fig. 1. Reflected and transmitted pulses due to an incident unit pulse.

interface into a reflected pulse of amplitude ε and a transmitted pulse of amplitude τ. Both the reflected and transmitted pulses occur at a split second (or split nanosecond or whatever) later than the incident pulse. Moreover, the reflected pulse is just above the interface while the transmitted pulse is just below the interface. In other words, the reflected pulse is in the first quadrant and the transmitted pulse is in the fourth quadrant. In Figure 1, the vectors show the space-time direction of these pulses and the small circles show their location. In summary, a downgoing incident pulse and the resulting reflected and transmitted pulses occur at a given interface at a given instant of time, but in actuality we must regard them as being within the proper quadrants formed by the interface and timing line. This convention is important because the energy in the incident pulse is divided into two parts, one part going into the reflected pulse and the other part going into the transmitted pulse. That is, the incident pulse appears at the interface slightly before the resulting reflected and transmitted pulses, even though the mathematical notation says that all three pulses appear at the interface at time t. More specifically, we should say that the incident pulse arrives at the interface at time $t - t_o$ and the resulting reflected and transmitted pulses emanate from the interface at time $t + t_o$, where t_o is a very small positive number.

All we have done so far is to describe the three pulses; our problem now is to work out the amplitudes of the reflected and transmitted pulses. The three pulses we have described satisfy the wave equation (or, alternatively, the telegrapher's equations) within their respective strata, but the telegrapher's equations must also be satisfied at the interface between the two strata. Thus, we must look at what happens right at the boundary. It will turn out that the telegrapher's equations demand that the three pulses fit together in a certain way.

From a purely physical argument, we can say that voltage must be the same on each side of an interface, which we describe by saying that voltage is continuous across the interface. Similarly, current must be the same on each side of the interface, or in other words, current must be continuous across the interface. In fact, continuity of voltage across an interface as well as continuity of current across the interface can be derived from the two telegrapher's equations, as we will now show.

Each telegrapher's equation holds in stratum 1 (above the interface) and in stratum 2 (below the interface). In addition, these equations also must hold in the interface, which we call region 1 - 2. Although we usually think of the interface as being sharply discontinuous, in reality it is not. The physical properties change very rapidly but not infinitely fast. In any case, we can imagine that there is a very rapid but continuous transformation of the characteristic impedance between stratum 1 and stratum 2 within the short distance we call region 1 - 2. Also, voltage and current will make a similar kind of transition in intermediate region 1 - 2. In this intermediate region, the telegrapher's equations will still be satisfied, and it is by following these equations in this region that we can arrive at the needed interface conditions.

Let us look at the first telegrapher's equation,

$$\frac{\partial v}{\partial x} = - L \frac{\partial i}{\partial t} .$$

The left-hand side involves the (partial) derivative of voltage with respect to distance. Suppose that the voltage jumped by some smooth but rapid transition from stratum 1 to stratum 2. (Our purpose here is to show that this supposition is wrong.) As a result of this supposition, the derivative $\partial v/\partial x$ will have some very large values in the intermediate region 1 - 2, because of the tremendous slope of v. In other words, the derivative $\partial v/\partial x$ will have a sharp peak at the interface. If we imagine squeezing the interface to even a thinner layer, the peak would get much highter. If the boundary is really sharp for the waves we are interested in, the magnitude of $\partial v/\partial x$ in the intermediate region will approach a Dirac delta function. As a result, the magnitude of $\partial v/\partial x$ in the intermediate region would be much greater than any contribution we might have from any other variation, so we ignore any variations other than those due to the boundary. We remember that voltage and current themselves do not get especially large at the interface; only the derivatives with respect to x can become so huge that they dominate the telegrapher's equations. But now let us look at the right-hand side of the above telegrapher's equation. The x-derivative is on the left-hand side. We have supposed that this x-derivative is huge. But there is nothing on the right-hand side to match the huge left-hand side. Therefore, our supposition is wrong; voltage cannot have any jump in going from stratum 1 to stratum 2, for if it did there would be a Dirac delta function on the left of the first telegrapher's equation but none on the right, and the equation would be false. As a result, we have the condition that voltages in stratum 1 and stratum 2 are equal at the interface: $v_1 = v_2$ at interface 1 - 2. By the same argument, the second telegrapher's equation,

$$\frac{\partial i}{\partial x} = - C \frac{\partial v}{\partial t} ,$$

shows that currents in stratum 1 and stratum 2 are equal at the interface: $i_1 = i_2$ at interface 1 - 2.

Let us now return to our incident, reflected, and transmitted pulses as shown in Figure 1. All three pulses occur at the interface, but in reality the unit incident pulse and the reflected pulse ε are just above the interface and so are in stratum 1; whereas, the transmitted pulse τ is just below the interface and so is in stratum 2. These pulses are all in terms of normalized units of (voltage times current)$^{1/2}$ so in order to apply our interface conditions, we must transform them to both voltage and current. In stratum 1, the incident unit pulse is the downgoing wave d_1 and the reflected pulse ε is the upgoing wave u_1, so the voltage is

$$v_1 = \sqrt{Z_1}\, \tilde{v}_1 = \sqrt{Z_1}(d_1 - u_1) = \sqrt{Z_1}(1 - \varepsilon)$$

and the current is

$$i_1 = \frac{\tilde{i}_1}{\sqrt{Z_1}} = \frac{1}{\sqrt{Z_1}} (d_1 + u_1) = \frac{1}{\sqrt{Z_1}} (1 + \varepsilon) .$$

In stratum 2, the transmitted pulse τ is the downgoing wave d_2, and the upgoing wave u_2 is zero, so the voltage is

$$v_2 = \sqrt{Z_2}\, \tilde{v}_2 = \sqrt{Z_2}\, d_2 = \sqrt{Z_2}\, \tau$$

and the current is

$$i_2 = \frac{\tilde{i}_2}{\sqrt{Z_2}} = \frac{d_2}{\sqrt{Z_2}} = \frac{\tau}{\sqrt{Z_2}} .$$

Here Z_1 and Z_2 are the characteristic impedances of strata 1 and 2, respectively. By continuity, we have $v_1 = v_2$ and $i_1 = i_2$ so

$$\sqrt{Z_1}\ (1 - \varepsilon) = \sqrt{Z_2}\ \tau$$

$$\frac{1}{\sqrt{Z_1}}\ (1 + \varepsilon) = \frac{\tau}{\sqrt{Z_2}} .$$

Solving for ε and τ, we find that

$$\varepsilon = \frac{Z_1 - Z_2}{Z_1 + Z_2} , \quad \tau = \frac{2\sqrt{Z_1 Z_2}}{Z_1 + Z_2} .$$

Because ε is the amplitude of the upgoing reflected pulse due to a downgoing incident pulse of unit amplitude, we call ε the (down-to-up) reflection coefficient of the interface between two strata with characteristic impedance Z_1 above and Z_2 below. Because τ is the amplitude of the downgoing transmitted pulse due to a downgoing incident unit pulse, we call τ the (down-to-down) transmission coefficient between the two strata.

Let us now consider an upgoing incident pulse of unit amplitude on the same interface, that is, with characteristic impedance Z_1 above and Z_2 below. Everything is the same as before, except now the roles down and up (and consequently Z_1 and Z_2 are interchanged) so the up-to-down reflection coefficient is $(Z_2 - Z_1)/(Z_2 + Z_1)$ which is $-\varepsilon$, and the (up-to-up) transmission coefficient is the same as before, namely τ.

Let us now consider energy. Again, we refer to Figure 1. We recall that amplitude is measured in units of $(\text{voltage} \cdot \text{current})^{1/2}$, so the square of the amplitude is in units of energy. The incident pulse of amplitude 1 is converted into a reflected pulse of amplitude ε and a transmitted pulse of amplitude τ. Because no energy is lost, the energy of the incident pulse must equal the sum of the energies of the reflected and transmitted pulses; that is, we have the law of conservation of energy given by

$$1 = \varepsilon^2 + \tau^2 .$$

Solving this equation for τ, we have

$$\tau = \sqrt{1 - \varepsilon^2} .$$

Hence, all the coefficients can be expressed as follows:

Down-to-up reflection coefficient $= \varepsilon$;
Down-to-down transmission coefficient $= \tau = \sqrt{1 - \varepsilon^2}$;
Up-to-down reflection coefficient $= -\varepsilon$;
Up-to-up transmission coefficient $= \tau = \sqrt{1 - \varepsilon^2}$.

Wave Propagation.

We now want to consider wave propagation in a stratified medium. Each wave is in fact a collection of pulses in space and/or time. Figure 2 shows the type of stratified model assumed, where the distance variable x varies in a positive direction downward and the time t variable varies in a positive direction to the right. Now let us divide the x coordinate into increments which define strata, and let us make the subdivision in such a way that all time events are synchronized. More specifically, we divide the x axis by means of discrete points x_n (where n is an integer) in the following way. The point x_n is called interface n, the point x_{n+1} is interface n+1, and the material between these two interfaces is called stratum n, with characteristic impedance Z_n and velocity c_n. Each stratum is chosen with such a thickness Δx_n so that the two-way travel time in the stratum is a constant Δt; that is,

$$\frac{2\Delta x_n}{c_n} = \Delta t = \text{constant} \quad (\text{where } \Delta x_n = x_{n+1} - x_n) .$$

In other words, given a fixed time increment Δt and the depth x_n of layer n, then the depth of layer n+1 is determined by the formula

$$x_{n+1} = x_n + \frac{c_n \, \Delta t}{2} .$$

It is convenient to choose a time scale so that our time interval is unity; that is, we choose $\Delta t = 1$.

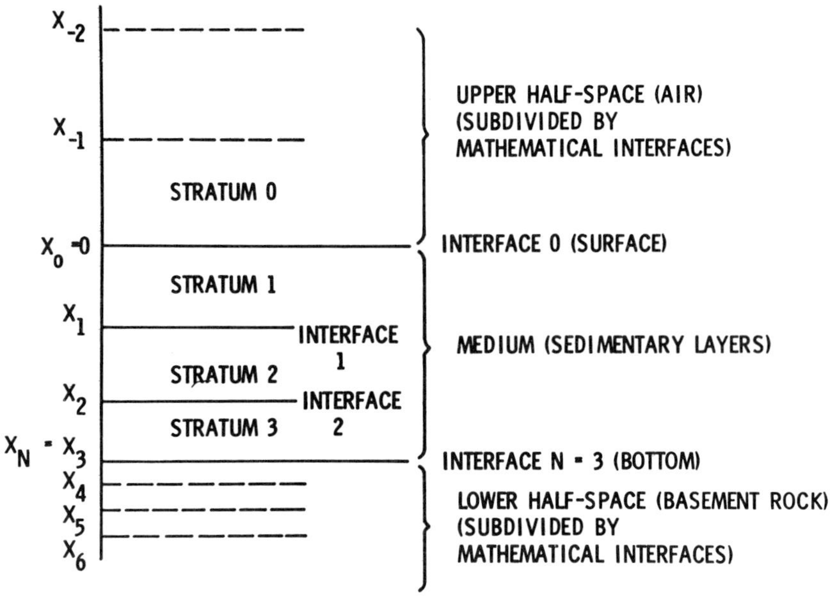

Fig. 2. Medium with three strata.

We want to consider a medium of finite thickness X which is sandwiched between two half-spaces. As we have just described, we divide the medium into N strata, where $x_0 = 0$ is the surface and $x_N = X$ is the bottom. The upper half-space has a certain characteristic impedance Z_0 and velocity c_0. With the velocity c_0, we can subdivide the upper half-space into strata. All of these upper half-plane strata have the same characteristic impedance Z_0 so the reflection coefficients of all the upper half-space interfaces are zero and transmission coefficients are one. Likewise, we can subdivide the lower half-space into such strata. We thus have the picture as shown in Figure 2.

As an input signal into the medium, we inject a unit downgoing pulse at zero time at the bottom of stratum 0. More precisely, the input pulse occurs just above interface 1 and just before time 0. The energy of this input spike is 1^2 or unity. This energy must be conserved. The energy in this input spike now propagates through time and space. Because it takes one-half time unit to travel through a layer, the energy must follow the ray paths as shown in Figure 3.

Upgoing waves are denoted by u and downgoing waves by d. Waves just above an interface are indicated by a prime; whereas, waves just below an interface are not tagged by a prime. A subscript indicates the stratum in which a wave occurs. Thus, the waves associated with interfaces n-1 and n are shown in Figure 4.

Let us now look at interface n at time t. Waves $d'_n(t)$ and $u_{n+1}(t)$ are inputs, and waves $u'_n(t)$ and $d_{n+1}(t)$ are outputs. Each output is the sum of transmitted and reflected parts of the inputs; that is,

$$\begin{aligned} d_{n+1}(t) &= \tau_n\, d'_n(t) - \varepsilon_n\, u_{n+1}(t) \\ u'_n(t) &= \varepsilon_n\, d'_n(t) + \tau_n\, u_{n+1}(t) \end{aligned} \tag{1}$$

where ε_n is the reflection coefficient and τ_n is the transmission coefficient.

Flow of Energy.

Let us now consider the time-flow of energy at interface n at time t as shown in Figure 4. Just before time t, the energy at the space-time point (n, t) is

$$d'^2_n(t) + u_{n+1}^2(t) \;;$$

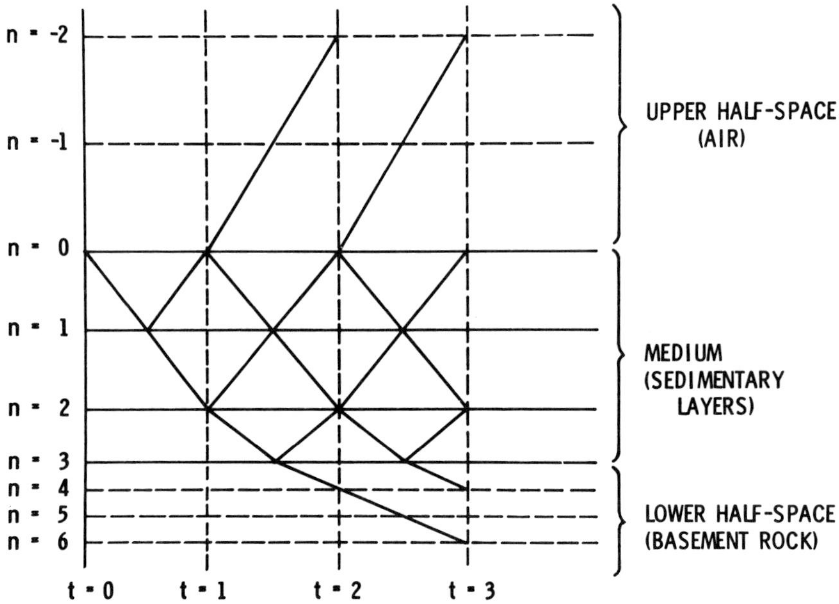

Fig. 3. Ray paths. Note that reflections can occur at real interfaces (shown by solid horizontal lines). The rays are synchronized at the interfaces, the rays crossing at integer times on even-numbered interfaces and at integer + 0.5 times on odd-numbered interfaces.

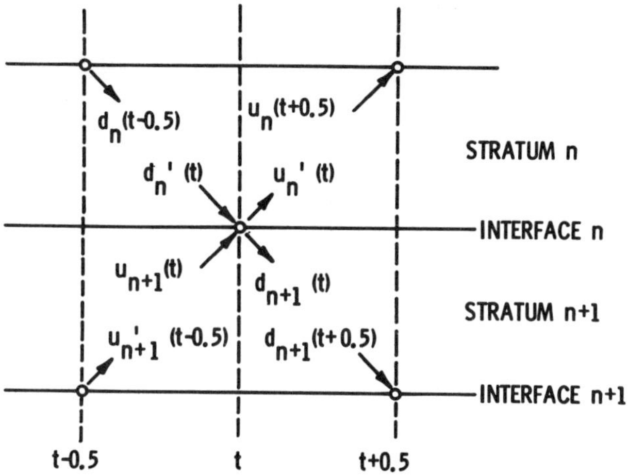

Fig. 4. Designation of waves. Note that t is an integer if n is even and t is an integer plus one-half if n is odd. The vectors of all the waves shown are in their correct quadrants with respect to interfaces and timing lines.

whereas, just after time t, the energy at space-time point (n, t) is

$$u_n'^2(t) + d_{n+1}^2(t) .$$

By the law of conservation of energy, the above two expressions should be equal, and indeed this is so, as can be directly verified by use of Eq. (1). The pulse $u_n'(t)$ travels up in stratum n without charge of form to become $u_n(t + 0.5)$; whereas, $d_{n+1}(t)$ travels down in stratum n+1, also without change of form to become $d_{n+1}'(t + 0.5)$. As a result, the above expressions for energy may be written as

$$d_n'^2(t) + u_{n+1}^2(t) = d_{n+1}'^2(t + 0.5) + u_n^2(t + 0.5) .$$

The left-hand side gives the total energy in strata n and n+1 just an instant before time t, and the right-hand side gives the total energy in strata n and n+1 just before time t + 0.5. If we sum each side of the above equation over all pairs of strata (including hypothetical strata in the upper and lower half-spaces), we get an equation that states that the total energy in all the strata just before time t is equal to the total energy in all the strata just before time t + 0.5. We have thus established the law of conservation of energy as energy flows with time. By recursive application of the law of conservation of energy, the energy at any time must be equal to the energy originally introduced into the system, namely one.

For ease of presentation, let us fix time t and the number N of strata, so we can get across the basic concept involved. Let t = 2 and N = 3, as shown in Figure 5.

The energy in the space waveform at time 2 is

$$u'^{2}_{-4}(2) + u'^{2}_{-2}(2) + u'^{2}_{0}(2) + d^{2}_{1}(2) + u'^{2}_{2}(2) + $$

$$+ d^{2}_{3}(2) + d^{2}_{5}(2) = 1$$

which is

$$[u'^{2}_{0}(0) + u'^{2}_{0}(1) + u'^{2}_{0}(2)] + [d^{2}_{1}(2) + u'^{2}_{2}(2) + $$

$$+ d^{2}_{3}(2)] + [d^{2}_{4}(1.5)] = 1 \ .$$

The three brackets in this expression show the division of the energy into three parts. The first bracket is the energy that has escaped by being reflected up and out of the medium as of time t. The second bracket is the energy that is still within the medium as of time t, so it is reactive energy. The third bracket is energy that has escaped by being transmitted down and out of the medium as of time t. The first bracket is the partial energy of the reflected time-waveform

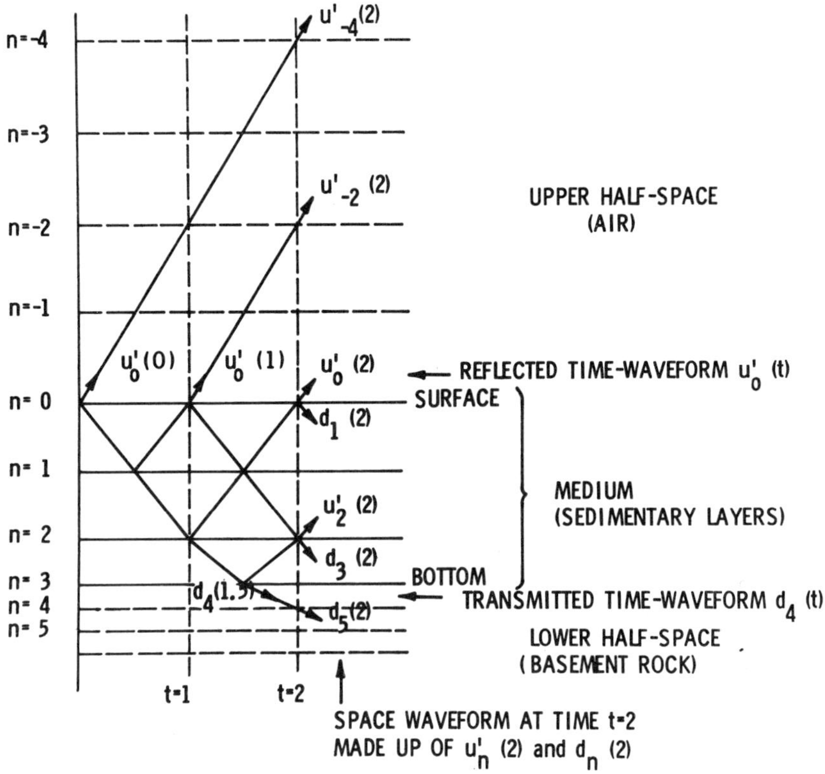

Fig. 5. The energy at $t = 2$ is the sum of squares of the pulses in the space-waveform. However, the space-waveform pulses above the surface are the same as the pulses in the reflected time-waveform up to and including time t, and the space-waveform pulses below the bottom are the same as the pulses in the transmitted time-waveform up to and including time t.

$u_0'(t)$ up to and including time t. The third bracket is the partial energy of the transmitted time-waveform $d_{N+1}(t)$ up to and including time t. As t tends to infinity, all the energy escapes and the second bracket tends to zero, the first bracket becomes the total energy in the reflected time-waveform, and the third bracket becomes the total energy in the transmitted time-waveform, i.e., for $t \to \infty$ we have

$$[u_0'^2(0) + u_0'^2(1) + u_0'^2(2) + u_0'^2(3) + \cdots] + [d_4^2(1.5) + $$

$$+ d_4^2(2.5) + \cdots] = 1 .$$

In the following sections, we further discuss the reflected wave $u_0'(t)$ and the transmitted wave $d_{N+1}(t)$ and introduce the special notation for them given by

$$u_0'(t) = y_t \qquad t = 0,1,2,...$$

$$d_{N+1}(t + \frac{N}{2}) = \tau\, b_t \qquad t = 0,1,2,...$$

where $\tau = \tau_0\tau_1\ldots\tau_N$ and $b_0 = 1$. Then the above equation for $t \to \infty$ is, in this new notation,

$$[y_0^2 + y_1^2 + y_2^2 + \cdots] + [\tau^2 + \tau^2\, b_1^2 +$$

$$+ \tau^2\, b_2^2 + \cdots] = 1 .$$

It can be established (Robinson and Treitel, 1976) that there is a constant net downgoing flow of energy in each layer which results in the fact that the downgoing wave in each layer is minimum delay.

Wave Propagation Across Interfaces.

A wave such as $d_{n+1}(t)$ considered as a function of time is actually a series of pulses of different amplitudes, each pulse in the time-wave occurring one time unit apart. A wave such as $d_{n+1}(t)$ considered a function of space is a series of pulses of different amplitudes, each pulse in the space-wave occurring at a distance of two strata apart (since a downgoing pulse in $d_{n+1}(t)$ travels through two strata per time unit).

The z-transform of the downgoing wave $d_{n+1}(t)$ is defined as

$$D_{n+1}(z) = \sum d_{n+1}(t)\ z^t$$

where the summation is for

$$t = \frac{n}{2},\ \frac{n}{2} + 1,\ \frac{n}{2} + 2,\ \ldots$$

Likewise, the z-transform of the upgoing wave $u_{n+1}(t)$ is

$$U_{n+1}(z) = \sum u_{n+1}(t)\ z^t$$

where the summation is for

$$t = \frac{n}{2} + 1,\ \frac{n}{2} + 2,\ \ldots$$

(since $u_{n+1}(n/2) = 0$). See Figure 6.

The wave $u_n'(t - 1/2)$ travels up in stratum n without change of form to become the wave $u_n(t)$. Since z is a unit delay operator (i.e., $zf(t) = f(t - 1)$ where $f(t)$ is any time function), we have

$$u_n(t) = u_n'(t - \frac{1}{2}) = z^{1/2}\ u_n'(t).$$

Thus, the corresponding z-transforms satisfy

$$U_n(z) = z^{1/2}\ U_n'(z)\ . \tag{2}$$

The wave $d_n(t)$ travels down in stratum n without change of form to become the wave $d_n'(t + 1/2)$. Thus,

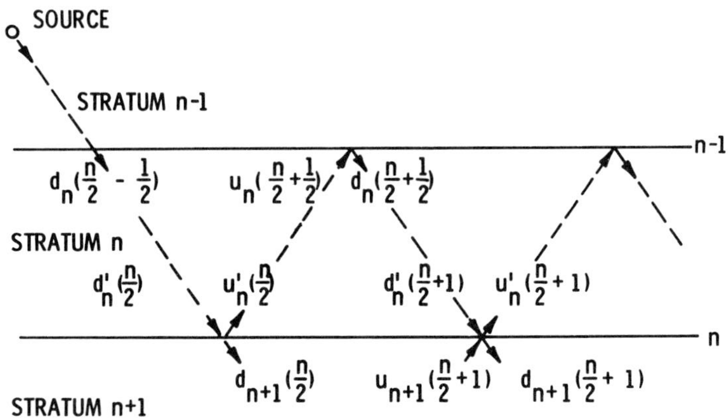

Fig. 6. Pulses associated with strata n and n+1.

$$d_n(t) = d'_n(t + \frac{1}{2}) = z^{-1/2}\, d'_n(t)$$

with the resulting equation

$$D_n(t) = z^{-1/2}\, D'_n(z)\ . \tag{3}$$

Let us now take the z-transforms of Eq. (1). We have

$$\begin{aligned} D_{n+1}(z) &= \tau_n\, D'_n(z) - \varepsilon_n\, U_{n+1}(z) \\ U'_n(z) &= \varepsilon_n\, D'_n(z) + \tau_n\, U_{n+1}(z). \end{aligned} \tag{4}$$

Using Eqs. (2) and (3) in Eq. (4), we obtain after some algebra the propagation equation,

$$\begin{bmatrix} D_{n+1} \\ \\ U_{n+1} \end{bmatrix} = \frac{z^{-1/2}}{\tau_n} \begin{bmatrix} z & -\varepsilon_n \\ \\ -\varepsilon_n z & 1 \end{bmatrix} \begin{bmatrix} D_n \\ \\ U_n \end{bmatrix} . \tag{5}$$

At the surface (i.e., interface 0), we have

$$\begin{bmatrix} D_1 \\ \\ U_1 \end{bmatrix} = \frac{1}{\tau_0} \begin{bmatrix} 1 & -\varepsilon_0 \\ \\ -\varepsilon_0 & 1 \end{bmatrix} \begin{bmatrix} D_0' \\ \\ U_0' \end{bmatrix} . \tag{6}$$

Use of Eq. (6) and recursive use of Eq. (5) allow us to relate the waves at the bottom and top of the medium by the equation

$$\begin{bmatrix} D_{N+1} \\ \\ U_{N+1} \end{bmatrix} = \frac{z^{-N/2}}{\tau_N \tau_{N-1} \cdots \tau_1 \tau_0} \begin{bmatrix} A^R & -E^R \\ \\ -E & A \end{bmatrix} \begin{bmatrix} D_0' \\ \\ U_0' \end{bmatrix} \tag{7}$$

where we define

$$\begin{bmatrix} A^R & -E^R \\ -E & A \end{bmatrix} = \begin{bmatrix} z & -\varepsilon_N \\ -\varepsilon_N z & 1 \end{bmatrix} \begin{bmatrix} z & -\varepsilon_{N-1} \\ -\varepsilon_{N-1} z & 1 \end{bmatrix} \cdots$$

$$\cdots \begin{bmatrix} z & -\varepsilon_1 \\ -\varepsilon_1 z & 1 \end{bmatrix} \begin{bmatrix} 1 & -\varepsilon_0 \\ -\varepsilon_0 & 1 \end{bmatrix} . \tag{8}$$

We see that A and E are polynomials of degree N and A^R and E^R are their respective reverse polynomials given by

$$A^R(z) = z^N A(z^{-1}) , \quad E^R(z) = z^N E(z^{-1}) .$$

That is, the coefficients of A^R are the same as those of A, but in reverse order. The coefficients of E^R are the same as those of E, but in reverse order. It is seen that the coefficients of A and of E depend only upon the reflection coefficients $\varepsilon_0, \varepsilon_1, \ldots, \varepsilon_N$. Explicit expressions may be obtained for the coefficients of A and E in terms of the reflection coefficients, but for our purposes we will make use of the following approximation. We assume that the reflection coefficients are small in magnitude, so that the product of three or more reflection coefficients may be neglected. Then Eq. (8) yields the following approximations for A and E:

$$\begin{aligned} A &\doteq 1 + \psi_1 z + \psi_2 z^2 + \cdots + \psi_N z^N \\ E &\doteq \varepsilon_0 + \varepsilon_1 z + \varepsilon_2 z^2 + \cdots + \varepsilon_N z^N \end{aligned} \tag{9}$$

where $\varepsilon_0, \varepsilon_1, \varepsilon_2, \ldots, \varepsilon_N$ is the reflection coefficient sequence and where $\psi_1, \psi_2, \ldots, \psi_N$ is the autocorrelation

$$\psi_t = \varepsilon_t \varepsilon_0 + \varepsilon_{t+1} \varepsilon_1 + \cdots + \varepsilon_N \varepsilon_{N-t} \quad (\text{for } t = 1,2,\ldots,N)$$

of the reflection coefficient sequence. Thus, the polynomial E(z) according to this approximation is the z-transform of the reflection coefficient sequence.

The Transmitted and Reflected Waves.

We recall our only input to the stratified medium is a downgoing unit surface pulse at time zero, so $D_0' = 1$ and $U_{N+1} = 0$. The resulting upgoing surface wave U_0' is designated by Y and is called the reflection response. The resulting downgoing bottom wave D_{N+1} is designated by T and is called the transmission response. Also, the product $\tau_N \tau_{N-1} \cdots \tau_1 \tau_0$ of transmission coefficients is simply denoted by τ. Thus, Eq. (7) becomes

$$\begin{bmatrix} T \\ \\ 0 \end{bmatrix} = \frac{z^{-N/2}}{\tau} \begin{bmatrix} A^R & -E^R \\ \\ -E & A \end{bmatrix} \begin{bmatrix} 1 \\ \\ Y \end{bmatrix} .$$

This matrix equation gives

$$T = \frac{z^{-N/2}}{\tau} (A^R - E^R Y)$$

$$0 = - E + AY$$

so the reflection response is

$$Y = \frac{E}{A}$$

and the transmission response is

$$T = \frac{z^{-N/2}}{\tau A} (A^R A - E^R E) .$$

But $A^R A - E^R E$ is the determinant of the matrix given in Eq. (8). Taking the determinant of each side of Eq. (8), we see that

$$A^R A - E^R E = (1 - \varepsilon_N^{\ 2}) \cdots (1 - \varepsilon_0^{\ 2}) z^N$$

$$= \tau_N^{\ 2} \cdots \tau_0^{\ 2} z^N = \tau^2 z^N .$$

As a result, the transmission response is

$$T = \frac{\tau\ z^{N/2}}{A} .$$

Let us look at this transmission response. The direct pulse travels from the surface source to the bottom (i.e., interface N). Since the first coefficient a_o of the polynomial A is unity, the direct pulse has amplitude τ, namely the product of the transmission coefficients of the interfaces through which the direct pulse travels. The travel time of the direct pulse in going through N strata is N/2; this travel time represents the time delay of the transmitted wave; this time delay is indicated by the factor $z^{N/2}$. Because the source contained unit energy, the transmitted wave must have finite energy and hence be stable, which means that the polynomial A(z) in the denominator of T must be minimum delay (i.e., A(z) must have no zeros within the unit circle). In fact, let us define B(z) as the inverse of A(z), i.e.,

$$B(z) = \sum_{t=0}^{\infty} b_t\ z^t = \frac{1}{A(z)}$$

and we call the one-sided waveform $b_0 = 1, b_1, b_2, \ldots$ the <u>reverberation wavelet</u>, or simply the reverberation. Because the inverse of a minimum-delay function also is minimum delay, the reverberation is minimum delay. In terms of the reverberation, the transmitted response is

$$T = \tau\ z^{N/2}\ B$$

and the reflection response is

$$Y = BE .$$

Let $y_0, y_1, y_2, \ldots$ denote the pulses of the reflection response. For example, in exploration seismology, $y_0, y_1, y_2, \ldots$ is the reflection seismogram. If we use the approximation for E given by Eq. (9), the above equation for Y is

$$\sum_{t=0}^{\infty} y_t z^t = (1 + b_1 z + b_2 z^2 + \cdots)(\varepsilon_0 + \varepsilon_1 z + \cdots + \varepsilon_N z^N).$$

As a result, the reflected waveform y_t is the convolution of the minimun-delay reverberation wavelet b_t with the reflection coefficient sequence ε_t; that is,

$$y_t = \sum_{s=0}^{\infty} b_s \varepsilon_{t-s} \quad (\text{for } t = 0,1,2,\ldots) . \tag{10}$$

Equation (10) for a stratified medium is the basic mathematical result of this paper. The purpose of this paper is the statistical interpretation which we give to Eq. (10) in the next section.

The Statistical Minimum-Delay Model and Predictive Deconvolution.

In order to fix ideas, let us consider a specific physical situation, namely the problem of seismic exploration for oil in the earth's sedimentary strata. The source is an explosion or another form of energy which is introduced into the ground at the surface. The reflection response y_t is the seismic reflection record (or trace or time series) which is digitally recorded at the surface. The reflection coefficient sequence ε_t is a digitized representation of the reflectivity of the earth as a function of depth. As a result, knowledge of the ε_t sequences for various geographic locations on the surface allows the seismic interpreter to make contour maps of the earth's sedimentary structure at depth. In other words, the reflection seismic method is a method of obtaining a picture of what the earth looks like thousands of feet below the surface from an experiment (i.e., a surface source of energy and a surface recording of the resulting reflection response to that source) carried out on the surface.

The nature of the geophysicist's problem can be imagined

by pretending that the earth's crust is a stack of Plexiglas sheets, all opaque with different properties and thicknesses, each having irregular surfaces, some of the plates being tilted, bent or deformed, and some containing pockets or traps of varying shapes and sizes. Some of these traps may contain accumulations of oil and natural gas. Now, the geophysicist taps the surface of this medium and he analyzes the resulting response at the surface to determine the thickness and physical properties of each plate, details of the interfaces between plates, and the locations and shapes of the traps in the plates. With this analogy in mind, the earth's crust may be described as an acoustically translucent, heterogeneous, multilayered medium. The traps for oil and natural gas may have an areal extent of only a half of a square mile, and be as much as four miles deep within the earth, with no geologic evidence at the surface that there is a trap deep below.

Let us now consider the magnitude of this problem. The United States consumes nearly 20 million barrels of oil each day and a correspondingly great amount of natural gas. An oil field with total reserves of 100 million barrels would be considered a good discovery, yet the United States would consume that much oil in five days. The only great oil discovery in the United States in the lifetime of the majority of Americans is the North Slope of Alaska, and if the United States had to rely entirely on the North Slope's known reserves, they would be consumed in two years. What makes matters really difficult is that not only in the United States but worldwide, the consumption of oil and natural gas is increasing at a geometric (i.e., exponential) rate.

In order to discover more oil and gas to replace the amount consumed, the oil exploration industry records about two million reflected seismic traces, y_t, each and every day, year after year. Each of these two million time series y_t has about 1000 data points (i.e., $t = 0,1,2,\ldots,1000$). These reflected traces, y_t, must be analyzed to yield a picture of the subsurface structure.

Before we describe the geophysical methods of today, we should first look at the history of exploration geophysics.

Before 1920, most oil was discovered by chance or by looking at surface indicators such as oil seeps or surface geological manifestations of subsurface traps. In the 1920s, great advances were made in geophysical methods of discovery, and since 1930 the reflection seismic method has been the chief method for discovering oil. Up until the digital revolution in seismology during the early 1960s, seismic interpretation was carried out on the basis of the reflected seismic trace, y_t. That is, up to about 1960, the recorded trace y_t (subject to only some electric filtering) was analyzed by eye in order to detect reflected events from the subsurface interfaces. From these observations, the seismic interpreter was able to draw subsurface contour maps. The amazing thing at the time was that the method worked at all; the interpretations on the whole were correct, and great oil fields were discovered. The question is why did the method work? But the question "why" is not important in comparison to the fact that the method did work; the seismic interpreter was able to detect primary reflected events by eye. The primary reflected event from interface t is defined as the pulse that travels directly down from the source to interface t and then directly back to the surface; its z-transform is

$$z^t \tau_0 \tau_1 \cdots \tau_{t-1} \varepsilon_t \tau_{t-1} \cdots \tau_1 \tau_0 .$$

However, for each primary reflected event, there are many multiple reflected events, and from a strictly mathematical point of view, these multiple events should overwhelm the primary events on the recorded trace y_t. That is, from a strictly mathematical point of view, the reverberations of this complex stratified system we call the earth should overwhelm our ability to pick distinct primary reflections on the seismic records.

By 1950, the search for oil extended into new areas all over the world, and it was realized that there were many areas where seismic record quality was poor, or in other words, the multiple reflections were overwhelming the primary reflections. In particular, in water-covered areas such as Lake Maracaibo in Venezuela and in the Persian Gulf, strong multiple waves in the water layer gave the observed seismic trace y_t a ringing character which made it impossible to detect the primary reflections coming up from the deep interfaces. Strong reverberations from surface layers masked the desired primary events in many land areas of exploration. In fact, as exploration pushed into more difficult areas, all types of multiple reflected energy began to be recognized on the traces and made conventional visual interpretation difficult to impossible. Moreover, conventional methods of electric filtering and field techniques could not overcome these difficulties.

From a theoretical point of view, it is important to develop models of the earth. In classical science, a great many models are known in which the behavior of the system is fully determined by a precisely fixed mathematical structure subject to idealized boundary-value or initial-value conditions. Such is the case for the earth model first treated in a classic paper of Lamb (1904) on the propagation of seismic waves in an elastic solid. Classical geophysics is concerned with the development of such deterministic models. However, because of the mathematical complexity of these models, only a very few simple geometric configurations within the earth are susceptible to analytic solutions. Because the field geophysicist historically was faced with a large number of different physical configurations, each of a complicated nature, the methods of classical geophysics were of no direct use and could only serve as a guide. The situation in exploration seismology in the early 1950s was summarized by one of the leading field geophysicists in this statement: "We have sharpened our tools as much as we can; what we need now are some new tools."

As it turned out, the new tool was the digital computer in conjunction with digital handling of seismic data from beginning to end. Use of digital computers to analyze geophysical data began in 1950 with an MIT research project which developed into the MIT Geophysical Analysis Group, an industry-supported project which ran from 1953 to 1957. Theoretical geophysical calculations made a significant advance when Norman Haskell, the geophysical advisor to this

MIT project, published his results on how seismic surface waves could be computed for a stratified earth model (Haskell, 1953). After extensive digital analysis of field records, Robinson (1954) proposed the statistical minimum-delay model of the earth together with the idea of deconvolution to separate the statistical component from the minimum-delay component. The present paper may be described as a synthesis of the stratified earth model and the statistical minimum-delay model. The stratified earth model for reflection seismograms has been extensively developed by Peterson, Fillippone and Coker (1955), Wuenschel (1960), Goupillaud (1961), Kunetz (1964) as well as many others.

Let us now explain the statistical minimum-delay model. The earth's crust is made out of a large number of strata in complex configurations. In the analysis of seismic waves from a physical point of view, we try to understand why the various waves that propagate through these strata behave the way they do. It is apparent that this is a difficult subject, and so we have to deal with it differently than with simpler situations. One way to analyze the properties of the strata is to start from a physical point of view, namely the point of view acquired over the years by the field geophysicists who analyze and interpret the seismic records. These geophysicists have a rough idea where they are going and they begin by making the right kind of approximations, knowing what is big and what is small in a given complicated situation. These interpretation problems are so complex that even an elementary understanding, although inaccurate and incomplete, is worth having. One of the things learned by these field geophysicists is that the analysis requires an understanding of the theory of probability. We do not want to know where every wave is actually moving, but rather how many move here and there on the average and the mutual buildup or cancelling of effects. Large numbers of seismic reflection records are needed to carry out an exploration program over a geographic area. The quantity of data necessarily requires the consideration of each record as a member of a larger group or ensemble of records. Thus, the reliability of a single record is considerably less than the reliability of the ensemble average in connection with the description of the geological conditions existing in that area. Also, from an economic standpoint, the amount of control in such an exploration program must be kept at a bare minimum consistent with worthwhile results. As a result, the working geophysicist must proceed to fit his empirical data into the larger overall framework from a statistical point of view. A major justification for using the statistical approach in seismology is due to the fact that large amounts of data must be processed; any data in large enough quantities takes on a statistical character, even though each individual piece of data is of a deterministic nature. Thus, the field geophysicist in working with the physical properties of the earth is faced with a situation which is essentially statistical. For example, a reflection which may be followed from trace to trace, record to record, usually has more value to the seismic interpreter, and hence is statistically more significant, than a reflection which appears only on a few traces.

Because the wavelet b_t in Eq. (10) is minimum-delay, the model represented by Eq. (10) is a minimum-delay model. Let us now give reasons why Eq. (10) may be regarded as a statistical model. According to the statistical point of view, the reflection coefficient sequence ε_t may be considered as a realization of a random (or stochastic) process. Equation (10), therefore, represents the reflection seismic trace as the composition of a minimum-delay wavelet and a random sequence. Equation (10) gives the entire reflected trace y_t from time zero to infinity. In practice, it is usually not feasible to analyze the whole trace in its entirety, and so instead the trace y_t is cut up into time sections (i.e., time gates) and then is analyzed section by section. Often these gates overlap to some extent, and then the results from the several sections are blended together so as to give a time-varying character to the overall analysis.

The part of the seismic trace within a given time gate contains the primary reflections from the interfaces within a certain section of the earth's crust. If we assume that the reflection coefficients ε_t within this earth section are mutually uncorrelated (i.e., that these ε_t represent a realization from a white-noise process), then the classical method of predictive deconvolution becomes applicable. At this point, let us note that the terms "minimum-delay" and "minimum-phase" are the same so they may be used interchangeably. Also, the terms "predictive deconvolution" and "predictive decomposition" may be used interchangeably.

The method of predictive deconvolution may be described as follows. Within the time gate, the ε_t are white; hence, all the spectral shape of the trace y_t within the time gate is due to b_t. As a result, computation of the autocorrelation of y_t within the gate averages out the white-noise elements ε_t and, therefore, gives us the autocorrelation of b_t. Since b_t is minimum-delay, its inverse a_t can be computed directly from its autocorrelation by solving the normal equations. The inverse operator a_t is the required deconvolution operator. Deconvolution is the process of convolving the inverse operator a_t with the trace y_t. The result of the deconvolution is the reflection coefficient sequence ε_t which is called the deconvolved trace. The deconvolved trace ε_t represents the earth's reflectivity as a function of depth within the earth section being analyzed. In practice, of course, many variations of the basic deconvolution method described here can be used. For example, the inverse operator a_t represents an autoregressive operator, that is, a prediction error operator with unit prediction distance; prediction error operators for other prediction distances also can be used. Adaptive and time-varying deconvolution methods as well as multitrace methods may be used.

By the early 1960s, the petroleum exploration industry had undergone what was called a digital revolution. Digital methods were used for recording, processing, and display of seismic data. This digital approach was based on the statistical point of view. Since 1965, it has been routine practice to deconvolve each seismic trace.

During these years, oil companies and geophysical contractors have made careful studies which shed light on the statistical minimum-delay model. For example, White and O'Brien (1974) of British Petroleum consider the assumption that the reflection coefficient sequence ε_t is white, that is, the assumption that the power spectrum of ε_t is flat and that the shape of the power spectrum of the reflected trace y_t is due solely to the wavelet b_t. They say: "This is a crucial assumption. An investigation of some reflection spike sequences estimated from well-log data has shown that it is usually valid. The only definite exceptions found occurred beyond the seismic frequency range, although there also may be a departure from a flat spectrum at low frequencies." In their study of sonic logs from oil wells, Schoenberger and Levin (1974) of Exxon find the assumption of a white reflection sequence ε_t is a good one for depth intervals of an oil well, but breaks down for certain intervals where a few strong reflections dominate.

Let us now look at the minimum-phase assumption. The fact that reverberating seismic energy is minimum-phase is well established (Kunetz, 1964; Treitel and Robinson, 1966). We must also consider the source waveform which is relatively narrow although it cannot be considered as narrow as a spike. Many studies have been made on source pulses. As stated by White and O'Brien (1974): "The source measurements already mentioned have shown that the pulse radiated by an explosion is minimum-phase or very close to it. It is possible for underwater sources to emit non-minimum-phase signals but operational conditions usually avoid this. Absorption in the earth is generally held to be minimum-phase, and so too is the response of most recording systems. Minimum-phase then is likely to be satisfied in practice." As a result, all these minimum-phase effects can be included in the minimum-delay wavelet b_t in the model as given by Eq. (10).

What then is the final test for the statistical minimum-delay model and the associated method of predictive decomposition or deconvolution? The answer lies in the fact that today as well as during the past 15 years, virtually every single seismogram taken in the exploration of oil and natural gas has been processed according to this model. These 10 billion seismic traces from every part of the world must speak for themselves, and they say the statistical minimum-delay model and the method of predictive deconvolution are physically valid.

Conclusions.

Let us now answer the question we asked in the last section, namely why did the seismic method work in so many areas before the advent of deconvolution and other forms of digital processing? The answer is that in these good seismic areas, the reflection coefficient sequence ε_t is white or nearly white for the entire crustal section (Treitel, 1976). Thus, the autocorrelations ψ_t of the reflection coefficient sequence are approximately zero, so Eq. (9) becomes $A \doteq 1$. As a result, B is also approximately equal to 1, so b_t becomes a unit pulse and Eq. (10) becomes simply $y_t \doteq \varepsilon_t$. In other words, in those areas where the reflection coefficients are small and uncorrelated, the reflection seismogram reduces simply to the sequence of reflection coefficients ε_t. The net effect of the white nature of the subsurface reflection coefficients is to cancel out all the multiplies and to amplify all the primaries so they have amplitude ε_t instead of amplitude $(\tau_0 \; \tau_1 \; \cdots \; \tau_{t-1})^2 \; \varepsilon_t$. Here is a case in nature where randomness has taken a very complex physical situation and made it extremely simple and beautiful. As we have seen in the application of deconvolution, we make use of those crustal sections in which we can consider the sequence ε_t uncorrelated.

The digital revolution in seismology as well as in other sciences is based on statistical (or data-oriented) models which make possible the unraveling (or deconvolution) of the observed time series. The essence of this statistical approach is that, in addition to its physical structure, a system has a structure of behavior within which the system operates. This structure of behavior can be deduced from field observations. Experiments with the statistical approach have confirmed the existence of this behavior and have brought us closer to understanding the workings of nature. The basis of this approach lies within the behavior of the physical media and in the interaction of their components. The underlying permanence is in the rules by which energy is exchanged between components, and in this respect the concept of the minimum-delay of energy becomes a central concept. We see that there is a subtle shift of emphasis here between the deterministic and the statistical approach. The statistical approach is concerned about the transactions rather than the things; that is it deals with the dynamics of what happens rather than with characteristics of the system as an object. The field data in the form of time series are the keys to this approach, for the analysis of the time series tells us what happens and in this way depicts the structure of the system. For example, in the statistical approach, we are interested in the correlation among reflection coefficients in a sequence of strata, and with the flow of energy within this structure. In other words, the statistical approach looks at the relationship between things rather than the things themselves.

References.

Goupillaud, P. (1961). An approach to inverse filtering of near surface layer effects from seismic records. *Geophysics* *26*, 754-760.

Haskell, N. A. (1953). The dispersion of surface waves on multilayered media. *Bull. Seismological Soc. Amer.* *43*, 17-34.

Kunetz, G. (1964). Généralisation des opérateurs d'anti-résonance à un nombre quelconque de réflecteurs. *Geophys. Prospecting* *12*, 283-289.

Peterson, R. A., W. R. Fillippone, and F. B. Coker (1955). The synthesis of seismograms from well log data. *Geophysics* *20*, 516-538.

Robinson, E. A. (1954). Predictive decomposition of time series with application to seismic exploration. MIT Geophysical Analysis Group, Report No. 7, Cambridge, Massachusetts. Reprinted in *Geophysics* *32*, 418-484 (1967).

Robinson, E. A., and S. Treitel (1976). Net downgoing energy and the resulting minimum-phase property of downgoing waves. *Geophysics* *41*, 1394-1396.

Schoenberger, M., and F. K. Levin (1974). Apparent attenuation due to intrabed multiples. *Geophysics 39*, 278-291.

Treitel, S. (1976). Personal communication.

Treitel, S., and E. A. Robinson (1966). Seismic wave propagation in layered media in terms of communication theory. *Geophysics 31*, 17-32.

White, R. E., and P. N. S. O'Brien (1974). Estimation of the primary seismic pulse. *Geophys. Prospecting 22*, 627-651.

Wuenschel, P. E. (1960). Seismogram synthesis including multiples and transmission coefficients. *Geophysics 25*, 106-129.

ESTIMATING THE INTENSITY OF A POISSON PROCESS

G. S. Watson
Princeton University

1. Introduction.

There is an immense and increasing literature on point processes, especially Poisson processes in which the intensity is a function $\lambda(t)$ of a scalar parameter t, usually time. Summaries are given in Cox and Lewis (1966), Lewis (1972), 1975). There are many motivations. Lewis has often been concerned with times of breakdown of computers. My own interest was aroused by two problems--the effects of the environment on the deaths per day in a large city and a discussion of wildcat oil strikes in Alberta. The latter was merely a numerical example in a paper by Clevenson and Zidek (1975) but it suggested that this topic would be appropriate for this Symposium. However, our discussion will be theoretical--we will speak of events, not deaths or strikes--and its aim is to show the simplicity and unity of a number of statistical and mathematical methods.

If the chances that an event occurs or does not occur in a smaller interval $(t,t+dt)$ are given by $\lambda(t)dt$, and $1 - \lambda(t)dt$, respectively, independently of what happened before t, then the number of events $N(t)$ in $(0,t)$ is said to follow a non-homogeneous Poisson process (N.H.P.P.). The numbers of events in non-overlapping intervals are independent and have Poisson distributions whose means are the integrals of $\lambda(t)$ over the relevant time intervals. Classical examples are deaths due to horse kicks in the Prussian Army and clicks of a Geiger Counter.

If the times of events are marked on a time axis, they will be densest, on the average, where $\lambda(t)$ is greatest. In this respect, $\lambda(t)$ is like a probability density. One can show that if n events occur in $(0,T)$, then the times of

ISBN 0-12-257250-5

the events may be treated as a random sample from a probability distribution with density

$$f(t) = \lambda(t)/\int_0^T \lambda(t)\,dt \quad . \tag{1.1}$$

Given a set of data consisting of the successive times of events from an N.H.P.P., the statistician will wish to estimate the <u>function</u> $\lambda(t)$ and perhaps test some hypotheses about it. In some circumstances, the form of $\lambda(t)$ may be given and the problem reduced to estimating some parameters in the formula for $\lambda(t)$.

Given a random sample from some distribution, precisely the same problems arise--estimating its density or parameters in a formula for it.

In both problems, the data are often grouped to begin with or later for convenience--deaths per day, strikes per month. The mathematical difficulties of estimating $\lambda(t)$ are greatly reduced by dividing the time axis into intervals of equal length δ. We will start with such data. Clevenson and Zidek discussed very similar topics in continuous time and were essentially forced by difficulties to make discrete approximations later.

The problem of estimating a function $\lambda(t)$, $0 < t < T$ or the function λ_t, $t = 0,1,2$ N-1 is clearly quite distinct, given a finite amount of data, from the more usual problem of estimating a few parameters. We will use the total mean square error to assess the quality of the estimator, i.e., we will want an estimator $\hat{\lambda}_t$ that makes

$$E \sum_0^{N-1} (\hat{\lambda}_t - \lambda_t)^2 \tag{1.2}$$

small. We will further consider a measure of smoothness of the sequence $\hat{\lambda}_t$ and require it to be small too, on the grounds that the true sequence λ_t is not rough, namely

$$E \sum_0^{N-1} (\Delta^P \hat{\lambda}_t)^2 \tag{1.3}$$

where $\Delta^p\hat{\lambda}_t$ is the p-th difference of the $\hat{\lambda}_t$ sequence.

By combining these two measures of variability, we will have something to optimize. It will turn out that the "best" estimator depends upon the unknown sequence so that, in practice, one needs to put in one's intuition. While this is familiar in time series analysis, it comes as surprise to some people since such results are concealed in elementary statistics by assuming a parametric form. More familiar is our restriction to a class of estimators linear in the counts of events.

Sometimes it is realistic to assume that $\lambda(t)$ is itself a random function. Thus, we will have two cases--fixed and random functions. Since $\lambda(t)$ is often the signal that something is present, it is often referred to as the signal, rather than as the function.

Of course, for the statistical analyses of particular point processes, the N.H.P.P. model may not be appropriate. It is beyond our scope to consider the general problem. However, in Section 6, we briefly mention another model. In Section 7, some ideas for simpler and more exploratory analyses are suggested.

2. A Time Series Approach to the Fixed Signal Case.

Suppose we have the number of events in N successive intervals and that these are denoted by $n_0, n_1, \ldots, n_{N-1}$ with mean values $\lambda_0, \lambda_1, \ldots, \lambda_{N-1}$ which we may think of first as constants--a "fixed signal". Since they are integrals of $\lambda(t)$ over intervals of length δ, if $\lambda(t)$ is a smooth function, the λ_t's will behave regularly, and be smoother than the n_t's. Therefore, it is reasonable to consider a *moving average* estimator of λ_t given by

$$\hat{\lambda}_t = \sum_{-K}^{K} a_k n_{t-k} \quad . \tag{2.1}$$

To avoid negative estimates, the a_k would usually be taken as non-negative. Since a constant record ought to be equal to its estimate, one usually supposes $\Sigma a_k = 1$. Then the

a_k's form a discrete probability distribution. In the absence of other ideas, it would be symmetric around zero, its support, determined by K, would be large if λ_t is a very smooth function, small if it is not. It is fairly standard to call $\hat{\lambda}_t$ a filtered version of the n_t series, and the set of a_k's, a filter. It is clear that smoothing reduces the variance but increases the bias.

To find an optimal filter, one might ask for the filter which minimizes the average mean square error

$$E \sum_{0}^{N-1} (\hat{\lambda}_t - \lambda_t)^2 \tag{2.2}$$

where, for this section we regard the λ_t's as constants and we make no assumptions about the a_k's. However, one cannot calculate $\hat{\lambda}_t$ for t's such that (2.1) calls for n_t's with t outside 0, ..., N-1. To overcome this, we put the n_t and λ_t equally spaced around the circumference of a circle. It is then natural to identify n_N with n_0, n_{N+1} with n_1 etc., and n_{-1} with n_{N-1} etc. The functions n_t, λ_t, $\hat{\lambda}_t$ then are defined for all integers and they have period N. If we define

$$\omega_j = \frac{2\pi j}{N}, \quad j=0,1,...,N-1 \tag{2.3}$$

$\exp i\omega_j t$ also has period N. Multiplying (2.1) by $\exp i\omega_j t$ and adding over t from 0 to N-1, we find

$$\hat{\lambda}(\omega_j) = A(\omega_j)n(\omega_j) \tag{2.4}$$

where

$$\hat{\lambda}(\omega_j) = \sum_{0}^{N-1} \hat{\lambda}_t \exp i\omega_j t,$$

$$A(\omega_j) = \sum_{-K}^{K} a_k \exp i\omega_j k, \tag{2.5}$$

$$n(\omega_j) = \sum_{0}^{N-1} n_t \exp i\omega_j t,$$

where we make no assumptions about K. Since

$$\sum_{t=0}^{N-1} \exp i\omega_j t = \begin{cases} N & \text{if} \quad j=0 \\ 0 & \text{if} \quad j=1,\ldots,N-1 \end{cases} \tag{2.6}$$

we can verify that

$$\hat{\lambda}_t = \frac{1}{N} \sum_0^{N-1} \hat{\lambda}(\omega_j) \exp -i\omega_j t \tag{2.7}$$

and

$$\sum_0^{N-1} \hat{\lambda}_t^{\,2} = \frac{1}{N} \sum_0^{N-1} |\hat{\lambda}(\omega_j)|^2 \tag{2.8}$$

with similar formulae for any other function of period N.

Thus

$$\sum_0^{N-1} (\hat{\lambda}_t - \lambda_t)^2 = \frac{1}{N} \sum_0^{N-1} |\hat{\lambda}(\omega_j) - \lambda(\omega_j)|^2$$

$$= \frac{1}{N} \sum_0^{N-1} |A(\omega_j)n(\omega_j) - \lambda(\omega_j)|^2 . \tag{2.9}$$

Hence, the criterion (2.2) may be rewritten: find $A(\omega)$ so that

$$E \sum_0^{N-1} |A(\omega_j)n(\omega_j) - \lambda(\omega_j)|^2 = \min ! \tag{2.10}$$

The j-th term in the left-hand side (l.h.s.) of (2.10) is

$$|A(\omega_j)|^2 E|n(\omega_j)|^2 - A(\omega_j)En(\omega_j)\bar{\lambda}(\omega_j)$$

$$- \bar{A}(\omega_j)E\bar{n}(\omega_j)\lambda(\omega_j) + |\lambda(\omega_j)|^2 . \tag{2.11}$$

Now

$$En(\omega_j) = \sum_0^{N-1} E(n_j) \exp i\omega_j t = \lambda(\omega_j) , \tag{2.12}$$

$$E|n(\omega_j)|^2 = E\, n(\omega_j)\bar{n}(\omega_j),$$

$$= \sum \sum E(n_t n_{t'}) \exp i\omega_j(t-t'),$$

$$= \sum_0^{N-1} \lambda_t + |\lambda(\omega_j)|^2 , \qquad (2.13)$$

since the n_t are independent for distinct t, $En_t = \lambda_t$, and $En_t^2 = \lambda_t + \lambda_t^2$. Now (2.11) may be rewritten as

$$E|n(\omega_j)|^2 \left| A(\omega_j) - \frac{En(\omega_j)\bar{\lambda}(\omega_j)}{E|n(\omega_j)|} \right|^2$$

$$+ |\lambda(\omega_j)|^2 - \frac{|En(\omega_j)\bar{\lambda}(\omega_j)|^2}{E|n(\omega_j)|^2} \qquad (2.14)$$

so that every term in the sum (2.10) is minimized if we choose

$$A(\omega_j) = \frac{|\lambda(\omega_j)|^2}{\sum_0^{N-1} \lambda_t + |\lambda(\omega_j)|^2} , \qquad (2.15)$$

giving (2.10) a minimum value of

$$\sum_{j=0}^{N-1} \frac{\sum_0^{N-1} \lambda_t}{\sum_0^{N-1} \lambda_t + |\lambda(\omega_j)|^2} |\lambda(\omega_j)|^2 . \qquad (2.16)$$

The optimal filter, defined by (2.15), has the following properties. We note that $0 \le A(\omega_j) \le 1$, all j. Since $A(\omega_j) = A(\omega_{-j})$ is real, $a_k = a_{-k}$ (i.e., the filter is symmetric about zero):

For

$$a_k = \frac{1}{N} \sum_0^{N-1} A(\omega_j) \exp -ik\omega_j,$$

$$= \frac{1}{N} \sum A(\omega_{-j}) \exp -ik\omega_j,$$

$$= \frac{1}{N} \sum A(\omega_{-j}) \exp -i(-k)\omega_{-j},$$

$$= a_{-k} \cdot$$

Since $\sum a_k = A(0)$, (2.15) gives $(\sum \lambda_t = \lambda(0) > 0)$

$$\sum a_k = \frac{\lambda(0)^2}{\lambda(0) + \lambda(0)^2} < 1,$$

so that for <u>all</u> filters considered here, there is a downwards bias. Finally, since $A(\omega_j) = A(\omega_{-j}) = A(\omega_{N-j})$, it will be convenient to arrange that $N = 2K + 1$ or $K = (N-1)/2$.

If $\lambda_t = \lambda$, $t = 0,\ldots,N-1$, then $\lambda(\omega_j) = 0$ unless $j = 0$ when it is

$$\sum_0^{N-1} \lambda_t = N\lambda$$

so that

$$A(0) = \frac{(N\lambda)^2}{N\lambda + (N\lambda)^2}, \quad A(\omega_j) = 0, \quad j = 1,\ldots,N-1.$$

Hence, the a_k sequence is constant, a_N say, and equal to $A(0)/N$. Thus, $Na_N \uparrow 1$ as $N\lambda \to \infty$ so that the downward bias disappears.

By contrast, if $\lambda_t = \lambda$ for all t except t_0 when $\lambda_{t_0} = \lambda + \delta$ with δ very large, then

$$\lambda(\omega_j) = \begin{cases} N\lambda + \delta & \text{if} \quad j=0 , \\ \delta \exp i\omega_j t_0 & \text{if} \quad j \neq 0 , \end{cases} \tag{2.17}$$

so that

$$A(\omega_j) = \begin{cases} \dfrac{(N\lambda + \delta)^2}{(N\lambda + \delta) + (N\lambda + \delta)^2} , & j = 0 \\ \dfrac{\delta^2}{(N\lambda + \delta) + \delta^2} , & j \neq 0 . \end{cases}$$

If $\delta > N\lambda$, $A(\omega_j) \simeq 1$ for all j and

$$a_k \simeq \begin{cases} 1 , & k = 0 , \\ 0 , & k \neq 0 . \end{cases}$$

Given the signal in the process, this is very reasonable--one would not want to smooth at all since only one λ_t is out of line. It is important to note that the optimal filter is unaffected by t_0. It will pick up this signal just as well wherever it is, i.e., the filter is <u>time-invariant</u> with respect to the position of the signal.

This is true in general since moving the signal on t_0 time units is the same here as multiplying $\lambda(\omega_j)$ by $\exp -i\omega t_0$ and $A(\omega_j)$ is unaffected since it depends only on the modulus of $\lambda(\omega_j)$.

Finally, the computation of the time domain filter coefficients given $A(\omega_j)$ can be done by using the Fast Fourier Transform to calculate

$$a_k = \frac{1}{N} \sum_{j=0}^{N-1} A(\omega_j) \exp -i\omega_j k . \tag{2.18}$$

Equally, given some idea of λ_t, it can be used to calculate the $\lambda(\omega_j)$ which will be needed to obtain the approximate $A(\omega_j)$. Naturally, in practice, the ideal filter cannot be used for the estimation of the quantity it depends on. In recurring problems, or in the testing or detection problems, discussed in Section 4, this <u>is</u> possible.

The circular device used above is an often exploited mathematical trick to make the mathematics easy. It will only lead to trouble if the a_k's do not fall off to zero rapidly as $|k|$ increases. The two examples show when we may expect it to work.

3. The Case of a Random $\lambda(t)$.

It was suggested in Section 1 that in many problems $\lambda(t)$ should be taken as a random function. In this case to the expectation in (2.2) and so (2.10) should be added another expectation $\underset{\lambda}{E}$ over all $\lambda(t)$. The analysis is otherwise the same. Looking at (2.11), (2.12), and (2.13), we need to evaluate

$$\underset{\lambda}{E}\ |\lambda(\omega_j)|^2 \quad \text{and} \quad \underset{\lambda}{E} \sum \lambda_t \ .$$

If $\lambda(t)$ is a second-order stationary process in continuous time, then λ_t will be second-order stationary in discrete time. As we cannot with grouped data do more than study the λ_t process, we will only make assumptions about it. These are the following:

$$E\ \lambda_t = \lambda$$

$$\mathrm{Cov}(\lambda_t, \lambda_{t+\tau}) = \rho(\tau), \quad \sum_{\tau} |\rho(\tau)| < \infty$$

$$f(\phi) = \frac{1}{2\pi} \sum_{-\infty}^{\infty} \rho(\tau) \exp i\tau\phi,$$

$$\lambda_t - \lambda = \int_{-\pi}^{\pi} e^{i\phi t}\, dZ(\phi), \tag{3.1}$$

$$\rho(\tau) = \int_{-\pi}^{\pi} e^{i\phi\tau} f(\phi)\, d\phi,$$

$$\underset{\lambda}{E}\ dZ(\phi) = 0,$$

$$\underset{\lambda}{E}\, dZ(\phi)\overline{dZ(\phi')} = 0 , \quad \phi \neq \phi',$$

$$\underset{\lambda}{E}\, |dZ(\phi)|^2 = f(\phi)d\phi .$$

These are the basic relations and quantities for such a process. $f(\phi)$ is called the (power) spectral density of the process. Readers unacquainted with this theory need only regard the integral formula for $\lambda_t - \lambda$ as an expression for this deviation from zero as a sum of sinusoidal oscillations with uncorrelated amplitudes whose average magnitudes vary with the frequency as described by $f(\phi)$, as seen in the last three expectations. The fact that the Fourier series for $f(\phi)$ has coefficients proportional to the covariances $\rho(\tau)$ will not be important here.

Then

$$\lambda(\omega_j) = \sum_0^{N-1} \lambda_t \exp i\omega_j t$$

$$= \lambda \sum_0^{N-1} \exp i\omega_j t + \int_{-\pi}^{\pi} \sum_0^{N-1} \exp(i(\phi+\omega_j)t)dZ(\phi) .$$

So

$$\lambda(0) = N\lambda + \int_{-\pi}^{\pi} \frac{1 - e^{iN\phi}}{1 - e^{i\phi}}\, dZ(\phi),$$

$$\lambda(\omega_j) = \int_{-\pi}^{\pi} \frac{1 - e^{iN(\phi+\omega_j)}}{1 - e^{i(\phi+\omega_j)}}\, dZ(\phi) , \quad j \neq 0 .$$

Hence

$$\underset{\lambda}{E}\, \lambda(0) = E \sum \lambda_t = N\lambda$$

$$\underset{\lambda}{E}\, \lambda(\omega_j) = 0 , \quad j \neq 0$$

$$\mathop{E}_{\lambda} |\lambda(0)|^2 = N^2\lambda^2 + \int_{-\pi}^{\pi} \left| \frac{1 - e^{iN\phi}}{1 - e^{i\phi}} \right|^2 f(\phi)d\phi, \tag{3.2}$$

$$\mathop{E}_{\lambda} |\lambda(\omega_j)|^2 = \int_{-\pi}^{\pi} \left| \frac{1 - e^{iN(\phi+\omega_j)}}{1 - e^{i(\phi+\omega_j)}} \right|^2 f(\phi)d\phi. \tag{3.3}$$

The multiplier of $f(\phi)$ when divided by $2\pi N$ in these last two integrals is called the Fejér kernel, i.e.,

$$g(\phi) = \frac{1}{2\pi N} \left| \frac{1 - e^{iN\phi}}{1 - e^{i\phi}} \right|^2 = \frac{1}{2\pi N} \left| \frac{e^{\frac{iN\phi}{2}} - e^{-\frac{iN\phi}{2}}}{e^{\frac{i\phi}{2}} - e^{-\frac{i\phi}{2}}} \right|^2$$

$$= \frac{1}{2\pi N} \left| \frac{\sin \frac{N\phi}{2}}{\sin \frac{\phi}{2}} \right|^2 . \tag{3.4}$$

It may be shown that $g(\phi)$ has an integral from $-\pi$ to π of unity and that it resembles more and more as $N\to\infty$ a Dirac delta function with singularity at $\phi = 0$. The same is true of $g(\phi+\omega_j)$ except that the peak is moved to $-\omega_j$.

Thus, for large N, (3.2) and (3.3) become approximately

$$\left.\begin{aligned} E|\lambda(0)|^2 &\simeq N^2\lambda^2 + 2\pi N\, f(0) \simeq N^2\lambda^2 \\ E|\lambda(\omega_j)|^2 &\simeq 2\pi N\, f(\omega_j) \end{aligned}\right\} \tag{3.5}$$

so that the analogue of (2.14) here reads

$$(N\lambda + N^2\lambda^2) \left| A(0) - \frac{N^2\lambda^2}{N\lambda + N^2\lambda^2} \right|^2 + N^2\lambda^2 - \frac{N^2\lambda^2}{N\lambda + N^2\lambda^2}$$

and

$$(N\lambda + 2\pi N\ f(\omega_j)) \left| A(\omega_j) - \frac{2\pi N\ f(\omega_j)}{N\lambda + 2\pi N\ f(\omega_j)} \right|^2 + 2\pi N\ f(\omega_j) - $$

$$- \frac{2\pi N\ f(\omega_j)}{N\lambda + 2\pi N\ f(\omega_j)}$$

so that for large N

$$\left. \begin{aligned} A(0) &\simeq 1\ , \\ A(\omega_j) &\simeq \frac{2\pi\ f(\omega_j)}{\lambda + 2\pi\ f(\omega_j)} \end{aligned} \right\} . \tag{3.6}$$

Since $f(\phi)$ has period 2π and is symmetrical about 0, it is symmetrical about π. If the realizations of λ_t are smooth, on the average $f(\phi)$ will be larger near the origin and smaller away from it. Thus, the interpretation of this filter is exactly as before.

4. Invariant Testing.

Watson (1974) studied the problem of detecting a peak of unknown position in the intensity function. Both the situations, times of events given and counts in equal intervals given, were discussed. The first case was reduced to testing for uniformity on a circle. To make the correspondence to testing probability densities for uniformity clearer, the data were divided by T to fit on a circle of unit perimeter. If the times are then $x_1 \ldots . x_M$, the locally most powerful invariant tests of Beran (1968) have the form: reject when T_M is large where

$$T_M = \sum_1^\infty |c_m|^2 \frac{2}{M} \left| \sum_{j=1}^{M} e^{2\pi i m x_j} \right|^2 \tag{4.1}$$

and the c_m are the Fourier coefficients of $g(x)$, a density with period unity. The family of alternatives is $(1 - \varepsilon) + \varepsilon g(x - \phi)$ where ϕ is any number in $(0,1)$ and ε is small. An alternative form of (4.1) is

$$T_M = \frac{1}{M} \int_0^1 [\Sigma\{g(x_i - \phi) - 1\}]^2 d\phi$$

$$= M \int_0^1 [\frac{1}{M} \Sigma g(x_i - \phi) - 1]^2 d\phi \qquad (4.2)$$

where

$$M^{-1} \sum_1^M g(x_i - \phi)$$

is a kernel-type estimator of the density of x, a test suggested by Watson (1967) on intuitive grounds. The test T_M is invariant with respect to the position of ϕ of the alternative density $g(x - \phi)$. In our case,

$$g(x - \phi) = \frac{\lambda(T(x - \phi))}{\int_0^T \lambda(t)dt)}$$

where $\lambda(\cdot)$ has been made to have period T. If only the counts in N equal intervals

$$(\frac{j}{N}, \frac{j+1}{N})$$

for $j = 0, \ldots, N-1$ are given, the natural approximation to (4.1) and (4.2) is

$$T_M' = \sum_{m=1}^{\infty} |d_m|^2 \frac{2}{M} \left| \sum_0^{N-1} n_j \exp \frac{2\pi ij}{N} \right|^2 \qquad (4.3)$$

where

$$d_m = c_m \frac{N(1 - \exp -2\pi im/N)}{2\pi im}, \quad m > 0 . \tag{4.4}$$

Then (4.3) was shown to reduce to

$$T_M' = \sum_{m=1}^{N-1} \left\{ \sum_{\ell=0}^{\infty} |d_{m+N\ell}|^2 \right\} \frac{2}{M} \left| \sum n_j \exp \frac{2\pi imj}{N} \right|^2 \tag{4.5}$$

because of aliasing. Experiments with a special case of T_M, called U_M^2 (Watson (1974)) with $N = 20$, showed that the null distributions of T_M and T_M' were practically the same.

Thus, for invariant testing, often little will be lost by grouping and the optimal statistic (then (4.5)) is easily computed using the F.F.T. In the notation of this paper, (4.5) is

$$T_M' \propto \sum_{m=1}^{N-1} \frac{|\lambda(\omega_m)|^2}{\Sigma\lambda_t} |n(\omega_m)|^2 \tag{4.6}$$

$$\simeq \sum_{m=1}^{N-1} |A(\omega_m)|^2 \; |n(\omega_m)|^2 \tag{4.7}$$

if $|\lambda(\omega_m)|^2/\Sigma\lambda_t$ is always small. This is so, for example, in the first special case of Section 2--a constant signal. The test in the second case, a short very strong signal, is based on

$$\sum_{m=1}^{N-1} |n(\omega_m)|^2$$

which is the discrete form of a test suggested in Watson (1967) for this situation.

Thus, the grouped form of invariant testing is very close to the grouped estimation problem of Section 2. Both benefit from being discussed in terms of finite Fourier transforms. If one ignored the test theory of Beran and just knew the results of Section 2, a natural test statistic, to test

λ_t = constant, λ say, would be

$$\sum_{t=0}^{N-1} (\hat{\lambda}_t - \lambda)^2 = \frac{1}{N} \sum_{0}^{N-1} |A(\omega_j)n(\omega_j) - \lambda'(\omega_j)|^2 \tag{4.8}$$

where

$$\lambda'(\omega_j) = \sum_{0}^{N-1} \lambda \exp i\omega_j t = \begin{cases} N\lambda & \text{if} \quad j = 0 \\ 0 & \text{if} \quad j \neq 0 . \end{cases}$$

But

$$A(0)n(0) = \frac{|\lambda(0)|^2}{\sum_{1}^{N} \lambda_t + |\lambda(0)|^2} \sum n_t \quad .$$

Thus, the first term in (4.8) merely tests the value chosen λ for the constant λ_t. The remaining terms test whether λ_t is constant so one would use only them for the test. But this is (4.7). In fact, $|\lambda(\omega_m)|^2/\Sigma\lambda_t$ small, all $m \neq 0$, really means that λ_t does not vary much and so defines a neighbourhood of λ_t = constant. Since Beran-type tests are only _locally_ most powerful invariant, it is not surprising that the two tests coincide in this case.

Alternatives may be local, intermediate or distant. The last concept is defined in Watson (1974) from probability distributions on the circle and corresponds to cases like the second example at the end of Section 2. The optimal invariant tests are based on suprema, the associated distributions are hard to find and Fourier analysis seems to be of no avail. In the present case of (2.15), this suggests the intuitive test--is the greatest of $n_0, n_1, \ldots, n_{N-1}$ too large to have occurred by chance? Since the n_t's are independent Poissons with the same unknown mean on the null hypothesis, the test which is independent of the mean uses the distribution of the largest frequency in a multinomial of N equal cells when the total frequency Σn_t is kept fixed. This is not an easy problem. It has been studied by Dudley (1971). In general then distant tests may be expected to have the form:

$$\max_{t=0,\ldots,N-1} \hat{\lambda}(t) > C$$

and to be difficult to use because of distributional troubles.

5. Spline Estimation.

In Sections 2 and 3, we have looked only at average mean square error of the estimator $\hat{\lambda}_t$ of λ_t. A method which has a Bayesian type of motivation arises if we insist also that $\hat{\lambda}_t$ be a smooth function of t. We treat here the situation of Section 2. Thus, to the criterion (2.2) we wish to add

$$c^2 \sum_1^N (\Delta^2 \hat{\lambda}_t)^2 \quad .$$

From (2.7)

$$\Delta^2 \hat{\lambda}_t = \frac{1}{N} \sum \hat{\lambda}(\omega_j) \Delta^2 \exp -i\omega_j t$$

$$= \frac{1}{N} \sum \hat{\lambda}(\omega_j) \exp -i\omega_j t \, (\exp i\omega_j - 1)^2 \, .$$

By the analogue of (2.8)

$$\sum_1^N (\Delta^2 \hat{\lambda}_t)^2 = \frac{1}{N} \sum_1^N |\hat{\lambda}(\omega_j)(\exp i\omega_j - 1)^2|^2 \, .$$

Hence, our criterion will now be: find $A(\omega_j)$ so that

$$E \sum_{1}^{N} |A(\omega_j)n(\omega_j) - \lambda(\omega_j)|^2 + c^2|A(\omega_j)n(\omega_j)(\exp i\omega_j - 1)^2|^2$$

$$= \min! \qquad (5.1)$$

In (5.1), c^2 is a smoothing parameter. As $c^2 \to \infty$, $\hat{\lambda}(t)$ must become flat.

Evaluating (5.1) by making use of the results that lead up to (2.13), we find that the j-th term in (5.1) is

$$|A(\omega_j)|^2(\sum \lambda_t + |\lambda(\omega_j)|^2)(1 + c^2|\exp i\omega_j - 1|^4)$$

$$- (A(\omega_j) + \bar{A}(\omega_j))|\lambda(\omega_j)|^2 + |\lambda(\omega_j)|^2$$

$$= (\sum \lambda_i + |\lambda(\omega_j)|^2)(1 + c^2|e^{i\omega_j} - 1|^4)\left|A(\omega_j) - \frac{|\lambda(\omega_j)|^2}{(\sum \lambda_i + |\lambda(\omega_j)|^2)(1 + c^2|e^{i\omega_j} - 1|^4)}\right|^2$$

$$+ \left\{|\lambda(\omega_j)|^2 - \frac{|\lambda(\omega_j)|^4}{(\sum \lambda_i + |\lambda(\omega_j)|^2)(1 + c^2|e^{i\omega_j} - 1|^4)}\right\} \qquad (5.2)$$

where the last term in (5.2) is positive. Thus, the optimal filter now is given by

$$A(\omega_j) = \frac{|\lambda(\omega_j)|^2}{(\sum \lambda_t + |\lambda(\omega_j)|^2)(1 + c^2|\exp i\omega_j - 1|^4)} . \qquad (5.3)$$

Clearly, the result (5.3) reduces to (2.15) when $c^2 = 0$. When $c^2 \to \infty$, $A(\omega_j) \to 0$ unless $j = 0$ so that the a_k become equal and the estimator becomes constant for $t = 1, \ldots, N$ as predicted above.

As in Section 2, it is easy to calculate the time filter

coefficients a_k using the Fast Fourier Transform.

The use of

$$\sum_{1}^{N} (\Delta^2 \hat{\lambda}_t)^2$$

to control smoothness is common but had we chosen

$$\sum_{1}^{N} (\Delta^p \hat{\lambda}_t)^2$$

the optimal filter would have been

$$A(\omega_j) = \frac{|\lambda(\omega_j)|^2}{(\sum \lambda_t + |\lambda(\omega_j)|^2)(1 + c^2 |\exp i\omega_j - 1|^{2p})} \tag{5.4}$$

and its properties for $p \geq 1$ are the same.

Splines in the discrete case were first suggested by Whittaker (1923) and in the continuous case by Schoenberg (1964). To obtain a Bayesian motivation for (5.1), it is usual to invent a process for λ_t which makes the likelihood function depend on $\sum(\Delta^p \lambda_t)^2$. This would be so if $\Delta^p \lambda_t$ for the various values of t were independently Gaussian with the same variance. In continuous time, this requires $\lambda(t)$ to be an integrated Brownian motion.

6. Another Model.

Given the above development, it seems worthwhile to show how the Weiner Filter may be easily derived. This means changing the model from the N.H.P.P. In fact, we have only assumed, in the Section 2 derivation of the optimal linear filter, that

$$E(n_t) = \lambda_t, \ \mathrm{var}(n_t) = \lambda_t, \mathrm{cov}(n_t, n_{t'}) = 0, \ t \neq t' . \tag{6.1}$$

Instead, suppose that

$$E(n_t) = \lambda_t \text{ , } \operatorname{cov}(n_t \text{ , } n_{t+\tau}) = \rho(\tau) \text{ .} \tag{6.2}$$

Thus, the counts in different intervals may be correlated but this correlation should depend only upon their spacing. In this sense, the assumptions are weaker. However, $\rho(0) = \operatorname{var}(n_t)$ must now be constant and not depend upon λ_t.

Then (2.13) becomes

$$E|n(\omega_j)|^2 = \sum\sum E(n_t n_{t'}) \quad \exp i\omega_j(t-t')$$

$$= |\lambda(\omega_j)|^2 + \sum\sum \rho(t-t') \exp i\omega_j(t-t'). \tag{6.3}$$

Given the fact that we have extended the data to make n_t of period N, $\rho(\tau)$ in (6.2) must have period N. Thus, $\rho(-1) = \rho(N-1)$, $\rho(-2) = \rho(N-2)$, etc., so that the covariance matrix C whose element in the t-th row, t'-th column is $\rho(t-t')$ is an N×N circulant. The eigenvectors of such matrices are $[\exp i\omega_j \cdot 0, \exp i\omega_j \cdot 1, \ldots, \exp i\omega_j(N-1)]$, $j = 0, \ldots, N-1$. Denoting the corresponding eigenvalues by $f(\omega_j)$, (6.3) becomes

$$E|n(\omega_j)|^2 = |\lambda(\omega_j)|^2 + f(\omega_j) \text{ .} \tag{6.4}$$

In this case, the analogue of (2.15) is seen to be

$$A(\omega_j) = \frac{|\lambda(\omega_j)|^2}{|\lambda(\omega_j)|^2 + f(\omega_j)} \text{ .} \tag{6.5}$$

This is nothing but the classical result of Wiener for filtering a stationary process--except that we have derived it on the circle.

In order to use (6.5) in practice to derive a good smoothing formula, one needs to make a good guess at the "noise" spectrum $f(\omega_j)$ as well as the signal spectrum $|\lambda(\omega_j)|^2$, which was all that was needed in Section 2. This is the price of the weaker assumption.

7. An Even More Applied Approach.

A data analyst, knowing that $E(n_t) = \text{Var}\,(n_t) = \lambda_t$, would probably consider $y_t = 2\sqrt{n_t}$ which will have approximately unit variance and mean $2\sqrt{\lambda_t}$, provided λ_t is not small. In passing a curve through the points (t, y_t), the analyst will not be distracted by variance changes. Further y_t will be approximately Gaussian so it makes sense to pass the curve through the "center" of the band of points. Finally, any smoothing method could be considered, without regard to positivity, since, given the smoothed values, $\hat{y}_t$ say, a positive smoothed estimate of λ_t is available from

$$\hat{\lambda}_t = \frac{1}{4}\hat{y}_t^{\,2} . \qquad (7.1)$$

If other transformations such as $2\sqrt{n_t + 3/8}$ or $\sqrt{n_t} + \sqrt{n_t+1}$ are used, the formulae replacing (7.1) are obtained similarly.

To counter criticisms, this analyst might object to (1.2) as a criterion on the grounds that the derivations $\hat{\lambda}(t) - \lambda(t)$ are weighted equally despite the fact that $\text{Var}(\hat{\lambda}(t) - \lambda(t)) = O(\lambda(t))$. The analyst might also criticize the notion of optimal smoothing by saying that, since the true $\lambda(t)$ is not known, one must try various smoothers on the data, possibly smoothing less in some stretches than in others to retain resolution.

It is, of course, possible to formalize smoothing with y_t, the square roots of the counts. A standard procedure would be to use a spline fit, e.g., find $\hat{\mu}_t = 2\sqrt{\hat{\lambda}_t}$ to minimize

$$\sum (y_t - \hat{\mu}_t)^2 + K \sum (\Delta^p \hat{\mu}_t)^2 .$$

The first term is more justifiable now since the y_t have approximately the same variance about μ_t. The second term insists on smoothness of the fitted μ_t's. The K is a tuning parameter, $K = 0$ implying no smoothing at all, i.e., $\hat{\mu}_t = y_t$.

References.

Beran, R. J. W. (1968). "Testing for uniformity on a compact homogeneous space," *J. Appl. Prob.* *5*, pp. 177-195.

Bloomfield, P. (1976). *Fourier Analysis of Time Series - An Introduction*. Wiley-Interscience, New York.

Brillinger, D. R. (1974). *Time Series, Data Analysis and Theory*. Holt, Rinehart and Winston, New York.

Clevenson, M. L., and Zidek, J. V. (1975). "Bayesian estimation of the intensity function of the non-stationary Poisson process," Technical Report No. 222, Department of Statistics, Stanford University.

Cox, D. R., and Lewis, P. A. P. (1966). *The Statistical Analysis of Series of Events*. Methuen, London.

Dudley, R. M. (1971). "Binomial tests of goodness of fit," Unpublished Manuscript.

Grandell, J. (1971). "On stochastic processes generated by a stochastic intensity," *Skand. Akt. Tid.* *54*, pp. 205-240.

Lewis, P. A. P. (1972). *Stochastic Point Processes: Statistical Analysis and Applications*. Wiley-Interscience, New York.

Lewis, P. A. P. (1975). "Statistical analysis of transacting processing in a data base system," IBM Report No. RJ1629.

Lundberg, O. (1940). *On Random Processes and Their Applications to Sickness and Accident Statistics*. Almquist and Wiksells, Uppsala. (Republished 1964).

Schoenberg, I. J. (1964). "Spline functions and the problem of graduation," *P. N. A. S.* *52* (4), pp. 947-950.

Watson, G. S. (1961). "Goodness of fit tests on a circle," *Biometrika* *48*, pp. 109-114.

Watson, G. S. (1967). "Some problems in statistics of directions," *Bull. I. S. I.*, 33rd Session, pp. 374-385.

Watson, G. S. (1974a). "Detecting signals in Poisson noise," in *Reliability and Biometry*, pp. 775-786, ed., F. Proschan and R. J. Serfling, S. I. A. M., Philadelphia

Watson, G. S. (1974b). "Optimal invariant tests for uniformity," *Studies in Probability and Statistics*. Jerusalem Academic Press, pp. 121-128.

Whittaker, E. T. (1923). "On a new method of graduation," *Proc. Edin. Math. Soc.* *41*, pp. 63-75.

A B C 8 D 9 E 0 F 1 G 2 H 3 I 4 J 5